The Dragonfly Nymphs of Thailand (Odonata: Anisoptera)

Rodolfo Novelo-Gutiérrez • Robert W. Sites

The Dragonfly Nymphs of Thailand (Odonata: Anisoptera)

An Identification Guide to Families and Genera

Springer

Rodolfo Novelo-Gutiérrez
Red de Biodiversidad y Sistemática
Instituto de Ecología AC
Xalapa, Veracruz, Mexico

Robert W. Sites
Enns Entomology Museum
University of Missouri
Columbia, MO, USA

ISBN 978-3-031-33711-6 ISBN 978-3-031-33712-3 (eBook)
https://doi.org/10.1007/978-3-031-33712-3

This Springer imprint is published by the registered company Springer Nature Switzerland AG
The registered company address is: Gewerbestrasse 11, 6330 Cham, Switzerland

Preface

This book is intended as a primer and first substantial effort to provide a comprehensive understanding of the Odonata nymphs of Thailand. Most of the studies devoted to this spectacular group of insects have been focused on the adults, leaving behind the knowledge of the immature stages. In many ways, this is understandable because of the charismatic nature of the adults, their apparency, and the great species richness in this tropical country. Many new species are discovered every year, and numerous nomenclatural changes have taken place in the last 25 years, since Hämäläinen and Pinratana's book "Atlas of the Dragonflies of Thailand" was published in 1999. On the other hand, an interest in the nymphs of Odonata in Thailand began just a very few years ago and is still nascent. Unfortunately, there are no resources available that bring together the widely scattered literature related to the Thai odonate nymphs, which makes it overwhelmingly difficult to begin working with this group even at the genus level. Further, there are no appropriate taxonomic keys for the local fauna, which makes accurate identification of most of the nymphs tenuous at best. For this reason, we tasked ourselves to eliminate this impediment by providing detailed descriptions, high-resolution photographs, and illustrated keys to genus level of the Anisoptera nymphs of Thailand, including distribution maps, references, and an updated list of the species recorded for this country.

Our primary motivation to prepare this book was to provide a basis from which naturalists, students, and professionals can easily and confidently identify Thai odonate nymphs to the genus level, so they will have a starting point from which to delve deeper into the fields of taxonomy, ecology, biomonitoring, and conservation involving Thai odonate nymphs. However, the reader should keep in mind that, as with all preliminary work, not all the variation in each genus has been included; thus, inconsistencies could exist in some of the descriptions and keys. Likewise, the keys are based on structures that are well or fully developed; thus, the keys are for late instars. Earlier instars might not key properly. Now that genus-level identifications are more easily possible, we hope that many enthusiastic people will take up working with Thai dragonfly nymphs; we encourage them to collect, rear, describe, and publish new findings. A more complete knowledge of the nymphs will enable us to then focus at the species level, which will in turn enable a deeper understanding of phylogenetic relationships among the taxa.

For each genus treated in this book, several montage photographs of the nymphs are provided. These photos were obtained with a Leica MZ16 stereomicroscope coupled with the Leica Application Suite V4.10 Extended Depth of Focus module, followed by image preparation with Photoshop CC2018 (Adobe Systems Inc., San Jose, California). Photographs that accompany the keys were obtained with a Nikon SMZ25 stereomicroscope (Chiyoda-ku, Tokyo, Japan) with Nikon DS-U3 camera and processed with the program NIS elements AR version 4.5 (Laboratory Imaging s.r.o., Praha, Czech Republic).

Xalapa, Veracruz, Mexico Rodolfo Novelo-Gutiérrez
Columbia, MO, USA Robert W. Sites

Acknowledgments

Many people contributed to making this endeavor possible with their invaluable help during field collections. We are grateful to Akekawat Vitheepradit (Kasetsart University, Bangkok), Taeng-On Prommi (Kasetsart University, Nakhon Pathom), Chaweewan Hutacharern (Royal Forestry Department), Porntip Chantaramongkol (Chiang Mai University), Surakrai Permkam (Prince of Songkla University, Hat Yai), Pongsak Laudee (Prince of Songkla University, Surat Thani), Kriengkrai Setaphan (Naresuan University), and Penkhae Thamsenanupap (Mahasarakham University) for assisting with collecting, logistical support, and sharing specimens and to the National Research Council of Thailand, the Royal Forestry Department, and the Department of National Parks, Wildlife, and Plant Conservation for permission to conduct fieldwork in Thailand. Also, thanks are due to the student teams of Chiang Mai University and the University of Missouri, Columbia, who collected many of the specimens used in this book. Kristin B. Simpson provided invaluable support sorting and preserving specimens. In addition, José Antonio Gómez-Anaya edited the photographs that accompany the keys and composed plates.

The following colleagues and friends provided literature of the Oriental region and/or invaluable comments: Che Salmah, Klaus-D. Dijkstra, Vincent Kalkman, Matti Hämäläinen, John Mitchalski, Milen Marinov, Alonso Ramírez, Frank Suhling, Albert Orr, Gunther Fleck, Mike Ferro, Enrique González-Soriano, and Bastiaan Kiauta. Special thanks are due to Dr. Akihiko Sasamoto for the donation of nymphs and exuviae of Aeshnidae and Gomphidae as well as for providing diverse difficult-to-obtain literature. Also, thanks are due to Dr. Fang-Shuo Hu for allowing us to use his original photographs of the nymph of *Cephalaeschna risi*.

I (RNG) particularly thank Dr. Takashi Aoki (Rokko Island High School) for the exchange of Odonata larvae from Japan, which were very important for the corroboration of several generic determinations of the Thai specimens. Also, I am especially indebted to Dr. Che Salmah (Universiti Sains Malaysia) who facilitated my first contact with the amazing world of the Odonata nymphs of Southeast Asia.

Finally, we thank the editors Dr. Florian Weihrauch (Odonatologica) and Dr. John Abbott (International Journal of Odonatology) for allowing us to use several drawings published in those journals. We also thank Magnolia Press for permission to reproduce here our own photographs of *Anotogaster gregoryi*, *Amphigomphus somnuki*, and *Phaenandrogomphus tonkinicus* published previously in Zootaxa. The corrections and suggestions of four anonymous reviewers substantially improved the last version of the manuscript.

Contents

1 Introduction 1
1.1 General 1
1.1.1 Life Cycle 2
1.1.2 Importance in Food Webs 4
1.1.3 Conservation 4
1.2 Study Area 5
1.3 A List of the Anisoptera of Thailand 9
References 16

2 Morphology of the Nymphs 23
References 32

3 Keys to Suborders and Families of Odonata of Thailand 33
3.1 Key to Families of Anisoptera 34

4 Aeshnidae 39
4.1 Key to the Genera of Aeshnidae 42
4.2 *Anaciaeschna* Selys, 1878 51
4.2.1 General Appearance 51
4.2.2 Description 51
4.2.3 Remarks 54
4.2.4 Distribution 54
4.3 *Anax* Leach, 1815 55
4.3.1 General Appearance 55
4.3.2 Description 55
4.3.3 Remarks 58
4.3.4 Distribution 58
4.4 *Cephalaeschna* Selys, 1883 59
4.4.1 General Appearance 60
4.4.2 Description 60
4.4.3 Remarks 63
4.4.4 Distribution 63

4.5 *Gynacantha* Rambur, 1842 64
4.5.1 General Appearance 64
4.5.2 Description 64
4.5.3 Remarks 67
4.5.4 Distribution 67
4.6 *Heliaeschna* Selys, 1882 68
4.6.1 General Appearance 68
4.6.2 Description 68
4.6.3 Remarks 69
4.6.4 Distribution 70
4.7 *Indaeschna* Fraser, 1926 71
4.7.1 General Appearance 71
4.7.2 Description 71
4.7.3 Remarks 72
4.7.4 Distribution 72
4.8 *Periaeschna* Martin, 1908 73
4.8.1 General Appearance 73
4.8.2 Description 73
4.8.3 Remarks 76
4.8.4 Distribution 76
4.9 *Petaliaeschna* Fraser, 1927 77
4.9.1 General Appearance 77
4.9.2 Description 78
4.9.3 Remarks 81
4.9.4 Distribution 82
4.10 *Planaeschna* McLachlan, 1896 83
4.10.1 General Appearance 83
4.10.2 Description 83
4.10.3 Remarks 86
4.10.4 Distribution 86
4.11 *Polycanthagyna* Fraser, 1933 87
4.11.1 General Appearance 87
4.11.2 Remarks 90
4.11.3 Distribution 90
4.12 *Sarasaeschna* Karube and Yeh, 2001 91
4.12.1 General Appearance 91
4.12.2 Description 92
4.12.3 Remarks 94
4.12.4 Distribution 94
4.13 *Tetracanthagyna* Selys, 1883 95
4.13.1 General Appearance 95
4.13.2 Description 96
4.13.3 Remarks 99
4.13.4 Distribution 99
References 102

5 Gomphidae 105
5.1 Key to Genera of Gomphidae 109
5.2 *Acrogomphus* Laidlaw, 1925 117
5.2.1 General Appearance 117
5.2.2 Description 117
5.2.3 Remarks 120
5.2.4 Distribution 120
5.3 *Amphigomphus* Chao, 1954 121
5.3.1 General Appearance 121
5.3.2 Description 121
5.3.3 Remarks 123
5.3.4 Distribution 124
5.4 *Anisogomphus* Selys, 1858 125
5.4.1 General Appearance 125
5.4.2 Description 125
5.4.3 Remarks 128
5.4.4 Distribution 128
5.5 *Asiagomphus* Asahina, 1985 129
5.5.1 General Appearance 130
5.5.2 Description 130
5.5.3 Remarks 133
5.5.4 Distribution 133
5.6 *Burmagomphus* Williamson, 1907 134
5.6.1 General Appearance 134
5.6.2 Description 134
5.6.3 Remarks 137
5.6.4 Distribution 137
5.7 *Davidius* Selys, 1878 138
5.7.1 General Appearance 138
5.7.2 Description 138
5.7.3 Remarks 141
5.7.4 Distribution 141
5.8 *Euthygomphus* Kosterin, 2016 142
5.8.1 General Appearance 142
5.8.2 Description 143
5.8.3 Remarks 145
5.8.4 Distribution 146
5.9 *Gomphidia* Selys, 1854 146
5.9.1 General Appearance 147
5.9.2 Description 147
5.9.3 Remarks 150
5.9.4 Distribution 150
5.10 *Gomphidictinus* Fraser, 1942 151
5.10.1 General Appearance 151
5.10.2 Description 152

5.10.3 Remarks 155
5.10.4 Distribution 155
5.11 *Heliogomphus* Laidlaw, 1922 156
5.11.1 General Appearance 156
5.11.2 Description 157
5.11.3 Remarks 160
5.11.4 Distribution 161
5.12 *Ictinogomphus* Cowley, 1934 162
5.12.1 General Appearance 162
5.12.2 Description 162
5.12.3 Remarks 165
5.12.4 Distribution 166
5.13 *Lamelligomphus* Fraser, 1922 167
5.13.1 General Appearance 167
5.13.2 Description 167
5.13.3 Remarks 170
5.13.4 Distribution 170
5.14 *Leptogomphus* Selys, 1878 171
5.14.1 General Appearance 171
5.14.2 Description 171
5.14.3 Remarks 174
5.14.4 Distribution 174
5.15 *Macrogomphus* Selys, 1857 175
5.15.1 General Appearance 175
5.15.2 Description 175
5.15.3 Remarks 178
5.15.4 Distribution 178
5.16 *Megalogomphus* Campion, 1923 179
5.16.1 General Appearance 179
5.16.2 Description 180
5.16.3 Remarks 183
5.16.4 Distribution 183
5.17 *Microgomphus* Selys, 1857 184
5.17.1 General Appearance 184
5.17.2 Description 184
5.17.3 Remarks 187
5.17.4 Distribution 187
5.18 *Nepogomphus* Fraser, 1934 188
5.18.1 General Appearance 188
5.18.2 Description 189
5.18.3 Remarks 191
5.18.4 Distribution 191
5.19 *Nihonogomphus* Oguma, 1926 192
5.19.1 General Appearance 192
5.19.2 Description 193

5.19.3 Remarks 195
5.19.4 Distribution 195
5.20 *Nychogomphus* Carle, 1986 196
5.20.1 General Appearance 196
5.20.2 Description 197
5.20.3 Remarks 199
5.20.4 Distribution 199
5.21 *Onychogomphus* Selys, 1854 200
5.21.1 General Appearance 200
5.21.2 Description 201
5.21.3 Remarks 203
5.21.4 Distribution 203
5.22 *Orientogomphus* Chao & Xu, 1987 204
5.22.1 General Appearance 204
5.22.2 Description 205
5.22.3 Remarks 207
5.22.4 Distribution 207
5.23 *Paragomphus* Cowley, 1934 208
5.23.1 General Appearance 208
5.23.2 Description 208
5.23.3 Remarks 210
5.23.4 Distribution 211
5.24 *Phaenandrogomphus* Lieftinck, 1964 212
5.24.1 General Appearance 212
5.24.2 Description 212
5.24.3 Remarks 215
5.24.4 Distribution 215
5.25 *Sieboldius* Selys, 1854 216
5.25.1 General Appearance 216
5.25.2 Description 217
5.25.3 Remarks 219
5.25.4 Distribution 220
5.26 *Sinictinogomphus* Fraser, 1939 221
5.26.1 General Appearance 221
5.26.2 Description 221
5.26.3 Remarks 224
5.26.4 Distribution 224
5.27 *Stylogomphus* Fraser, 1922 225
5.27.1 General Appearance 225
5.27.2 Description 226
5.27.3 Remarks 228
5.27.4 Distribution 228
5.28 Genus Undetermined 229
5.28.1 General Appearance 229
5.28.2 Description 230

5.28.3 Remarks . . . 232
5.28.4 Distribution . . . 233
References . . . 235

6 **Chlorogomphidae** . . . 239
6.1 Key to Genera of Chlorogomphidae of Thailand . . . 240
6.2 *Chlorogomphus* Selys, 1854 . . . 241
6.2.1 General Appearance . . . 241
6.2.2 Description . . . 241
6.2.3 Remarks . . . 245
6.2.4 Distribution . . . 245
References . . . 247

7 **Cordulegastridae** . . . 249
7.1 *Anotogaster* Selys, 1854 . . . 250
7.1.1 General Appearance . . . 250
7.1.2 Description . . . 251
7.1.3 Remarks . . . 254
7.1.4 Distribution . . . 254
References . . . 255

8 **Macromiidae** . . . 257
8.1 Key to Genera of Macromiidae of Thailand . . . 258
8.2 *Epophthalmia* Burmeister, 1839 . . . 259
8.2.1 General Appearance . . . 259
8.2.2 Description . . . 259
8.2.3 Remarks . . . 262
8.2.4 Distribution . . . 263
8.3 *Macromia* Rambur, 1842 . . . 264
8.3.1 General Appearance . . . 264
8.3.2 Description . . . 264
8.3.3 Remarks . . . 267
8.3.4 Distribution . . . 268
References . . . 269

9 **Corduliidae** . . . 271
9.1 Key to Genera of Corduliidae of Thailand . . . 272
9.2 *Hemicordulia* Selys, 1870 . . . 273
9.2.1 General Appearance . . . 273
9.2.2 Description . . . 273
9.2.3 Remarks . . . 276
9.2.4 Distribution . . . 277
9.3 *Procordulia* Martin, 1907 . . . 277
9.3.1 General Appearance . . . 277
9.3.2 Description . . . 278
9.3.3 Remarks . . . 279
9.3.4 Distribution . . . 279

9.4 *Somatochlora* Selys, 1871 280
9.4.1 General Appearance 280
9.4.2 Description 280
9.4.3 Remarks 283
9.4.4 Distribution 283
References 284

10 Libellulidae 287
10.1 Key to Genera of Libellulidae 292
10.2 *Acisoma* Rambur, 1842 303
10.2.1 General Appearance 303
10.2.2 Description 303
10.2.3 Remarks 306
10.2.4 Distribution 306
10.3 *Aethriamanta* Kirby, 1889 307
10.3.1 General Appearance 307
10.3.2 Description 307
10.3.3 Remarks 310
10.3.4 Distribution 310
10.4 *Agrionoptera* Brauer, 1864 311
10.4.1 General Appearance 311
10.4.2 Description 312
10.4.3 Remarks 314
10.4.4 Distribution 315
10.5 *Brachydiplax* Brauer, 1868 316
10.5.1 General Appearance 316
10.5.2 Description 316
10.5.3 Remarks 319
10.5.4 Distribution 319
10.6 *Brachythemis* Brauer, 1868 320
10.6.1 General Appearance 321
10.6.2 Description 321
10.6.3 Remarks 324
10.6.4 Distribution 324
10.7 *Bradinopyga* Kirby, 1893 325
10.7.1 General Appearance 325
10.7.2 Description 326
10.7.3 Remarks 327
10.7.4 Distribution 327
10.8 *Camacinia* Kirby, 1889 328
10.8.1 General Appearance 328
10.8.2 Description 328
10.8.3 Remarks 330
10.8.4 Distribution 330

10.9 *Cratilla* Kirby, 1900 . . . 331
10.9.1 General Appearance . . . 331
10.9.2 Description . . . 332
10.9.3 Remarks . . . 334
10.9.4 Distribution . . . 335
10.10 *Crocothemis* Brauer, 1868 . . . 336
10.10.1 General Appearance . . . 336
10.10.2 Description . . . 336
10.10.3 Remarks . . . 339
10.10.4 Distribution . . . 340
10.11 *Diplacodes* Kirby, 1889 . . . 341
10.11.1 General Appearance . . . 341
10.11.2 Description . . . 341
10.11.3 Remarks . . . 344
10.11.4 Distribution . . . 345
10.12 *Hydrobasileus* Kirby, 1889 . . . 346
10.12.1 General Appearance . . . 346
10.12.2 Description . . . 346
10.12.3 Remarks . . . 350
10.12.4 Distribution . . . 350
10.13 *Hylaeothemis* Ris, 1909 . . . 351
10.13.1 General Appearance . . . 351
10.13.2 Description . . . 351
10.13.3 Remarks . . . 354
10.13.4 Distribution . . . 354
10.14 *?Indothemis* Ris, 1909 . . . 355
10.14.1 General Appearance . . . 356
10.14.2 Description . . . 356
10.14.3 Remarks . . . 359
10.14.4 Distribution . . . 359
10.15 *Lathrecista* Kirby, 1889 . . . 360
10.15.1 General Appearance . . . 360
10.15.2 Description . . . 360
10.15.3 Remarks . . . 361
10.15.4 Distribution . . . 361
10.16 *Lyriothemis* Brauer, 1868 . . . 362
10.16.1 General Appearance . . . 362
10.16.2 Description . . . 362
10.16.3 Remarks . . . 366
10.16.4 Distribution . . . 366
10.17 *Macrodiplax* Brauer, 1868 . . . 367
10.17.1 General Appearance . . . 367
10.17.2 Description . . . 367
10.17.3 Remarks . . . 371
10.17.4 Distribution . . . 371

10.18 *Nannophya* Rambur, 1842 372
10.18.1 General Appearance 372
10.18.2 Description 372
10.18.3 Remarks 376
10.18.4 Distribution 376
10.19 *Nannophyopsis* Lieftinck, 1935 377
10.19.1 General Appearance 377
10.19.2 Description 377
10.19.3 Remarks 378
10.19.4 Distribution 378
10.20 *Neurothemis* Brauer, 1867 379
10.20.1 General Appearance 379
10.20.2 Description 380
10.20.3 Remarks 383
10.20.4 Distribution 384
10.21 *Onychothemis* Brauer, 1868 385
10.21.1 General Appearance 385
10.21.2 Description 385
10.21.3 Remarks 389
10.21.4 Distribution 389
10.22 *Orchithemis* Brauer, 1878 390
10.22.1 General Appearance 390
10.22.2 Description 390
10.22.3 Remarks 391
10.22.4 Distribution 392
10.23 *Orthetrum* Newman, 1833 392
10.23.1 General Appearance 393
10.23.2 Description 393
10.23.3 Remarks 397
10.23.4 Distribution 397
10.24 *Palpopleura* Rambur, 1842 398
10.24.1 General Appearance 398
10.24.2 Description 399
10.24.3 Remarks 402
10.24.4 Distribution 402
10.25 *Pantala* Hagen, 1861 403
10.25.1 General Appearance 403
10.25.2 Description 403
10.25.3 Remarks 407
10.25.4 Distribution 407
10.26 *Potamarcha* Karsch, 1890 408
10.26.1 General Appearance 408
10.26.2 Description 409
10.26.3 Remarks 413
10.26.4 Distribution 413

10.27 *Rhodothemis* Ris, 1909 414
10.27.1 General Appearance 414
10.27.2 Description 414
10.27.3 Remarks 419
10.27.4 Distribution 419
10.28 *Rhyothemis* Hagen, 1867 420
10.28.1 General Appearance 420
10.28.2 Description 420
10.28.3 Remarks 424
10.28.4 Distribution 424
10.29 *Tetrathemis* Brauer, 1868 425
10.29.1 General Appearance 426
10.29.2 Description 426
10.29.3 Remarks 430
10.29.4 Distribution 430
10.30 *Tholymis* Hagen, 1867 431
10.30.1 General Appearance 431
10.30.2 Description 431
10.30.3 Remarks 435
10.30.4 Distribution 435
10.31 *Tramea* Hagen, 1861 436
10.31.1 General Appearance 436
10.31.2 Description 437
10.31.3 Remarks 441
10.31.4 Distribution 441
10.32 *Trithemis* Brauer, 1868 442
10.32.1 General Appearance 442
10.32.2 Description 442
10.32.3 Remarks 446
10.32.4 Distribution 446
10.33 *Urothemis* Brauer, 1868 447
10.33.1 General Appearance 447
10.33.2 Description 448
10.33.3 Remarks 452
10.33.4 Distribution 452
10.34 *Zygonyx* Hagen, 1867 453
10.34.1 General Appearance 453
10.34.2 Description 454
10.34.3 Remarks 458
10.34.4 Distribution 458
10.35 *Zyxomma* Rambur, 1842 459
10.35.1 General Appearance 459
10.35.2 Description 460
10.35.3 Remarks 463
10.35.4 Distribution 464
References 468

11 Incertae Sedis 471
11.1 Key to *Incertae Sedis* Genera of Thailand 472
11.2 *Idionyx* Hagen, 1867 472
11.2.1 General Appearance 472
11.2.2 Description 472
11.2.3 Remarks 476
11.2.4 Distribution 476
11.3 *Macromidia* Martin, 1907 477
11.3.1 General Appearance 478
11.3.2 Description 478
11.3.3 Remarks 481
11.3.4 Distribution 481
References 482

Glossary 485

Index 489

About the Authors

Rodolfo Novelo-Gutiérrez was educated at the National Autonomous University of Mexico. Following graduation at this institution, he taught general entomology courses for more than a decade. In 1989, he joined the Institute of Ecology, A.C. in Xalapa city. Since then, his studies have focused on the taxonomy of tropical Odonata, although he has also published several papers on aquatic Coleoptera, Hemiptera, and Trichoptera. In 2010, he studied odonate nymphs at the University of Sains Malaysia at the invitation of Dr. Che Salmah. From this experience, he published several papers on the odonate fauna of Southeast Asia. In 2012, he began collaborating with Prof. Sites on samples of Odonata nymphs of Thailand that had been collected over the previous 25 years.

Robert W. Sites has had extensive involvement with Thailand over the past 25 years, including in both education and research. He has taught various entomology courses to Thai students, served as major advisor for several Thai graduate students, and is known to his Thai colleagues and students as "Ajarn Bob." For nearly 20 years, he taught an annual University of Missouri study abroad course to Thailand for American students. His research on aquatic insects has been conducted in collaboration with Thai colleagues and has included field work in nearly all Thai provinces. The taxonomic focus of his studies has been primarily on Odonata and Heteroptera but also has included studies on Trichoptera and several families of aquatic Coleoptera. In an ecological study, he studied the recovery of the freshwater lentic insect community in waterbodies along the coastline that were inundated by seawater during the tsunami of 2004.

Introduction

1

Contents

1.1 General ... 1
1.1.1 Life Cycle ... 2
1.1.2 Importance in Food Webs ... 4
1.1.3 Conservation ... 4
1.2 Study Area ... 5
1.3 A List of the Anisoptera of Thailand ... 9
References ... 16

1.1 General

Few groups of invertebrates, other than butterflies, have captured the attention of both scientists and amateur naturalists as have the dragonflies. Their attraction to so many people is because of a myriad of reasons. They are common, colorful, and occur in many terrestrial habitats, not being restricted to areas near their nymphal aquatic habitats. They are among the most ancient winged insects, with three-foot wingspan Protodonata ancestors dating back to the Carboniferous period approximately 320 million years ago, long before dinosaurs roamed the earth. Their flight capabilities are unmatched, as they can hover, fly straight up or down, dart quickly, and stop instantly. With more than 10,000 facets per eye, their visual acuity is superb, which is one reason why they are so difficult to catch. Their biology and complex behavior are amazing considering their unique mating behavior, levels of territoriality, migrations, and more. Think about the ecological pursuit of mosquitoes as a food source in which mosquito larvae are preyed on by dragonfly nymphs in the water, and then after the mosquitoes leave the water as winged adults, the

R. Novelo-Gutiérrez, R. W. Sites, *The Dragonfly Nymphs of Thailand (Odonata: Anisoptera)*, https://doi.org/10.1007/978-3-031-33712-3_1

dragonflies are still preying on them, but now also as winged adults. In fact, adult dragonflies can eat over a hundred mosquitoes per day, and they can catch and eat them while in flight. Understandably, most people are interested in dragonfly adults for these reasons and more. However, the seldom-studied nymphs living in water also have amazing attributes worthy of attention. Because they live beneath the water surface, dragonfly nymphs are less apparent and thus have not been the subject of attention, in comparison with the adults.

Damselflies and dragonflies belong to the order Odonata (from the Greek "odonto," tooth, and "gnathos," mandibles), referring to the sharp teeth of their mouthparts. The Odonata itself dates back to the Permian period, approximately 250–280 million years ago. They have an amphibiotic life cycle with nymphs (also called *larvae* or *naiads*) being aquatic and adults terrestrial/aerial. The nymphs develop in practically any type of standing or running freshwater, although each species is highly specific to a particular aquatic habitat; some species occupy specialized microhabitats such as seepages and phytotelmata (bromeliads and tree holes filled with rain water), a few live on the vertical rock surfaces of waterfalls, some are salt-tolerant and live in brackish waters, and just a handful of species are terrestrial or semi-terrestrial. The nymphs and adults of all species around the world are voracious predators, feeding exclusively on living prey.

Adults of the closely related damselflies have a delicate, thin, slight body with both pairs of wings similar in size, shape, and venation, which is the reason they are classified into the suborder Zygoptera (Greek *zygos*, even, and *pteron*, wing). Adult dragonflies are more robust and stronger fliers than are damselflies, with both pairs of wings of different shape and venation, so they are classified into the suborder Anisoptera (Greek *anisos*, uneven, and *pteron*, wing). A third suborder, the Anisozygoptera, has three relict species of a single extant genus, *Epiophlebia*, restricted to the Indo-Oriental region, with all other representatives of this suborder extinct. The nymphs of damselflies and dragonflies also are strikingly different in morphology, most notably by the presence of three caudal gills in Zygoptera. In contrast, the nymphs of Anisoptera exhibit a short anal pyramid consisting of five acute appendages, with gills housed internally in a rectal branchial chamber.

To date, 6404 species of Odonata have been described worldwide (Paulson et al. 2023). Although the adults are quite well known, the nymphs are still poorly known, especially in the tropical regions where diversity is the highest.

1.1.1 Life Cycle

Odonata are hemimetabolous insects; thus, they have three stages of life, egg, nymph, and adult, with the first two stages developing usually in water.

Eggs: Females lay their eggs (oviposition) in a wide range of aquatic habitats, selecting sites using cues such as type and height of riparian vegetation, length of shoreline, and presence/absence of aquatic macrophytes, among others. Three modes of oviposition are used by dragonfly females: endophytic (eggs are inserted into plant tissue), epiphytic (eggs are attached to the outside of a plant), and exophytic (eggs are dropped into the water). In general, eggs deposited in the first two modes are elongate and spindle-shaped, whereas eggs laid exophytically usually are elliptical or subspherical. Eggs are laid in successive batches of 100–400 eggs in endophytic species to several hundreds to thousands per batch in exophytic species. Freshly deposited eggs usually are yellow to orange, becoming darker in a few hours. The embryos develop in a period of 7–80 days in the tropics; in temperate and cold regions, eggs may enter an obligatory diapause for overwintering, lasting several months. Eggs are vulnerable to the attack of mold, mites, predators, and parasitoids.

Nymph: After the embryo has developed, eclosion of the pronymph (first instar) takes place; the pronymph can last from a few seconds to 1–3 min, and then a rapid molting occurs to give rise to the second instar. At this point, the nymph has only three antennomeres and one tarsomere in the legs, and no wing buds are visible. The nymph continues growing by periodic moltings; the number of nymphal instars ranges from 8 to 16, completing its development in a few weeks to several years depending on the species and environmental conditions. The nymph is a voracious carnivore that behaves generally as an ambush predator. In this stage, the remarkable labium is modified to become a raptorial grabbing device, which is rapidly fired in a centesimal of a second when prey is detected, pulling the victim to the mouth to be consumed with the aid of the mandibles and maxillae. In order to avoid being preyed on themselves, they exhibit cryptic coloration while climbing among submerged vegetation, clinging to underwater objects, burrowing in the substrate, covering themselves with mud, sand, or debris, or employing jet propulsion.

At the end of the nymphal stage, the fully grown nymph stops feeding, and an internal metamorphosis occurs. The pharate adult features are visible through the nymphal cuticle, indicating the nymph is ready to climb up emergent vegetation or a rock where the final molt will occur. The newly emerged, teneral adult must wait before it can fly. During this time, it is vulnerable to predation since it is not capable of much evasive movement; thus, eclosion to the adult stage usually occurs at night. By morning, it is sufficiently sclerotized and ready for its maiden flight. In some groups (e.g., Gomphidae), emergence can occur in the morning or at noon.

Adult: Newly emerged adults fly away from the emergence site, and their cuticle continues to harden as they feed. Male and female juvenile adults exhibit similar coloration and lustrous wings. The maturation period may last from a few weeks to several months. When mature, males are usually more colorful than are females. Colors and patterns in the wings and body may play an important role in territoriality and courtship. At this time, they are sexually mature and ready for reproduction, which takes place usually near waterbodies. Males of some species hold territories near the water and defend them aggressively against conspecific males, and patrol in search of females. When a female arrives looking for a mate or attempting to oviposit, she is quickly seized by a male and copulation takes place. One mating provides enough sperm to fertilize several batches of eggs. Afterward, the female will attempt to oviposit while under the guard of the male, either in tandem or without contact with the male.

1.1.2 Importance in Food Webs

Odonates are one of the most dominant and important groups of invertebrate predators. Both nymphs and adults have important roles in natural population control of other macroinvertebrates, while at the same time serving as food for other animals, contributing to food web complexity. Adult odonates feed on mosquitoes, midges, small butterflies, moths, bees, and other odonates; some members of Coenagrionidae (e.g., Pseudostigmatinae) and Aeshnidae (e.g., *Aeschnophlebia longistigma*) are specialized in consuming web-making spiders. In contrast, adult odonates are frequently preyed upon by other invertebrates such as robber flies (Asilidae), spiders, larger odonates, and by some vertebrates such as frogs and birds. Also, they are attacked by external parasites such as gnats (Ceratopogonidae) and mites (Arrenuridae), as well as some internal parasites as viruses (Circoviridae) (Rosario et al. 2011). In the same way, odonate nymphs capture many kinds of small invertebrates such as mosquito larvae or even other odonate nymphs; large odonate nymphs can prey on tadpoles and small fish. They also serve as prey for other animals such as giant water bugs, saucer bugs, shrimp, fish, turtles, and even young crocodiles.

Odonates provide a significant service to humans and their domestic animals by consuming harmful insects that can act as vectors of pathogens such as bloodsucking flies, or insect pests that attack crops in agroecosystems.

1.1.3 Conservation

Natural aquatic ecosystems are complex and exposed to a wide variety of simultaneous stressors with cumulative effects. In recent times, habitat destruction and fragmentation have increased to alarming levels, threatening many sensitive species that require a certain amount of habitat integrity and good water quality to maintain breeding populations. In addition, recent climate change and desiccation, as well as the use of pesticides such as neonicotinoids in developing countries, may further exacerbate the situation.

Several studies have demonstrated the value of using odonate communities or assemblages as indicators of ecosystem integrity, especially in riparian corridors. Degraded habitats tend to maintain a community of very common, generalist, and widely distributed odonate species, whereas well-preserved areas sustain a more diverse community with many specialized odonate species.

The great diversity of Odonata is found in the tropical areas of the world, with much of its biodiversity still unknown to us. It is precisely in these areas where the more intensive and irrational human impacts are destroying the most fragile ecosystems. Thus, it is necessary to highlight the importance of habitat conservation to avoid the loss of more aquatic habitats, and consequently the extinction of species. It is also necessary to raise awareness among politicians, resource managers, and other decision makers before we arrive at the point of environmental no-return, after which communities cannot rebound or be recovered.

1.2 Study Area

The Kingdom of Thailand is a special country for many reasons. It has some of the most spectacular scenery, delicious food, rich culture, friendly people, and exotic biota. It is not surprising that nearly 18% of the country's GDP is from services associated with tourism. Thailand is situated in Southeast Asia and borders Myanmar, Cambodia, Laos, and Malaysia. The coastlines are over 3000 km in length and border the Andaman Sea to the west and the Gulf of Thailand to the east. At 513,120 square kilometers, Thailand is 25% larger in landmass than California in the United States and slightly smaller than France. The country is divided into six regions (Northern, Western, Northeast, Central, Eastern, and Southern) and 77 political units as provinces (Fig. 1.1).

Thailand was known as Siam until 1939, when the name was changed to reflect that it had never been brought under control by any of the European colonizing countries; Thailand means land of the free. Cambodia, Laos, and Vietnam were colonized by France; Myanmar, Malaysia, and Singapore by Britain. Thailand is the only Southeast Asian country to remain independent during the European colonial period.

A chain of mountains extends along the western border of the country from the north near Chiang Rai and Mae Hong Son south to Malaysia (Fig. 1.2). The mountains are higher and more extensive throughout the north and become lower and less continuous southward. Doi Inthanon in Chiang Mai Province is the highest mountain in the country at 2565 m elevation, and only nine other mountains exceed 2000 m, all in the north. Smaller mountain ranges, some of which are isolated, occur in the east near Sakhon Nakhon and southeast near Nakhon Ratchasima and Chantaburi. Most of eastern Thailand is flat and used for agriculture.

Being a tropical country, Thailand experiences a great amount of precipitation, but rainfall varies depending on region and time of year. The Southwest Monsoon is responsible for the rainy season; it brings moisture from the Indian Ocean northeasterly to most parts of Thailand and generally lasts from May to October. In contrast, the Northeast Monsoon passes southwesterly over China, Vietnam, and Laos, where most of its rain falls; thus, by the time this monsoon reaches Thailand from November to March, the air is relatively dry. Just as rainfall amounts vary among regions and time of year, so do temperatures. In general, the lowest temperatures occur in December and January and the highest in April and May. It might seem counterintuitive, but the best collecting for odonate nymphs and other aquatic insects is toward the end of the dry season because the water levels are lower and the insects are able to organize to their preferred habitats in the absence of scouring floods. Also, drier conditions enable easier and safer access to waterbodies.

Agriculture represents 8–10% of Thailand's GDP. Thailand is the world's biggest rice producer and exporter. This is the predominant crop in the lower flat regions, including the Korat Plateau where the fragrant jasmine rice is produced. Other major crops in Thailand include rubber, sugarcane, corn, cassava, and tree fruits. Rubber production can have a detrimental effect on aquatic insect communities because in the post-harvest treatment of rubber, sulfuric acid is used to coagulate the latex. In some unscrupulous operations, the used acid is discarded in streams, which obviously has catastrophic effects on the environment and biota.

Approximately 70 million people live in Thailand with most people in rural areas, and specifically in the rice-growing areas in the northern half of the country. Most of those who live in urban areas are concentrated in Bangkok and the surrounding area. As with most developing countries and with so many people living in rural areas, conflict with a host of environmental concerns is inevitable. In 1961, 53% of Thailand was covered by forests, whereas today that number is 31%. Despite the 1989 logging ban, tree poaching continues to occur, including within national parks and other protected areas. Rural farmers also burn the forest undergrowth in the dry season to remove old vegetation ostensibly to make way for new growth. This practice results in a myriad of detrimental effects, including widespread air pollution, the loss of forest nutrients, and it denudes the forest floor of vegetation, which exacerbates flooding. Deforestation and burning result in the predictable problems of soil erosion, sedimentation in rivers, and associated loss of habitat and biota. An estimate 36 years ago reported that 47 million tons of sediment, mostly eroded topsoil, was carried by Thai rivers annually into the Gulf of Thailand (Arbhabhirama et al. 1988). Further affecting aquatic insect communities is that coastal wetlands and mangroves have been impacted by the construction of a dramatic number of unsustainable commercial shrimp ponds. The ponds have a relatively short life span because of disease and a gradual reduction in productivity, after which the pond is relocated to a newly cleared area of coastline.

Thailand is perhaps more proactive about conservation efforts than most other developing countries. Single species conservation programs are in place that focus on protection for elephants, tigers, clouded leopards, and other species of the charismatic megafauna. Broader conservation programs protect entire ecosystems, including mangroves, coral reefs, and forests. Although laws are in place for protection, penalties can be light and enforcement negligible.

Two major river systems drain most of the country. The Mekong River and associated tributaries drain the eastern part of Thailand toward the southeast and empty into the South China Sea. Most of the land included in this watershed are lowlands, including the Korat Plateau. The Chao Phraya River watershed is the largest in Thailand and includes approximately one third of the country, draining the mountains in the north toward Bangkok and emptying into the Gulf of Thailand. In addition to these lotic systems, numerous standing waterbodies occur throughout the landscape, from small roadside ponds and check-dam ponds to larger wetlands, reservoirs, and lakes.

South of Bangkok is the Isthmus of Kra, which is a narrow neck of land only 40 km wide in Chumphon Province. It separates the Andaman Sea in the west from the Gulf of Thailand to the east. The isthmus also presents a limit to distribution for certain biota, with some taxa on one side or the other, occurring either in the Sundaic zoogeographic region to the south, or the Indochinese region to the north. Most of the biogeographic work has involved mammals, birds, and plants, although a few investigations have considered insects, arachnids, and prawns.

Faunistic studies of Thai aquatic insect taxa have spanned many of the major groups. Most recently, the Trichoptera have been studied by Porntip Chantaramongkol and her many students in Chiang Mai (e.g., Malicky and Chantaramongkol 1999), Ephemeroptera (e.g., Sites et al. 2001; Boonsoong and Braasch 2010; Tungpairojwong and Boonsoong 2011; Boonsoong and Sartori 2016; Auychinda et al. 2020; Boonsoong 2022), several families of Coleoptera by William Shepard (e.g., Shepard and Sites 2016, 2019), Diptera (e.g., Rattanarithikul and Panthusiri 1994; Cranston 2007; Plant et al. 2011), and Heteroptera (e.g., Sites and Vitheepradit 2011). Although a sizable literature exists for Odonata, this predominantly treats adults (e.g., Pinratana et al. 1988; Asahina 1993; Hämäläinen and Pinratana 1999, 2000; Hämäläinen 2002; Yeh 1999; Pinratana 2003; Kitagawa and Katatani 2002, 2003, 2005; Katatani et al. 2004; Katatani et al. 2010; Thipaksorn et al. 2003; Ferro et al. 2009; Kosterin and Vikhrev 2009; Kosterin et al. 2012; Sasamoto 2015; Yanagisawa and Sasamoto 2017; Makbun 2017a, b, 2022; Sribal et al. 2018; Farrell and Makbun 2020; Pierce and Makbun 2020; Buppachat et al. 2020; Kompier et al. 2021) to the near exclusion of the nymphs. Studies in which nymphs occurring in Thailand were described include Matsuki and Kitagawa (1986), Ferro and Sites (2006), Boonsoong and Chainthong (2014a, b), Chainthong and Boonsoong (2016), Makbun and Fleck (2018), Novelo-Gutiérrez and Sites (2019a, b, c), Chainthong et al. (2020), Fleck (2020), Saetung et al. (2020), and Keetapithchayakul et al. (2022), most concerning the Anisoptera family Gomphidae, except Novelo-Gutiérrez and Sites (2019b) which focused on the nymph of a cordulegastrid. Most of the knowledge on the nymphs inhabiting Thailand has been gleaned from descriptions of taxa from other parts of Asia.

Attributes of the aquatic insect community can be used to evaluate water quality, and this inexpensive technique has been applied worldwide, including in Thailand. To provide a broad understanding of the use of aquatic insects in biological monitoring, Mustow (2002) evaluated its use in rivers throughout Thailand, Parnrong (2002) reported on its application in the Mekong River watershed, and in northeastern Thailand, biotic indices (Uttaruk et al. 2011; Sirisinthuwanich et al. 2016) and a multimetrix index (Rattanachan et al. 2016) have been used to evaluate stream quality. Reports by Taeng-On Prommi and her colleagues have evaluated the entire aquatic insect community in particular river systems in northern Thailand (e.g., Payakka and Prommi 2014; Maneechan and Prommi 2015; Prommi and Payakka 2015), whereas others have focused on the response of certain taxa to water quality conditions (e.g., Mustow et al. 1997; Thani and Phalaraksh 2008; Jaihao and Phalaraksh 2013). Use of this technique requires the accurate identification of the insects to various taxonomic levels. Biological signal improves with finer taxonomic resolution, and Odonata and other groups have been evaluated only at the family level. This limitation is in part because of the lack of comprehensive taxonomic literature on odonate nymphs and that the nymphs of many species have not been described. Association of known adults with unknown nymphs can be achieved by matching DNA sequences or by rearing nymphs to adults in the laboratory. More faunistic work on aquatic insects in all groups and production of comprehensive resources for identification are needed to enable resource managers and others to use these communities to assess water quality as well as to enable scientists to address other important questions.

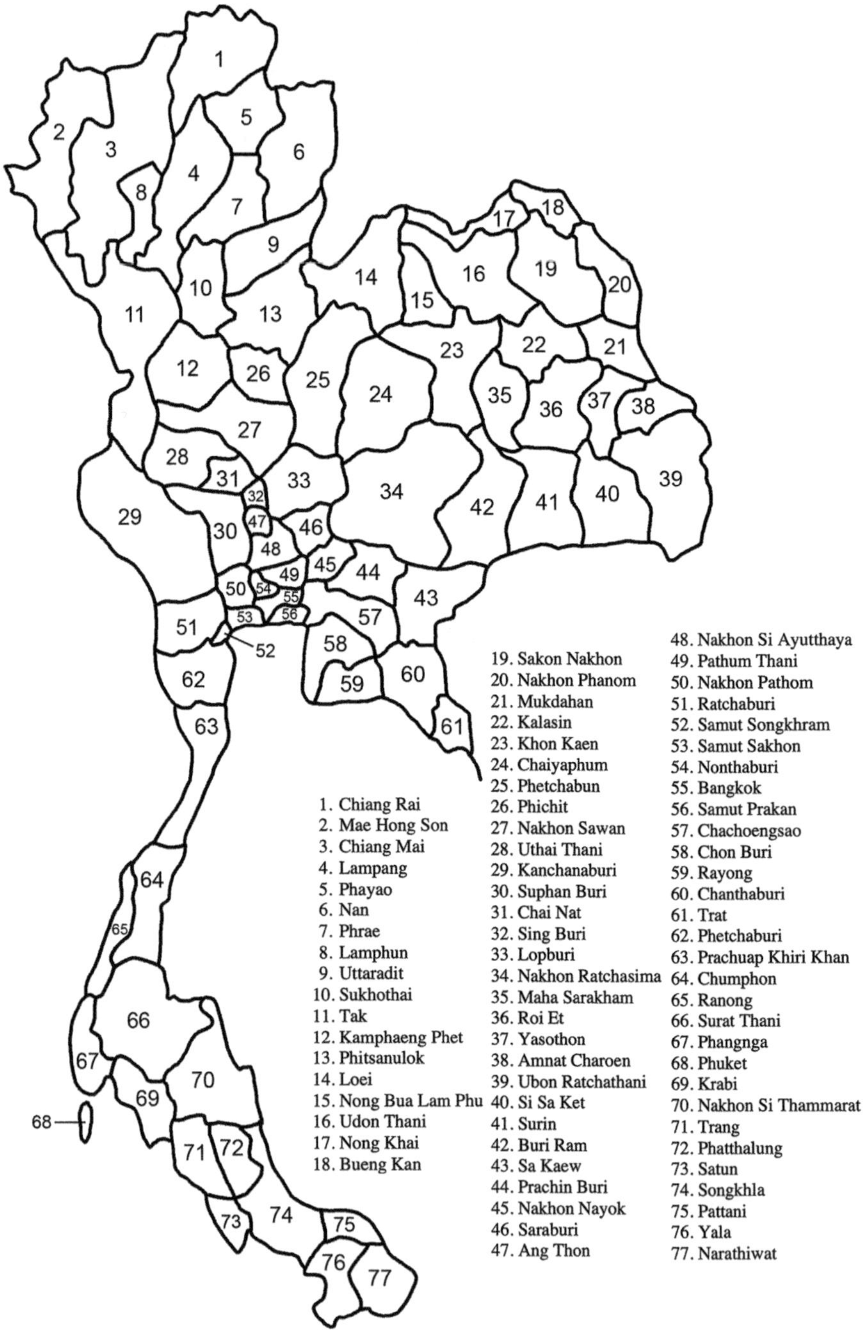

Fig. 1.1 The provinces of Thailand

Fig. 1.2 Relief map of Thailand. Note the northern mountains that extend to the south along the western border and the flat agricultural regions to the east

1.3 A List of the Anisoptera of Thailand (Modified From Hämäläinen and Pinratana 1999) (221 Species Recorded to Date). N = Nymph Described [Citation(s)]

AESHNIDAE

Anaciaeschna jaspidea (Burmeister, 1839)—**N** [Asahina 1957; Lieftinck 1962; Matsuki and Obana 1985; Ishida 1996]

Anaciaeschna martini (Selys, 1897)—**N** [Asahina 1957; Ishida 1996; Tze-wai et al. 2011]

Anax aurantiacus Makbun et al., 2022 [**N** included]

Anax ephippiger (Burmeister, 1839)—**N** [Gerken and Sternberg 1999; Sasamoto et al. 2022]

Anax guttatus (Burmeister, 1839)—**N** [Asahina 1957; Sangal and Kumar 1970; Matsuki 1987; Ishida 1996]

Anax indicus Lieftinck, 1942

Anax nigrofasciatus Oguma, 1915—**N** [Asahina 1974; Ishida 1996; Sasamoto and Ushijima 2000]

Anax panybeus Hagen, 1867—**N** [Asahina 1957, 1974; Matsuki 1987; Ishida 1996]

Cephalaeschna asahinai Karube, 2011

Gynacantha basiguttata Selys, 1882

Gynacantha bayadera Selys, 1891

Gynacantha corbeti Lempert, 1999

Gynacantha demeter Ris, 1911

Gynacantha incisura Fraser, 1935

Gynacantha limbalis Karsch, 1892

Gynacantha phaeomeria Lieftinck, 1960

Gynacantha saltatrix Martin, 1909—**N** [Tze-wai et al. 2011]

Gynacantha subinterrupta Rambur, 1842—**N** [Tze-wai et al. 2011]

Heliaeschna crassa Krüger, 1899

Heliaeschna simplicia (Karsch, 1891)—**N** [Butler and Orr 2013]

Heliaeschna uninervulata Martin, 1909—**N** [Orr and Ngiam 2011]

Indaeschna grubaueri (Förster, 1904)—**N** [Orr 1994, 2005]

Oligoaeschna sirindhornae Ngiam and Orr, 2017

Periaeschna nocturnalis Fraser, 1927

Petaliaeschna flavipes Karube, 1999—**N** [Kawashima and Karube 2006]

Planaeschna chiengmaiensis Asahina, 1981

Planaeschna intersedens (Martin, 1909)

Planaeschna sp.

Polycanthagyna erythromelas (McLachlan, 1896)—**N** [Matsuki and Lien 1985]

Polycanthagyna ornithocephala (McLachlan, 1896)—**N** [Wong et al. 2012]

Sarasaeschna minuta (Asahina, 1986)

Sarasaeschna pramoti (Yeh, 2000)

Tetracanthagyna brunnea McLachlan, 1898

Tetracanthagyna plagiata (Waterhouse, 1877)—**N** [Orr et al. 2010]

Tetracanthagyna waterhousei McLachlan, 1898—**N** [Matsuki 1988a; Wan et al. 2011]

GOMPHIDAE

Acrogomphus malayanus? Laidlaw, 1925—**N** [Lieftinck 1941; Orr 2005]

Amphigomphus somnuki Hämäläinen, 1996—**N** [Novelo-Gutiérrez and Sites 2019a]

Anisogomphus yanagisawai Sasamoto, 2015

Anisogomphus yingsaki Makbun, 2017

Asahinagomphus insolitus (Asahina, 1986)

Asiagomphus xanthenatus (Williamson, 1907)

Burmagomphus arboreus Lieftinck, 1940

Burmagomphus arthuri Lieftinck, 1953
Burmagomphus asahinai Kosterin, Makbun and Dawwrueng, 2012
Burmagomphus chiangmaiensis Makbun, 2017
Burmagomphus divaricatus Lieftinck, 1964—**N** [Lieftinck 1964; Yang and Davies 1996]
Burmagomphus johnseni Lieftinck, 1966
Burmagomphus williamsoni Förster, 1914—**N** [Lieftinck 1964; Orr 2005]
Davidius fruhstorferi Martin, 1904—**N** [Chao 1990]
Davidius sp.
Euthygomphus yunnanensis (Zhou and Wu, 1992)
Gomphidia abbotti Williamson, 1907
Gomphidia kruegeri Martin, 1904—**N** [Chao 1990]
Gomphidia maclachlani Selys, 1873
Gomphidictinus perakensis (Laidlaw, 1902)—**N** [Matsuki 1990; Ferro and Sites 2006]
Heliogomphus selysi Fraser, 1925—**N** [Boonsoong and Chainthong 2014a]
Heliogomphus sp.
Ictinogomphus decoratus (Selys, 1854)—**N** [Orr 2005]
Ictinogomphus rapax (Rambur, 1842)—**N** [Begum et al. 1980; Kumar 1985; Chao 1990; Butler 2007]
Lamelligomphus castor (Lieftinck, 1941)—**N** [Lieftinck 1941]
Lamelligomphus risi (Fraser, 1922)
Lamelligomphus sp. (aff. *castor* Lieftinck, 1941)
Leptogomphus gestroi Selys, 1891
Leptogomphus inclitus Selys, 1878
Leptogomphus risi Laidlaw, 1932
Leptogomphus sp.
Macrogomphus albardae Selys, 1878
Macrogomphus kerri Fraser, 1932
Macrogomphus matsukii Asahina, 1986
Macrogomphus parallelogramma (Burmeister, 1839)
Macrogomphus thoracicus McLachlan, 1884
Mattigomphus pinratani (Hämäläinen, 1991)
Megalogomphus sumatranus (Krüger, 1899)—**N** [Lieftinck 1941 as *M.icterops*]
Megalogomphus sp.
Merogomphus pavici Martin, 1904
Microgomphus chelifer Selys, 1859—**N** [Ngiam 2010 as *Heliogomphus* cf. *retroflexus*]
Microgomphus jurzitzai Karube, 2000
Microgomphus farrelli Makbun and Fleck, 2018 [**N** included]
Microgomphus svihleri (Asahina, 1969)—**N** [Boonsoong and Chainthong 2014b as *M. thailandicus*]
Nepogomphus walli (Fraser, 1924)—**N** [Yang and Davies 1993; Sasamoto and Kawashima 2009a]
Nihonogomphus pulcherrimus (Fraser, 1927)

Nihonogomphus sp.
Nychogomphus duaricus (Fraser, 1924)—**N** [Chainthong and Boonsoong 2016 as *Onychogomphus*]
Nychogomphus lui Zhou, Zhou, and Lu, 2005
Onychogomphus kerri Fraser, 1933
Onychogomphus (Siriusonychogomphus) louissiriusi Fleck, 2020—**N** [Novelo-Gutiérrez and Che Salmah 2013 as *Onychogomphus* sp.; Fleck 2020]
Orientogomphus minor (Laidlaw, 1931)—**N** [Chainthong and Boonsoong 2022]
Paragomphus capricornis (Förster, 1914)—**N** [Wilson 1995]
Phaenandrogomphus asthenes Lieftinck, 1964—**N** [Lieftinck 1964; Novelo-Gutiérrez and Sites 2019c]
Phaenandrogomphus dingavani Fraser, 1924
Phaenandrogomphus tonkinicus (Fraser, 1926)—**N** [Novelo-Gutiérrez and Sites 2019c]
Sieboldius japponicus Selys, 1854—**N** [Lieftinck 1932; Orr 2005]
Sieboldius nigricolor (Fraser, 1924)
Sinictinogomphus clavatus (Fabricius, 1775)—**N** [Chao 1990; Ishida 1996]
Stylogomphus malayanus Sasamoto, 2001 [**N** included]
Stylogomphus thongphaphumensis Chaithong, Sartori and Boonsoong, 2020 [**N** included]

CHLOROGOMPHIDAE
Chlorogomphus arooni Asahina, 1981—**N** [Orr 2005]
Chlorogomphus miyashitai Karube, 1995
Chlorogomphus piaoacensis Karube, 2013
Chlorogomphus yokoii Karube, 1995—**N** [Zhang and Tong 2013]
Chlorogomphus sp. 1
Chlorogomphus sp. 2
Watanabeopetalia atkinsoni (Selys, 1878)—**N** [Asahina 1961]

CORDULEGASTRIDAE
Anotogaster gregoryi Fraser, 1924—**N** [Novelo-Gutiérrez and Sites 2019b]

MACROMIIDAE
Epophthalmia frontalis Selys, 1871—**N** [Fraser 1919]
Epophthalmia vittigera (Rambur, 1842)—**N** [Lieftinck 1931; Orr 2005]
Macromia aculeata Fraser, 1927
Macromia arachnomima Lieftinck, 1953—**N** [Lieftinck 1953; Orr 2005]
Macromia callisto Laidlaw, 1922
Macromia chaiyaphumensis Hämäläinen, 1985
Macromia cupricincta Fraser, 1924
Macromia cydippe Laidlaw, 1922—**N** [Lieftinck 1950, 1955; Orr 2005]
Macromia erato Lieftinck, 1950—**N** [Lieftinck 1950, 1955]
Macromia flavocolorata Fraser, 1922—**N** [Xu 2010]
Macromia moorei Selys, 1874—**N** [Lieftinck 1950, 1955; Kumar 1973; Orr 2005]

Macromia pinratani Asahina, 1983
Macromia septima Martin, 1904—**N** [Xu 2010]
Macromia siamensis Makbun, 2022
Macromia vangviengensis Yokoi and Mitamura, 2002
Macromia spp. (six undescribed species)

CORDULIIDAE

Hemicordulia tenera tenera Lieftinck, 1930
Hemicordulia tenera donnellyi Kosterin, Karube and Futahashi, 2015
Procordulia artemis Lieftinck, 1930 [**N** included]
Somatochlora sp.

LIBELLULIDAE

Acisoma panorpoides Rambur, 1842—**N** [Kumar 1984; Ishida 1996; Orr 2005; Tze-wai et al. 2011]
Aethriamanta aethra Ris, 1912—**N** [Tze-wai et al. 2011]
Aethriamanta brevipennis (Rambur, 1842)
Aethriamanta gracilis (Brauer, 1878)
Agrionoptera insignis (Rambur, 1842)—**N** [Lieftinck 1962; Ishida 1996]
Agrionoptera sexlineata Selys, 1879
Amphithemis curvistyla Selys, 1891
Atratothemis reelsi Wilson, 2005
Brachydiplax chalybea Brauer, 1868—**N** [Ishida 1996; Orr 2005]
Brachydiplax farinosa Krüger, 1902
Brachydiplax sp.
Brachygonia oculata (Brauer, 1878)
Brachythemis contaminata (Fabricius, 1793)—**N** [Kumar 1973; Ishida 1996]
Bradinopyga geminata (Rambur, 1842)—**N** [Sangal and Kumar 1970]
Camacinia gigantea (Brauer, 1867)—**N** [van Tol 1992; Theischinger and Hawking 2006]
Camacinia harterti Karsch, 1890
Chalybeothemis fluviatilis Lieftinck, 1933
Cratilla lineata (Brauer, 1878)—**N** [Lien and Matsuki 1988]
Cratilla metallica (Brauer, 1878)—**N** [Orr 2005]
Crocothemis servilia (Drury, 1770)—**N** [Sangal and Kumar 1970; Ishida 1996]
Diplacodes nebulosa (Fabricius, 1793)
Diplacodes trivialis (Rambur, 1842)—**N** [Lieftinck 1962; Kumar 1984; van Tol 1992; Ishida 1996]
Hydrobasileus croceus (Brauer, 1867)—**N** [van Tol 1992; Ishida 1996; Orr 2005]
Hylaeothemis clementia Ris, 1909—**N** [Sasamoto and Kawashima 2009b]
Indothemis carnatica (Fabricius, 1798)
Indothemis limbata (Selys, 1891)
Lathrecista asiatica (Fabricius, 1798)—**N** [Adambukulam and Kakkassery 2013]
Lyriothemis biappendiculata (Selys, 1878)
Lyriothemis bivittata (Rambur, 1842)—**N** [Matsuki 1993]

Lyriothemis cleis Brauer, 1868—**N** [Orr 2005]
Lyriothemis elegantissima Selys, 1883—**N** [Ishida 1996]
Lyriothemis magnificata (Selys, 1878)—**N** [Lieftinck 1962]
Lyriothemis mortoni Ris, 1909
Lyriothemis pachygastra (Selys, 1878)—**N** [Ishida 1996]
Lyriothemis pallidistigma Kompier, Holden and Makbun, 2021
Macrodiplax cora (Brauer, 1867)—**N** [Lieftinck 1962; Ishida 1996; Orr 2005]
Nannophya pygmaea Rambur, 1842—**N** [Ishida 1996; Orr 2005; Tze-wai et al. 2011]
Nannophyopsis clara (Needham, 1930)—**N** [Yeh and Lien 1995]
Nesoxenia lineata (Selys, 1879)
Neurothemis fluctuans (Fabricius, 1793)
Neurothemis fulvia (Drury, 1773)
Neurothemis intermedia (Rambur, 1842)—**N** [Suri Babu 2000; Tze-Wai et al. 2011]
Neurothemis ramburii (Kaup in Brauer, 1866)—**N** [Ozono et al. 2013; Watanabe et al. 2013]
Neurothemis tullia (Drury, 1773)—**N** [Kumar 1988; Matsuki 1988b]
Onychothemis culminicola Förster, 1904—**N** [Orr 2005]
Onychothemis testacea Laidlaw, 1902
Orchithemis pulcherrima Brauer, 1878—**N** [Butler 2013]
Orchithemis pruinans (Selys, 1878)
Orthetrum chrysis (Selys, 1891)—**N** [Tze-Wai et al. 2011]
Orthetrum glaucum (Brauer, 1865)—**N** [Ishida 1996; Tze-Wai et al. 2011]
Orthetrum internum McLachlan, 1894—**N** [Ishida 1996]
Orthetrum luzonicum (Brauer, 1868)—**N** [Ishida 1996; Tze-Wai et al. 2011]
Orthetrum pruinosum (Hangenber in Burmeister, 1839)—**N** [Kumar 1972; Ishida 1996]
Orthetrum sabina (Drury, 1770)—**N** [Lieftinck 1962; Kumar 1973; Ishida 1996; Orr 2005]
Orthetrum schneideri Förster, 1903
Orthetrum testaceum (Burmeister, 1839)—**N** [Chowdhury and Akhteruzzaman 1981]
Orthetrum triangulare (Selys, 1878)—**N** [Khaliq et al. 1995; Ishida 1996; Tze-Wai et al. 2011]
Palpopleura sexmaculata (Fabricius, 1787)—**N** [Tze-Wai et al. 2011]
Pantala flavescens (Fabricius, 1798)—**N** [Cabot 1890; Geijskes 1934; Lieftinck 1962; Kumar 1973]
Phyllothemis eltoni Fraser, 1935
Potamarcha congener (Rambur, 1842)—**N** [Kumar 1973; van Tol 1992]
Pseudothemis jorina Förster, 1904
Pseudothemis zonata (Burmeister, 1839)—**N** [Ishida 1996]
Pseudotramea prateri Fraser, 1920

Rhodothemis rufa (Rambur, 1842)—**N** [Chowdhury and Akhteruzzaman 1981; van Tol 1992]

Rhyothemis obsolescens Kirby, 1889

Rhyothemis phyllis (Sulzer, 1776)—**N** [Lieftinck 1962]

Rhyothemis plutonia Selys, 1883

Rhyothemis triangularis Kirby, 1889—**N** [Orr 2005; Tze-Wai et al. 2011]

Rhyothemis variegata (Linnaeus, 1763)—**N** [Chowdhury and Akhteruzzaman 1981; Ishida 1996; Hussain and Riaz 1999]

Risiophlebia dohrni (Krüger, 1902)

Risiophlebia guentheri Kosterin, 2015

Sympetrum hypomelas (Selys, 1884)

Sympetrum thailandensis Makbun, 2023

Tetrathemis irregularis Brauer, 1868—**N** [Orr 2005 as *Tetrathemis* sp.]

Tetrathemis platyptera Selys, 1878—**N** [van Tol 1992]

Tholymis tillarga (Fabricius, 1798)—**N** [Fraser 1919; Lieftinck 1962; Chowdhury and Akhteruzzaman 1981; Ishida 1996]

Tramea basilaris (Palisot de Beauvois, 1817)—**N** [Kumar 1973]

Tramea transmarina Brauer, 1867—**N** [Lieftinck 1962; Ishida 1996]

Tramea virginia (Rambur, 1842)—**N** [Kumar and Prasad 1977; Ishida 1996; Tze-Wai et al. 2011]

Trithemis aurora (Burmesiter, 1839)—**N** [Kumar 1973; Ishida 1996; Orr 2005]

Trithemis festiva (Rambur, 1842)—**N** [Kumar 1972]

Trithemis pallidinervis (Kirby, 1889)—**N** [Khaliq et al. 1994]

Tyriobapta torrida Kirby, 1889

Urothemis abbotti Laidlaw, 1927

Urothemis signata (Rambur, 1842)—**N** [Chowdhury and Akhteruzzaman 1981; Tze-Wai et al. 2011]

Zygonyx ida Selys, 1869

Zygonyx iris Selys, 1869—**N** [St. Quentin 1973; Matsuki 1988c; Matsuki and Kitagawa 1986; Orr 2005]

Zyxomma breviventre (Martin, 1921)

Zyxomma petiolatum Rambur, 1842—**N** [Lieftinck 1962; Kumar 1973; Ishida 1996]

INCERTAE SEDIS

Idionyx iida Hämäläinen, 2002

Idionyx optatus Selys, 1878

Idionyx selysi Fraser, 1926—**N** [Zhang and Tong 2010]

Idionyx thailandicus Hämäläinen, 1985

Idionyx sp. (*yolanda* species group)—**N** [Lieftinck 1971; Orr 2005 (*I. montana*)]

Macromidia genialis Laidlaw, 1923—**N** [Matsuki 1989; Orr 2005]

Macromidia rapida Martin, 1907

References

Adambukulam SP, Kakkassery FK. Taxonomic studies of the last instar nymph of *Lathrecista asiatica asiatica* (Fabricius 1798) (Family: Libellulidae, Order: Odonata) by using its exuvia. J Entomol Zool Stud. 2013;1(5):103–9.

Arbhabhirama A, Phantumvanit D, Elkington J, Ingkasuwan P. Thailand Natural Resources Profile. Oxford: Oxford University Press; 1988.

Asahina S. Material for the fauna of Japanese Odonata, XII. Shin-Kontyu. 1957;10(8):49–55.

Asahina S. Descriptions of some dragonfly larvae from Darjeeling. Kontyu. 1961;29(4):240–6.

Asahina S. Diagnostic notes on the ultimate instar larvae of some *Anax* species. Tombo. 1974;17:10–6.

Asahina S. A list of the Odonata from Thailand (parts I–XXI). Bangkok: Brothers of Saint Gabriel in Thailand; 1993.

Auychinda C, Sartori M, Boonsoong B. Vietnamellidae (Insecta, Ephemeroptera) of Thailand. ZooKeys. 2020;902:17–36.

Begum A, Bashar MA, Biswas BR. Life history of *Ictinogomphus rapax* (Rambur) (Anisoptera: Gomphidae). Bangladesh J Zool. 1980;8(1):53–60.

Boonsoong B. Mayfly larvae (Ephemeroptera) in Thailand. Bangkok: Pre-One Limited Company; 2022.

Boonsoong B, Braasch D. Heptageniidae (Insecta, Ephemeroptera) of Thailand. ZooKeys. 2010;272:61–93.

Boonsoong B, Sartori M. *Sangpradubina*, an astonishing new mayfly genus from Thailand (Ephemeroptera: Leptophlebiidae: Atalophlebiinae). Zootaxa. 2016;4169(3):587. https://doi.org/10.11646/zootaxa.4169.3.11.

Boonsoong B, Chainthong D. Description of the final-instar larva of *Heliogomphus selysi* Fraser (Odonata: Gomphidae). Zootaxa. 2014a;3764(4):482–8.

Boonsoong B, Chainthong D. Description of the last stadium larva and female of *Microgomphus thailandica* Asahina, 1981 (Odonata: Gomphidae). Zootaxa. 2014b;3811(2):271–9.

Buppachat N, Inkaew P, Thitiarchagul T, et al. Additions to the Odonata fauna of Thailand (Aeshnidae, Devadattidae, Euphaeidae, Gomphidae, Libellulidae, Platycnemididae). Agrion. 2020;24(3):184–91.

Butler SG. The larva of *Gomphidia t-nigrum* Selys from Nepal (Anisoptera: Gomphidae). Odonatologica. 2007;36(4):399–403.

Butler SG. Description of the last instar larva of *Orchithemis pulcherrima* Brauer from Sarawak, Malaysia (Anisoptera: Libellulidae). Odonatologica. 2013;42(3):247–51.

Butler SG, Orr AG. The larva of *Heliaeschna simplicia* Karsch, 1891 (Anisoptera: Aeshnidae). Odonatologica. 2013;42(2):151–6.

Cabot L. The immature state of the Odonata. Part 3. Subfamily Cordulina. Mem Mus Comp Zool Harvard Coll. 1890;17:1–52.

Chainthong D, Boonsoong B. Description of two final stadium *Onychogomphus* larvae from Thailand (Odonata: Gomphidae). Zootaxa. 2016;4066(5):561–70.

Chainthong D, Boonsoong B. Taxonomy and Distribution of the Gomphid Dragonfly *Orientogomphus minor* (Laidlaw, 1931) (Odonata: Gomphidae) in Thailand. Diversity. 2022;14(291):1–12. https://doi.org/10.3390/d14040291.

Chainthong D, Sartori M, Boonsoong B. *Stylogomphus thongphaphumensis* (Odonata: Anisoptera: Gomphidae), a new gomphid dragonfly and the first record of *S. malayanus* Sasamoto, 2001 from Thailand. Zootaxa. 2020;4763(2):231–45.

Chao X-F. The Gomphid Dragonflies of China (Odonata: Gomphidae). Fuzhou, Fujian: The Science and Technology Publishing House; 1990.

Chowdhury SH, Akhteruzzaman MD. Dragonfly (Odonata: Anisoptera) larvae from Chittagong. Bangladesh J Zool. 1981;9(2):131–44.

Cranston PS. The Chironomidae larvae associated with the tsunami-impacted waterbodies of the coastal plain of southwestern Thailand. Raffles Bull Zool. 2007;55(2):231–44.

Farrell D, Makbun N. First records of three Lestidae species in Thailand, with a checklist of the dragonflies known from Nam Nao National Park and Phu Khieo Wildlife Sanctuary (Odonata). Faunistic Stud South-East Asian and Pacific Islands Odonata. 2020;32:1–12.

Ferro ML, Sites RW. Description of the larva of *Gomphidictinus perakensis* (Laidlaw) (Odonata: Gomphidae), with distributional notes. Proc Ent Soc Washington. 2006;108(1):76–81.

Ferro ML, Sites RW, Vitheepradit A. Contributions to the faunistics of Odonata in Thailand. Insecta Mundi. 2009;104:1–24.

Fleck G. *Onychogomphus (Siriusonychogomphus) louissiriusi*, a new species and new subgenus from Thailand (Odonata: Anisoptera: Gomphidae). Faunitaxys. 2020;8(7):1–9.

Fraser FC. Descriptions of new Indian odonate larvae and exuviae. Rec Indian Mus (B). 1919;16(7):459–67.

Geijskes DC. Notes on the Odonata fauna of the Dutch west Indian Islands Aruba, Curaçao and Bonaire with an account on their nymphs. Int Rev Gesam Hydrobiol Hydrogr. 1934;31:284–311.

Gerken B, Sternberg K. The exuviae of European dragonflies. Höxter und Jena: Arnika and Eisvogel; 1999.

Hämäläinen M. The species list of Thai dragonflies increase steadily—an update. Malangpo. 2002;19:176–9.

Hämäläinen M, Pinratana BA. Atlas of the dragonflies of Thailand. Distribution maps by provinces. Bangkok: Brothers of Saint Gabriel in Thailand; 1999.

Hämäläinen M, Pinratana BA. Additions and corrections to dragonfly lists of five protected areas in Thailand. Malangpo. 2000;17:156–7.

Hussain R, Riaz M. Description of last instar naiads of *Rhyothemis variegata variegata* (Linnaeus) and *Pantala flavescens* (Fabricius) (Libellulidae: Odonata). Int J Agric Biol Eng. 1999;1:145–6.

Ishida K. Monograph of Odonata larvae in Japan. Sapporo: Hokkaido University Press; 1996.

Jaihao R, Phalaraksh C. Relationship between water quality and distribution of stonefly larvae in pristine areas at Huai Nam Dung National Park, Thailand. Khon Kaen University Sci J. 2013;41(3):709–22.

Katatani N, Kitagawa K, Muraki A, Matuzawa T. Notes on Thai Odonata, Part 3. The Odonata from Chiang Rai. Aeschna. 2004;41:59–70.

Katatani N, Kitagawa K, Muraki A, et al. Notes on Thai Odonata, Part 5. The Odonata from Nan and Phrae. Aeschna. 2010;46:48–58.

Kawashima I, Karube H. External morphology of the last instar larva of probable *Petaliaeschna flavipes* Karube (Anisoptera, Aeshnidae, Brachytoninae) from Laos, Indochina. Tombo Matsumoto. 2006;48(1/4):7–11.

Keetapithchayakul TS, Makbun N, Phan QT, et al. Description of the larva of *Indocnemis orang* (Förster in Laidlaw, 1907) (Odonata: Platycnemididae: Calicnemiinae) from Thailand, with larval key to the known genera of the family Platycnemididae in Asia. Zootaxa. 2022;5134(4):504–20.

Khaliq A, Aslam S, Anjum SA. The naiads of *Orthetrum pruinosum neglectum* (Rambur) and *Trithemis pallidinervis* (Kirby) (Libellulidae: Odonata). Pakistan Entomol. 1994;16:51–3.

Khaliq A, Murtaza HG, Khan MR. Description of the naiads of three species of genus *Orthetrum* (Libellulidae: Odonata). Pakistan J Zool. 1995;27(3):245–7.

Kitagawa K, Katatani N. Notes on Thai Odonata, Part 1. The Odonata from Chanthaburi in March 1998. Aeschna. 2002;39:33–42.

Kitagawa K, Katatani N. Notes on the Odonata of Thailand, Part 2. The Odonata from Trang Prefecture in March 1998. Aeschna. 2003;40:13–23.

Kitagawa K, Katatani N. Notes on Thai Odonata, Part 4. The Odonata from Kanchanaburi. Aeschna. 2005;42:23–30.

Kompier T, et al. A new species of *Lyriothemis* Brauer, 1868 from Vietnam and Thailand (Odonata: Libellulidae). Zootaxa. 2021;4933(4):586–94.

Kosterin OE, Vikhrev NE. Some new provincial records of Odonata made in Thailand in 2007–2009 and a new record for Vietnam. Agrion. 2009;13(2):75–9.

Kosterin OE, Karube H, Futahashi R. Two new subspecies of *Hemicordulia tenera* Lieftinck, 1930 (Corduliidae) from Cambodia and Thailand. Int Dragonfly Fund Rep. 2015;82:1–19.

Kosterin OE, Makbun N, Dawwrueng P. *Burmagomphus asahinai* sp. nov., a new species from Cambodia and Thailand, with a description of the male of *B. gratiosus* Chao, 1954. Int J Odonatol. 2012;15(4):275–92.

Kumar A. Bionomics of Orthetrum pruinosum neglectum (Rambur) (Odonata: Libellulidae). Bull Ent. 1972;11(1):85–93.

Kumar A. Description of the last instar larvae of Odonata from Dehra Dun Valley (India), with notes on biology, II (Suborder: Anisoptera). Orient Insects. 1973;7(2):291–331.

Kumar A. Studies on the life history of Indian dragonflies, *Acisoma panorpoides panorpoides* Rambur, 1842 (Odonata: Libellulidae). Rec Zool Surv India. 1984;81(3–4):203–13.

Kumar A. Studies on the life history of Indian dragonflies, *Ictinogomphus rapax* (Rambur) (Gomphidae: Odonata). Ann Entomol. 1985;3(1):29–3.

Kumar A. Studies on the life history of *Neurothemis t. tullia* (Drury) from Dehra Dun, India (Odonata: Libellulidae). Indian J Odonatol. 1988;1:35–44.

Kumar A, Prasad M. On the larvae of *Tramea virginia* (Rambur, 1842) from India, with notes on the larvae of Indian representatives of the genus *Tramea* Hagen, 1861 (Libellulidae: Odonata). J Bombay Nat Hist Soc. 1977;74(1):199–202.

Lieftinck MA. Contributions of the dragonfly-fauna of the Dutch East indies. Treubia. 1930;12(2):135–66.

Lieftinck, MA. A revision of the genus *Epophthalmia* Burm. (Odonata: Corduliinae), with notes on habits and larvae. Treubia. 1931;13(1):21–80.

Lieftinck MA. Notes on the larvae of two interesting Gomphidae (Odon.) from the Malay peninsula. Bull Raffles Mu. 1932;7:102–15.

Lieftinck MA. Studies on Oriental Gomphidae (Odon.), with descriptions of new and interesting larvae. Treubia. 1941;18:233–53.

Lieftinck MA. On South East Asiatic species of *Macromia*. Treubia. 1950;20:677–89.

Lieftinck MA. New dragonflies (Odonata) from Borneo, with notes on their habits and larvae. Treubia. 1953;22:381–406.

Lieftinck MA. Further inquiries into the Old World species of *Macromia* Rambur (Odonata). Zool Med. 1955;33:251–77.

Lieftinck MA. Odonata. Insects Micronesia. 1962;5(1):1–95.

Lieftinck MA. Some Gomphidae and their larvae, chiefly from the Malay Peninsula (Odonata). Zool Ver. 1964;69:1–38.

Lieftinck MA. Studies in Oriental Corduliidae (Odonata) I. Tijdschr Ent. 1971;114(1):1–63.

Lien JC, Matsuki K. Description of the larva of *Cratilla lineata assidua* Lieftinck from Taiwan (Libellulidae: Odonata). Tombo. 1988;31(1–4):35–6.

Makbun N. A new species of the genus *Burmagomphus* Williamson (Odonata: Gomphidae) from Northern Thailand. Zootaxa. 2017a;4269(1):133–6.

Makbun N. *Anisogomphus yingsaki* (Odonata: Gomphidae) sp. nov., a new gomphid species from Thailand. Zootaxa. 2017b;4306(3):437–43.

Makbun N. *Macromia siamensis*, a new species from North Thailand (Odonata: Macromiidae). Zootaxa. 2022;5133(3):346–54.

Makbun N. The genus *Sympetrum* Newman, 1833 in Thailand, with description of *S. thailandensis* sp. nov. (Odonata: Libellulidae). Zootaxa. 2023;5296(4):569–81.

Makbun N, Fleck G. Description of *Microgomphus farrelli* sp. nov. (Odonata: Anisoptera: Gomphidae) based on adults of both sexes and larvae from Northern Thailand. Zootaxa. 2018;4422(3):442–50.

Makbun N, Wongkamhaeng K, Keetapithchayakul TS. *Anax aurantiacus* sp. nov., a new dragonfly from mainland Southeast Asia (Odonata: Aeshnidae). Odonatologica. 2022;51(3/4):301–39.

Malicky H, Chantaramongkol P. A preliminary survey of the caddisflies (Trichoptera) of Thailand. Proceedings of the 9th International Symposium of Trichoptera, 205–216. 1999.

Maneechan W, Prommi T-O. Diversity and distribution of aquatic insects in streams of the Mae Klong Watershed, western Thailand. Psyche. 2015;2015:1–7.

Matsuki K. Description of the larvae of four species of the genus *Anax* in Taiwan (Odonata: Aeshnidae). Tombo. 1987;30:25–32.

Matsuki K. Description of the larva of *Tetracanthagyna waterhousei* (McLachlan) from Hongkong and Thailand (Aeshnidae; Odonata). Tombo. 1988a;31:37–40.

Matsuki K. Description of the larva of *Neurothemis tullia tullia* obtained from Taiwan (Libellulidae) Tombo. Tokyo. 1988b;41:29–30.

Matsuki K. Description of the larva of *Zygonyx iris insignis* (Kirby) from Hongkong (Libellulidae, Odonata). Gekkan-Mushi. 1988c;210:24–5.

Matsuki K. Description of the larva of *Macromidia genialis shanensis* Fraser from Thailand (Corduliidae, Odonata). Gekkan-Mushi. 1989;219:28–9.

Matsuki K. Description of the larva of *Gomphidia perakensis* Laidlaw from Thailand (Gomphidae, Odonata.). Gekkan-Mushi. 1990;228:32–3.

Matsuki K. Description of the larva of *Lyriothemis bivittata* (Rambur, 1842) from Thailand (Libellulidae: Odonata). Tombo. 1993;36:25–8.

Matsuki K, Lien JC. Description of the larva of *Polycanthagyna erythromelas paiwan* Asahina from Taiwan (Aeshnidae: Odonata). Gekkan-Mushi. 1985;173:9–11.

Matsuki K, Kitagawa K. Description of the possible larva of *Zygonyx iris malayanus* (Laidlaw) from Thailand (Libellulidae: Odonata). Gekkan-Mushi. 1986;183:24–5.

Matsuki K, Obana S. Description of the larva of *Anaciaeschna jaspidea* (Burmeister) from Japan and Taiwan (Odonata: Aeshnidae). Gekkan-Mushi. 1985;175:29–31.

Mustow SE. Biological monitoring of rivers in Thailand: use and adaptation of the BMWP score. Hydrobiologia. 2002;479:191–229.

Mustow SE, Wilson RS, Sannarm G. Chironomid assemblages in two Thai watercourses in relation to water quality. Nat Hist Bull Siam Soc. 1997;45:53–64.

Novelo-Gutiérrez R, Che Salmah MR. Two interesting larvae of *Onychogomphus* from Malaysia (Anisoptera: Gomphidae). Odonatologica. 2013;42(1):31–8.

Novelo-Gutiérrez R, Sites RW. The larva of *Amphigomphus somnuki* Hämäläinen, 1996 and the first records of the genus *Stylogomphus* Fraser, 1922 for Thailand (Odonata: Gomphidae). Zootaxa. 2019a;4555(1):121–6.

Novelo-Gutiérrez R, Sites RW. The probable larva of *Anotogaster gregoryi* Fraser, 1923, with new distributional records of the genus from northern Thailand (Odonata: Cordulegastridae). Zootaxa. 2019b;4565(1):138–44.

Novelo-Gutiérrez R, Sites RW. The larvae of *Phaenandrogomphus* Lieftinck, 1964 in Thailand, including the description of *P. tonkinicus* (Fraser, 1926) with larval diagnosis and new province records of *P. asthenes* Lieftinck, 1964 (Odonata: Gomphidae). Zootaxa. 2019c;4700(3):377–84.

Orr AG. Life histories and ecology of Odonata breeding in phytotelmata in Bornean rainforest. Odonatologica. 1994;23(4):365–77.

Orr AG. Dragonflies of peninsular Malaysia and Singapore. A pocket guide. Kota Kinabalu, Sabah: Natural History Publications (Borneo); 2005.

Orr AG, Ngiam RWJ. A description of the larva of *Heliaeschna uninervulata* Martin (Odonata: Aeshnidae) from Singapore, with notes on its relationships. Int J Odonatol. 2011;14(2):163–9.

Orr AG, Ngiam RWJ, Leong TM. The larva of *Tetracanthagyna plagiata*, with notes on its biology and comparisons with congeneric species (Odonata: Aeshnidae). Int J Odonatol. 2010;13(2):153–66.

Ozono A, Kawashima I, Futahashi R. Dragonflies of Japan. 2nd ed. Tokyo: Bun-Ichi-Sôgô-Shuppan Publ. Co.; 2013.

Parnrong S. A review of the biological assessment of freshwater ecosystems in Thailand. Report to Mekong River Commission Environment Program. 2002: 4–31.

Paulson DR, Schorr M, Abbott, J et al. (Coordinators). World Odonata List. OdonataCentral, University of Alabama. 2023. https://www.odonatacentral.org/app/#/wol/. Accessed 18 July 2023.

Payakka A, Prommi T-O. Aquatic insects biodiversity and water quality parameters of receiving water body. Current World Environ. 2014;9:53–8.

Pierce AJ, Makbun N. First record of *Gynacantha limbalis* Karsch, 1892 from Thailand (Odonata: Anisoptera: Aeshnidae). Songklanakarin J Sci Technol. 2020;42(5):965–6.

Pinratana A, Kiauta B, Hämäläinen M. List of the Odonata of Thailand and annotated bibliography. Bangkok: Brothers of St. Gabriel in Thailand; 1988.

Pinratana A. Editorial. Malangpo. 2003;20:195.

Plant AR, Surin C, Saokhod R, Srisuka W. Higher taxon diversity, abundance and community structure of Empididae, Hybotidae and Brachystomatidae (Diptera: Empidoidea) in tropical forests—results of mass-sampling in Thailand. Stud Dipterol. 2011;18:121–49.

Prommi T-O, Payakka A. Aquatic insect biodiversity and water quality parameters of streams in northern Thailand. Sains Malaysiana. 2015;44:707–17.

Rattanachan K, Boonsoong B, Getwongsa P, Uttaruk Y, Sangpradub N. A benthic macroinvertebrate multimetric index for assessment of the ecological integrity of northeast streams, Thailand. Environ Asia. 2016;9(2):186–94.

Rattanarithikul R, Panthusiri P. Illustrated keys to the medically important mosquitos of Thailand. Southeast Asian J Trop Med Public Health. 1994;25 suppl 1:66.

Rosario K, Marinov M, Stainton D, et al. Dragonfly cyclovirus, a novel single-stranded DNA virus discovered in dragonflies (Odonata: Anisoptera). J Gen Virol. 2011;92(6):1302–8.

Saetung T, Makbun N, Sartori M, Boonsong B. The subfamily Platycnemidinae (Zygoptera: Platycnemididae) in Thailand, with description of the final stadium larva of *Copera chantaburii* Asahina, 1984. Int J Odonatol. 2020;23(3):219–37.

Sangal SK, Kumar A. Studies on the taxonomy of larvae of Doon Valley Odonata 1. *Crocothemis servilia servilia* (Drury) and *Bradinopyga geminata* (Rambur) (family Libellulidae). J Nat Hist. 1970;4:33–8.

Sasamoto A. Description of a new subspecies of *Stylogomphus lawrenceae* Yang et Davies, 1996 from the Malay Peninsula (Anisoptera: Gomphidae). Tombo. 2001;43:14–8.

Sasamoto A. *Anisogomphus yanagisawai* sp. nov., a new gomphid dragonfly from northern Thailand (Odonata: Anisoptera: Gomphidae). Zootaxa. 2015;3904(3):421–6.

Sasamoto A, Kawashima I. Description of the last instar larva of *Nepogomphus walli* (Fraser, 1924) (Anisoptera, Gomphidae, Onychogomphinae) from West Malaysia, southeastern Asia. Tombo Osaka. 2009a;52:7–12.

Sasamoto A, Kawashima I. Description of the last instar larva of *Hylaeothemis clementia* Ris from Laos (Anisoptera: Libellulidae). Odonatologica. 2009b;38(4):369–74.

Sasamoto A, Ushijima K. Notes on the larva of *Anax nigrofasciatus nigrolineatus* Fraser. Tombo. 2000;42:46–8.

Sasamoto A, Kotabe A, Takuma Y, et al. Description of larva of *Anax ephippiger* (Burmeister, 1839) from Japan, including changes and developments in external morphology (Odonata: Aeshnidae). Tombo. 2022;65:28–37.

St. Quentin D. Results of the Austrian Ceylonese Hydrobiological Mission 1970 of the 1st Zoological Institute of the University of Vienna (Austria) and the Department of Zoology of the Vidyalankara University of Ceylon, Kelaniya. Part XII: Contributions to the Ecology of the Larvae of some Odonata from Ceylon. Bull Fish Res Stat Sri Lanka. 1973;24(1–2):113–24.

Shepard WD, Sites RW. Aquatic beetles of the families Dryopidae and Elmidae (Insecta: Coleoptera: Byrrhoidea) of Thailand: Annotated list and illustrated key to genera. Nat Hist Bull Siam Soc. 2016;61(2):89–126.

Shepard WD, Sites RW. Larval Psephenidae (Coleoptera: Byrrhoidea) of Thailand: annotated list and illustrated key to genera. Coleop Bull. 2019;73(2):259–82.

Sirisinthuwanich K, Sangpradub N, Hanjavanit C. Development of biotic index to assess the Phong and Cheon rivers' healths based on benthic macroinvertebrates in northeastern Thailand. 2016

Sites RW, Vitheepradit A. The *Heleocoris* (Heteroptera: Naucoridae: Laccocorinae) of Thailand, with the description of a new species. Zootaxa. 2011;2736:1–16.

Sribal U, Paweenpermsuk Y, Thitiarchagul T, et al. New national records of Odonata from Thailand based mostly on photographs (Odonata: Argiolestidae, Philosinidae, Aeshnidae, Libellulidae). Agrion. 2018;22(1):30–6.

Sites RW, Wang TQ, Permkam S, Hubbard M. The mayfly genera (Ephemeroptera) of southern Thailand. Nat Hist Bull Siam Soc. 2001;49(2):243–68.

Suri Babu B. Description of the larva of *Neurothemis intermedia* (Rambur), with notes on biology. Odonatologica. 2000;29:341–6.

Thani I, Phalaraksh C. A preliminary study of aquatic insect diversity and water quality of Mekong River, Thailand. Khon Kaen University Sci J Suppl. 2008;36:95–106.

Theischinger G, Hawking J. The complete field guide to dragonflies of Australia. Collinwood: CSIRO Publishing; 2006.

Thipaksorn A, Kittayapong P, Milne JR, et al. Records of rice field Odonata from Thailand. Notul Odonatol. 2003;6(2):20–4.

Tungpairojwong N, Boonsoong B. New records of *Isonychia formosana*, *Prosopistoma annamense* and *Prosopistoma sinense* (Ephemeroptera) from Thailand. Ent Res. 2011;41(2):66–9.

Tze-wai T, Ka-king L, Kwan BSP, et al. The dragonflies of Hong Kong. Hong Kong: Friends of the Country Parks, Cosmos Books Ltd.; 2011.

Uttaruk P, Voharndee P, Jusanit P, et al. Development of biotic index based on rapid bioassessment approaches using benthic macroinvertebrates for Chi and Mun headwater streams, northeastern Thailand. AES Bioflux. 2011;3(1):29–43.

van Tol J. An annotated index to names of Odonata used in publications by M. A. Lieftinck. Zool Verh. 1992;279:3–263.

Wan FX, Jiang YH, Wan J. Descriptions of *Anax immaculifrons* Rambur and *Tetratcanthagyna waterhousei* McLachlan exuviae from China (Anisoptera: Aeshnidae). Odonatologica. 2011;40(4):339–45.

Watanabe K, Kawashima I, Sasamoto A. Notes on the larva of *Neurothemis ramburii ramburii* (Kaup in Brauer, 1866) obtained from Iriomote-jima Island, Yaeyama Islands, southern Ryukyus, Japan (Anisoptera: Libellulidae). Tombo Fukui. 2013;55:83–7.

Wilson KDP. The gomphid dragonflies of Hong Kong, with descriptions of two new species (Anisoptera: Gomphidae). Odonatologica. 1995;24(3):319–40.

Wong KC, Yeh W-C, Chan T-W. Description of the final stadium larvae of *Polycanthagyna ornithocephala* (McLachlan, 1896) from Taiwan, with a key to the known larvae of the genus (Odonata: Anisoptera: Aeshnidae). Zootaxa. 2012;3238(1):64–8.

Xu Q. The larvae of *Macromia flavocolorata* and *M. septima* from Fujian, China (Odonata: Macromiidae). Int J Odonatol. 2010;13(1):145–52.

Yanagisawa T, Sasamoto A. New records of *Chlorogomphus miyashitai* Karube, 1995 and *Ch. piaoacensis* Karube, 2013 from northern Thailand, with notes on the characteristics of *Ch. piaoacensis*. Tombo. 2017;59:71–3.

Yang B, Davies DAL. Gomphid dragonflies of Yunnan, China, with descriptions of new species and some views on the origin of the group (Anisoptera: Gomphidae). Odonatologica. 1993;22(1):45–62.

Yang B, Davies DAL. Two new species and one new subspecies of Gomphidae from southwestern China, with descriptions of larvae and distribution records (Anisoptera). Odonatologica. 1996;25(3):283–96.

Yeh W-C. Description of *Petaliaeschna pinratanai* spec. nov. from northern Thailand (Anisoptera: Aeshnidae). Odonatologica. 1999;28(3):283–8.

Yeh W-C, Lien J-C. Proven distribution of *Nannophyopsis clara* in Taiwan and morphological description of the ultimate instar larva (Odonata: Libellulidae). Tombo. 1995;38:24–6.

Zhang H, Tong X. Descriptions of the final instar larvae of three Chinese *Idionyx* species (Odonata: Anisoptera: Corduliidae). Zootaxa. 2010;2716(1):53–63.

Zhang H, Tong X. Descriptions of the final instar larvae of seven Chinese Chlorogomphidae species, with taxonomic notes on adults (Odonata: Anisoptera). Zootaxa. 2013;3620(2):223–44.

Morphology of the Nymphs

2

Contents

References.. 32

The Odonata nymphs have the typical insect body sections (tagmata) of head, thorax, and abdomen (Fig. 2.1). They are distinguished from other aquatic insects by the presence of an enlarged, extendable, labium that is longer than the head. The integument of the nymphs can be glabrous, rugose, setose, granulose, spinulose, or any combination of these.

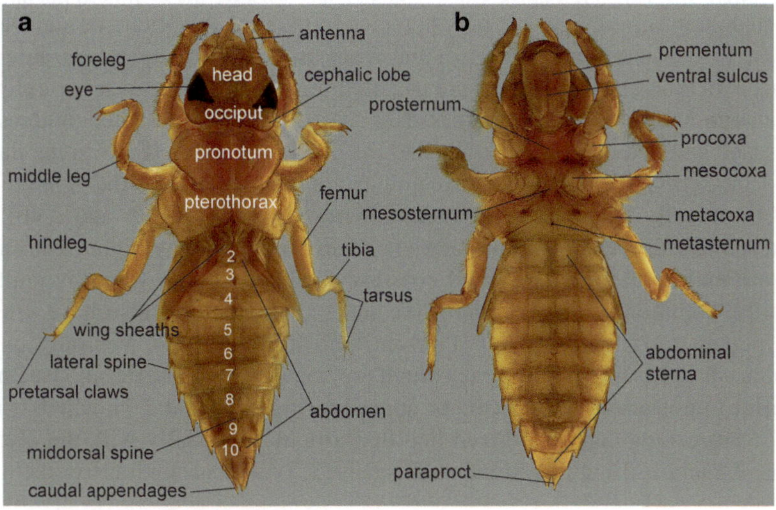

Fig. 2.1 Dorsal (**a**) and ventral (**b**) habitus of the nymph of *Acrogomphus* sp. (Gomphidae) showing important morphological characters

R. Novelo-Gutiérrez, R. W. Sites, *The Dragonfly Nymphs of Thailand (Odonata: Anisoptera)*, https://doi.org/10.1007/978-3-031-33712-3_2

The head. The head (Figs. 2.1 and 2.2) is the anteriormost part of the body, which bears the sense organs and mouthparts. The head capsule can be globose or slightly compressed dorsoventrally (Tennessen 2019). Anterodorsally is the clypeus (Fig. 2.3). The dorsal surface is mostly occupied by the frons, which comprises the antefrons and postfrons, behind which is the vertex bearing the ocelli. Posterior to the epicranial suture is the occiput, and the lobes behind the compound eyes are the cephalic lobes (Fig. 2.3).

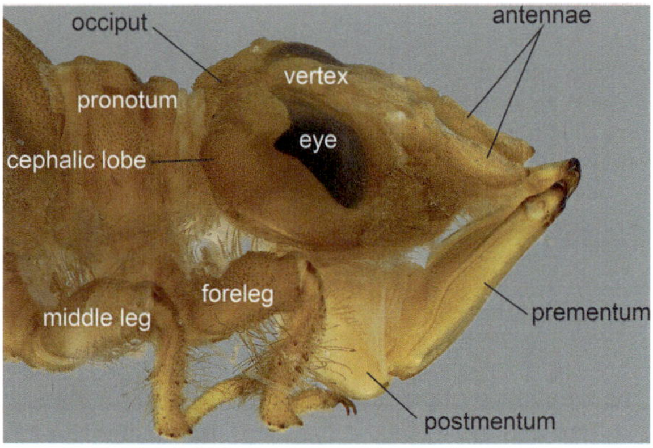

Fig. 2.2 The head and prothorax of *Lamelligomphus* sp. (Gomphidae) in lateral view

Dorsolaterally, the compound eyes (Figs. 2.1a and 2.3) are composed of many small facets that extend internally as ommatidia, which collectively form an image. On the top of the head (vertex) are usually three ocelli (Fig. 2.3), which do not form images but only perceive light intensity (Richards and Davies 1977). In the anterolateral part of the frons is a pair of segmented antennae (Figs. 2.1a, 2.2, and 2.3), which are usually long and typically of seven cylindrical antennomeres of different sizes (Fig. 2.4a); the first two antennomeres are the scape and pedicel, respectively, and the other antennomeres form the flagellum (Snodgrass 1935). Notable exceptions are the antennae of the Gomphidae, which typically bear only four antennomeres (Figs. 2.1a, 2.2, and 2.4b); the third is markedly larger and the fourth is greatly reduced or vestigial. The antennae are mainly tactile organs and possibly chemosensory covered by multiple sensilla, commonly including hair-like setae. The antennae also have sensilla that perceive olfactory stimuli (Piersanti and Rebora 2018). Ventrally on the head are the chewing mouthparts, which are composed of a series of structures: anterior to posterior are a labrum or upper lip (Figs. 2.3 and 2.7), which is usually a subrectangular, sclerotized flap that articulates basally with the clypeus (Figs. 2.3 and 2.4b); a pair of mandibles (Fig. 2.5) with strong incisive teeth and

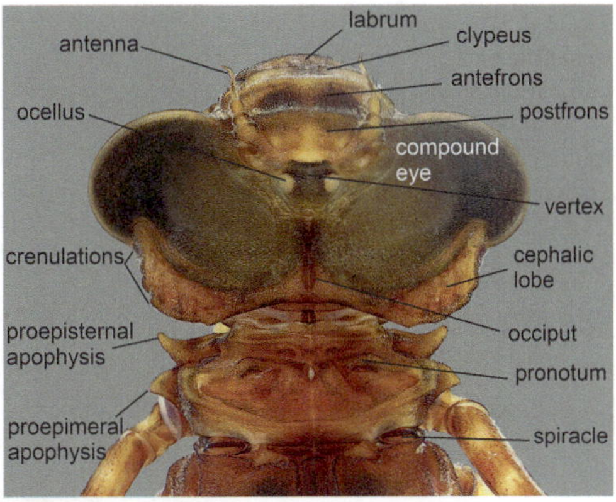

Fig. 2.3 The head and prothorax of *Periaeschna* sp. (Aeshnidae) in dorsal view

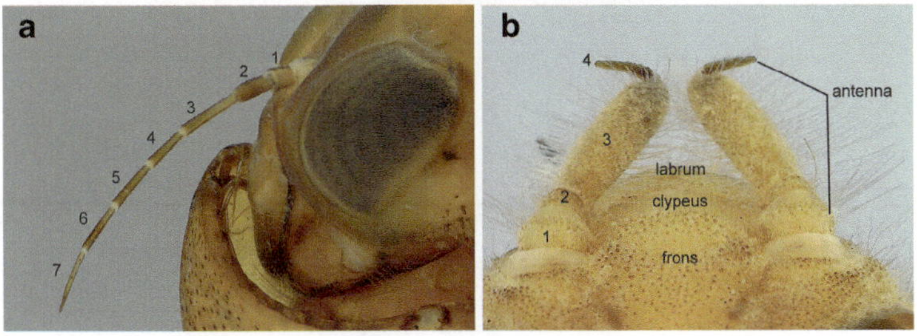

Fig. 2.4 Antenna of *Hydrobasileus croceus* (Libellulidae), lateral (**a**); antennae of *Paragomphus capricornis* (Gomphidae), dorsal (**b**)

highly variable molar teeth, which sometimes form a molar crest (Fig. 2.6); a pair of maxillae; the hypopharynx (Fig. 2.7); and the labium or lower lip (Fig. 2.2) with highly variable labial palps (Fig. 2.8).

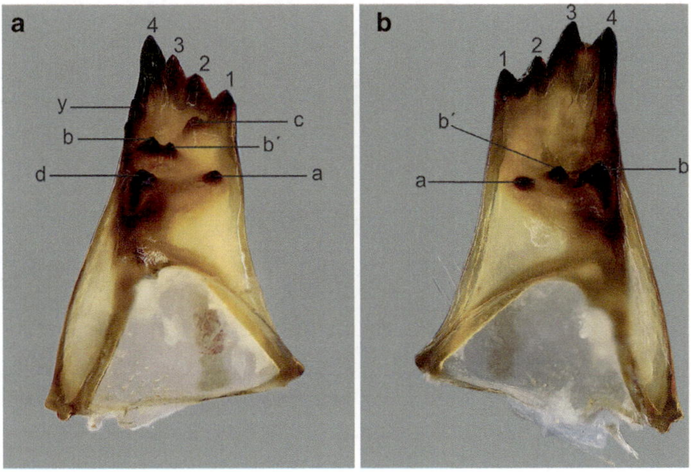

Fig. 2.5 The right (**a**) and left (**b**) mandibles of *Hydrobasileus croceus* (Libellulidae) showing the mandibular formula (Watson 1956) in mesal view. The number, size, and position of teeth is important for nymph taxonomy

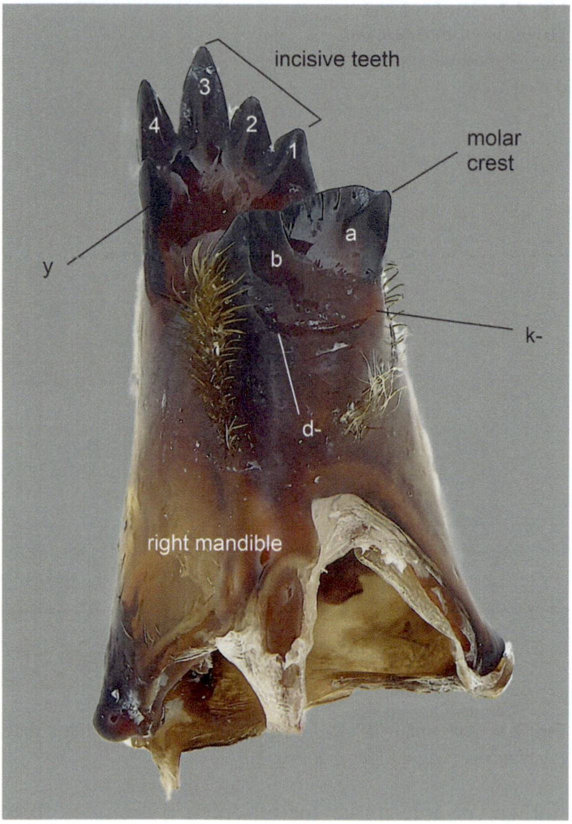

Fig. 2.6 The right mandible of *Tetracanthagyna* sp. (Aeshnidae) showing the molar crest in ventromesal view

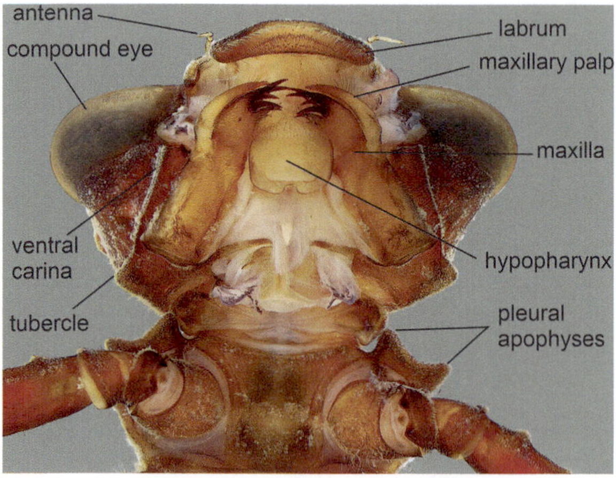

Fig. 2.7 The head and prothorax of *Tetracanthagyna* sp. (Aeshnidae) in ventral view (labium removed)

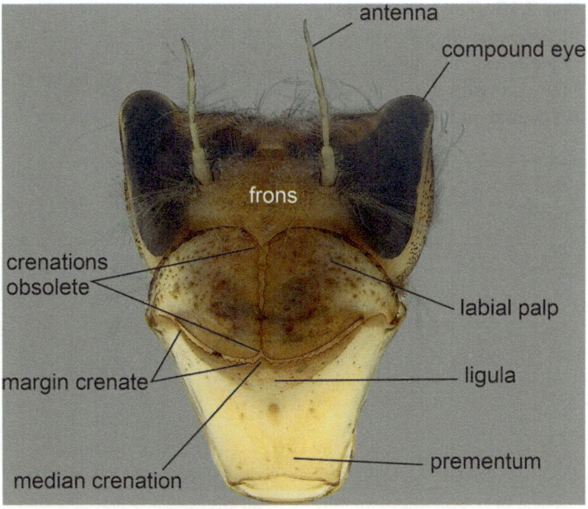

Fig. 2.8 The head and prementum of *Orthetrum* sp. (Libellulidae) in frontal view

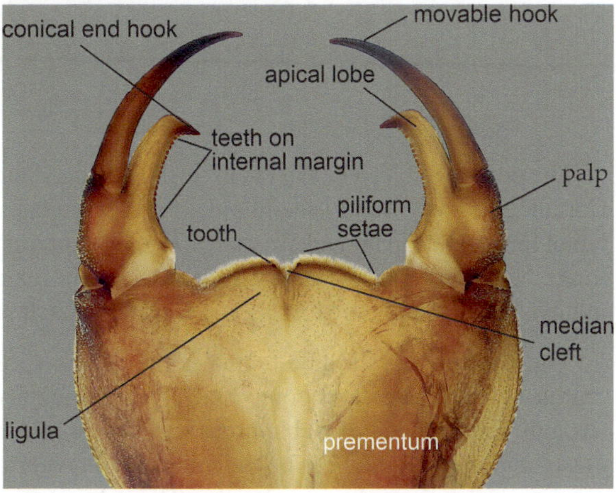

Fig. 2.9 Detail of the prementum of *Tetracanthagyna* sp. (Aeshnidae) in ventral view, showing the ligula and palps

The labium covers the ventral part of the head and frequently part of the thorax and comprises two large segments: a basal postmentum, which articulates with the head capsule, and a distal prementum (Snodgrass 1954) (Fig. 2.2), which bears a pair of labial palps on the distolateral corners, and a distomedian ligula which can be entire (Fig. 2.8) or with a median cleft; the width/length ratio of the ligula varies among genera and species (Figs. 2.9 and 2.10).

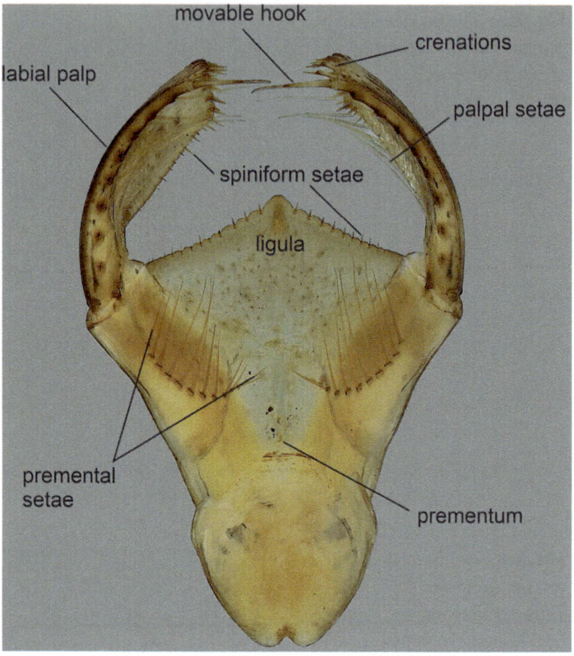

Fig. 2.10 The prementum of *Potamarcha congener* (Libellulidae) in dorsal view, showing the premental and palpal setae

The prementum can be flat (as in Aeshnidae and Gomphidae) or scoop-like (as in Cordulegasteroidea and Libelluloidea) and articulates with the postmentum by a labial hinge (Needham and Westfall 1954). The prementum and labial palps sometimes bear long setae dorsally (Fig. 2.10). The extendable labium of Odonata nymphs is unique among insects and is used to catch prey.

The thorax. The middle tagma is the thorax (Fig. 2.1); it is the locomotory region of the insect and composed of three heavily sclerotized segments: the strongly reduced and movable prothorax bearing the anterior pair of legs, and the meso- and metathorax bearing the middle and posterior pairs of legs as well as the anterior and posterior wing sheaths. These last two segments together compose the bulky, fused pterothorax.The pronotum is the dorsal surface of the prothorax with the raised area known as the pronotal disc (Figs. 2.1a and 2.2); ventrolaterally are the supracoxal processes (Fig. 2.3), which are projections of the ventral margin of the highly reduced sclerites of the propleuron: the proepisternal and proepimeral apophyses. Dorsally is a pair of mesospiracles on the boundary between the pro- and mesothorax (Fig. 2.3), with one spiracle on each side of the midline. Most of the pterothorax is composed of large pleural sclerites, the meso- and metepisterna, and the meso- and metepimera. The legs are usually unmodified and used for walking, but some anterior legs are modified to dig into the substrate (Fig. 5.98); in some cases they are very long and used to cling onto aquatic plants or attach to the substrate (e.g., some Libellulidae, Macromiidae). The legs are composed of six segments (Fig. 2.1): the basal coxa; the apparently subdivided trochanter; the femur, which is usually the stoutest segment; the elongate tibia; the tarsus, which is usually subdivided into three tarsomeres, although some gomphid nymphs have only two; and the last segment is the pretarsus, which is represented predominantly as a pair of well-developed claws which sometimes bears rows of setae. The medial pair of wing sheaths correspond to the mesothorax, and the lateral pair is metathoracic (Fig. 2.1a). The wing sheaths can be parallel or strongly divergent. In the last instar (F-0) two noticeable morphological changes occur: wing expansion and the onset of wing sheath melanization (Okude et al. 2021). The thoracic sternum (Figs. 2.1b and 2.11) is usually shallowly concave and mostly fused with the mes- and metinfraepimera. The prosternum is small and connects to the mesosternum by intersegmental membrane. In some nymphs, tubercles are at the anterolateral corners of the mesosternum and to a lesser extent on the metasternum; these tubercles are usually beset with stiff setae.

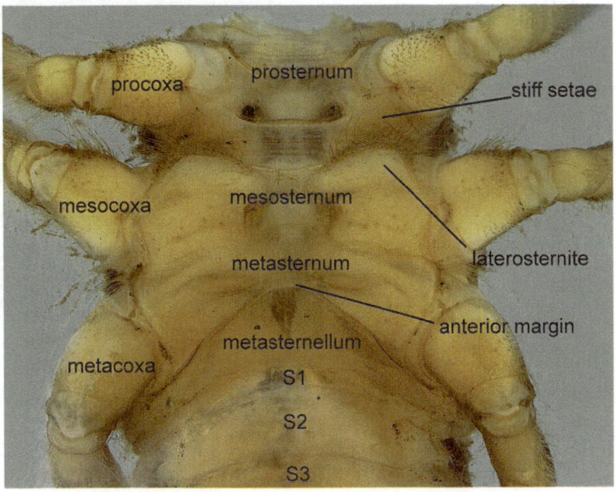

Fig. 2.11 Thoracic sternum of *Orthetrum* sp. (Libellulidae)

On the meso- and metasterna, furcal pits connected by a transverse suture (the sternacostal suture) are quite obvious and separate the basisternum (anterior) from the sternellum. The posteriormost sclerite of the metasternum is the metasternellum (Fig. 2.11).

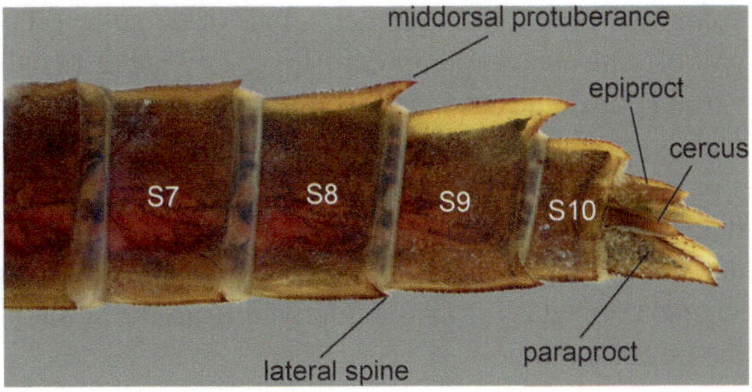

Fig. 2.12 The abdomen of *Tetracanthagyna* sp. (Aeshnidae) in lateral view

The abdomen. The posterior tagma is the abdomen (Fig. 2.1), which contains most of the viscera and the reproductive organs. It is composed of ten segments (with remnants of a primitive eleventh segment [see explanation below]) numbered S1 to S10 from base to apex. The segments are joined by intersegmental membrane. The dorsum of the abdomen is the tergum and can be shallowly convex to strongly arched. The venter of the abdomen is the sternum and can be flat or shallowly concave. Between the tergum and sternum is a very narrow membranous area. The shape of the abdomen is variable, but is most commonly cylindrical or subcylindrical, and sometimes can be flattened.

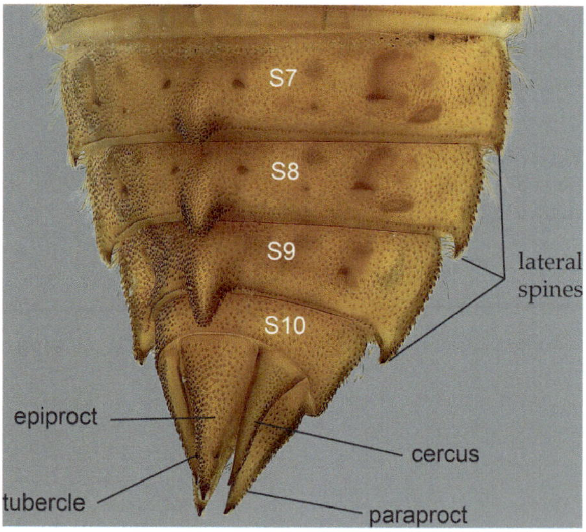

Fig. 2.13 Abdominal segments 7–10 and caudal appendages of *Lamelligomphus* sp. (Gomphidae) in dorsolateral view

The posterior end of the abdomen can taper gradually or be abruptly truncate. Dorsal protuberances may be present or absent on one, several, or almost all abdominal segments (Figs. 2.12 and 2.13); they can be shaped as low humps, moderately developed blunt protuberances, vertical or arched spines, or large hooks. The posterior margins of the terga can be bare, setose, or spiny. The lateral margins of the terga can be carinated or keeled and are usually beset with delicate or spiniform setae or both; the posterolateral corners sometimes terminate in variously sized and shaped spines (Fig. 2.13). At the posterior end of the abdomen is the anal pyramid composed of a dorsal epiproct, a pair of dorsolateral cerci (singular cercus), and a pair of ventrolateral paraprocts (Figs. 2.13 and 2.14); their conditions are extremely variable in shape, size, and texture that make them taxonomically instructive. The male's epiproct presents a more or less rounded basal tubercle (e.g., Aeshnidae) or a pair of small subapical tubercles (Fig. 2.13); in Libelluloidea they are usually difficult to see. Females lack such tubercles. The epiproct and paraprocts are either derived from the 11th segment or a postsegmental telson, whereas the cerci are 11th segment appendages that are derived from the serially homologous appendages of the insect ancestor.

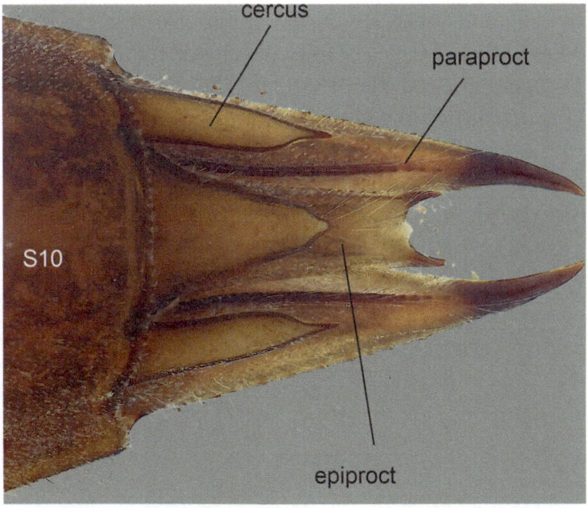

Fig. 2.14 The anal pyramid of *Planaeschna* sp. (Aeshnidae) in dorsal view

In all the genera descriptions, we refer to the epiproct, paraprocts, and cerci collectively as "caudal appendages" for convenience. The sternum is usually paler and smoother than the tergum, although it can be setose, granulose, or spiny.

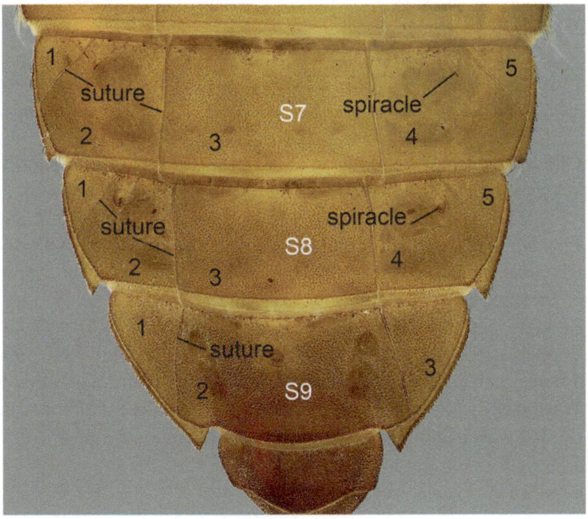

Fig. 2.15 Sterna 7–10 of an undetermined gomphid nymph, showing the five sternites on sterna 7–8 and three sternites on sternum 9

The sterna each can be divided into three or five sternites separated by longitudinal sutures (Fig. 2.15). When five sternites are present, the anterolateral sternites are the shortest. The sexes of Aeshnidae and Cordulegastridae can be easily determined in middle-aged to fully developed nymphs by the presence of a well-developed ovipositor in S8–S9 of females. In many Gomphidae, well-developed nymphs of females and males can be recognized by the presence of rudimentary or vestigial gonapophyses on S8 of females, and a low hump on the sterna of S2–S3 and dorsolateral tubercles on the epiproct of males. In Corduliidae, Libellulidae, and Macromiidae, males of fully developed nymphs can be distinguished by the secondary genitalia on the sterna of S2–S3 visible through the transparent cuticle (Fig. 2.16a) or in Chlorogomphidae by the presence of low humps on the sternum of S2 (Fig. 2.16b).

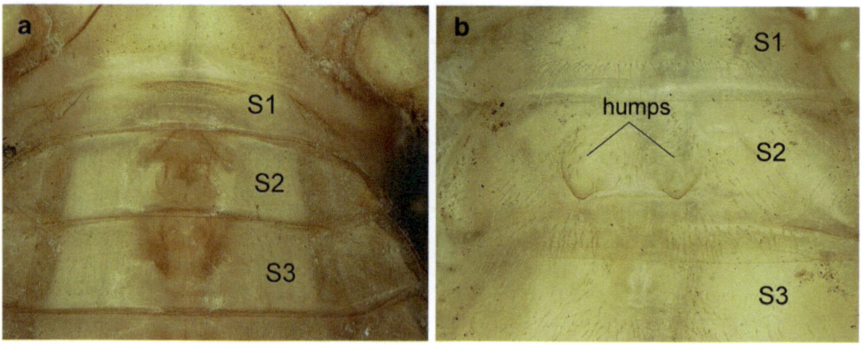

Fig. 2.16 Sterna S1–S3 of males of *Neurothemis* (Libellulidae) (**a**) and *Chlorogomphus* (Chlorogomphidae) (**b**)

References

Needham JG, Westfall MJ. A manual of the dragonflies of North America (Anisoptera) including the Greater Antilles and the provinces of the Mexican border. Berkeley, CA: University of California Press; 1954.

Okude G, Fukatsu T, Futahashi R. Comprehensive comparative morphology and developmental staging of final instar larvae toward metamorphosis in the insect order Odonata. Sci Rep. 2021;11(5164):1–14. https://doi.org/10.1038/s41598-021-84639-2.

Piersanti S, Rebora M. The antennae of damselfly larvae. Arthropod Struct Dev. 2018;47(1):36–44.

Richards OW, Davies RG. IMMS' General textbook of entomology. structure, physiology and development. London: Chapman and Hall; 1977.

Snodgrass RE. Principles of insect morphology. New York, NY: McGraw-Hill; 1935.

Snodgrass RE. The dragonfly larva. Smithsonian Misc Coll. 1954;123(2):1–38.

Tennessen KJ. Dragonfly nymphs of North America. An identification guide. Cham: Springer; 2019.

Watson MC. The utilization of mandibular armature in taxonomic studies of anisopterous nymphs. Trans Am Entomol Soc. 1956;81:155–205.

Keys to Suborders and Families of Odonata of Thailand

3

Contents

3.1 Key to Families of Anisoptera 34

1. Body slender, posterior end of abdomen with three (sometimes two) well-developed, caudal gills of several forms (Fig. 3.1) .... ZYGOPTERA

Fig. 3.1 Different types of caudal gills in Zygoptera: (**a**) inflated (*Bayadera* sp., Euphaeidae); (**b**) lamellar (*Rhinagrion* sp., Philosinidae); (**c**) triquetral (Chlorocyphidae)

1' Body robust, posterior end of abdomen without caudal gills, but with five, short, hard, usually sharply pointed, caudal appendages which together form the anal pyramid (Figs. 2.13 and 2.14) .... ANISOPTERA

R. Novelo-Gutiérrez, R. W. Sites, *The Dragonfly Nymphs of Thailand (Odonata: Anisoptera)*, https://doi.org/10.1007/978-3-031-33712-3_3

3.1 Key to Families of Anisoptera

1. Antennae 7-segmented (Fig. 3.2a); tarsi 3-segmented 2

1' Antennae 4-segmented (fourth segment reduced or vestigial) (Fig. 3.2b); tarsal formula 2–2–2 or 2–2–3 . . . . GOMPHIDAE

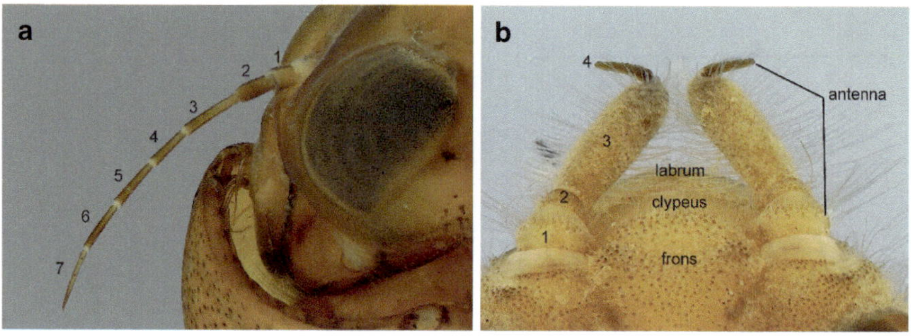

Fig. 3.2 Antenna 7-segmented (Libellulidae) (**a**); antennae 4-segmented (Gomphidae) (**b**)

2(1) Prementum flat or nearly so, without premental setae, ligula with median cleft, labial palps nearly flat and longer than wide (Fig. 3.3a), not covering the anterior part of head . . . . AESHNIDAE

2' Prementum dorsally concave, usually with premental setae (Fig. 3.3b), ligula with median cleft (Fig. 3.4a) or cleft lacking (Figs. 3.3b and 3.4b), labial palps spoon-shaped and covering the anterior part of head (Fig. 2.9) 3

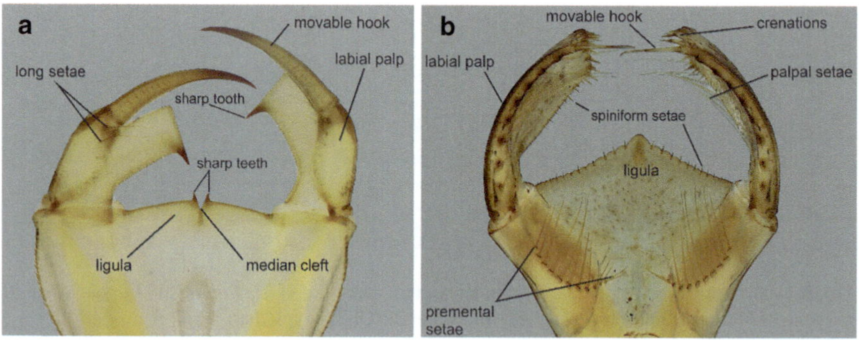

Fig. 3.3 Dorsal view of ligula and palps of *Gynacantha* sp. (Aeshnidae) (**a**) and *Potamarcha congener* (Libellulidae) (**b**)

3(2') Ligula with median cleft; distal margin of labial palp cut into large, irregular teeth (Fig. 3.4a) 4

3' Ligula without median cleft (Figs. 3.4b and 3.6); distal margin of labial palp not as above and variable 5

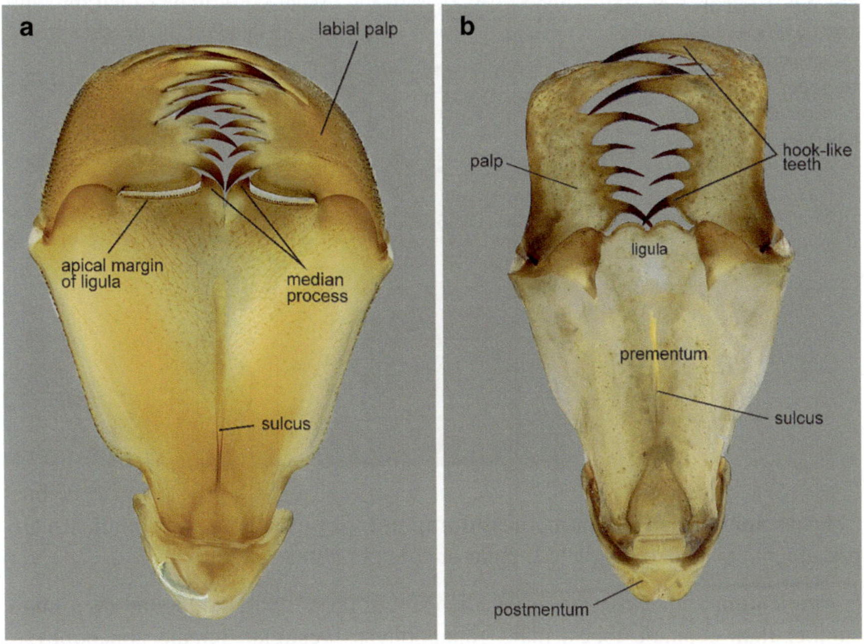

Fig. 3.4 Ventral view of prementa of *Chlorogomphus* sp. (Chlorogomphidae) (**a**) and *Epophthalmia* sp. (Macromiidae) (**b**)

4(3) Terga 2–10 covered with spiniform and scale-like setae including on posterior margins (Fig. 6.5); S8–S9 lacking posterolateral spines; sterna S3–S7 each divided into five sternites (lateralmost sternite very thin on S7), sternum S8 into three sternites, sterna of S9–S10 each forming a continuous ring with respective terga; female gonapophyses vestigial . . . . CHLOROGOMPHIDAE

4' Terga 2–10 covered with short and long fine setae on sides (Fig. 7.7), posterior margins of S2–S9 with a row of intermingled short and long fine setae; S8–S9 with short posterolateral spines; sterna 3–8 each divided into five sternites, sternum 9 undivided and not forming a continuous ring with its respective tergum, sternum 10 forming a continuous ring with its respective tergum; female gonapophyses well-developed . . . . CORDULEGASTRIDAE, ***Anotogaster***

5(3') Anterior margin of frons reflexed as a horn or with papilliform tubercles on different parts of the head or both (Fig. 3.5); ligula poorly (as Fig. 3.4b) to well-developed (Fig. 8.11), its apical margin convex or biconvex (Fig. 3.4b); sterna 3–9 each divided into three sternites . . . . MACROMIIDAE

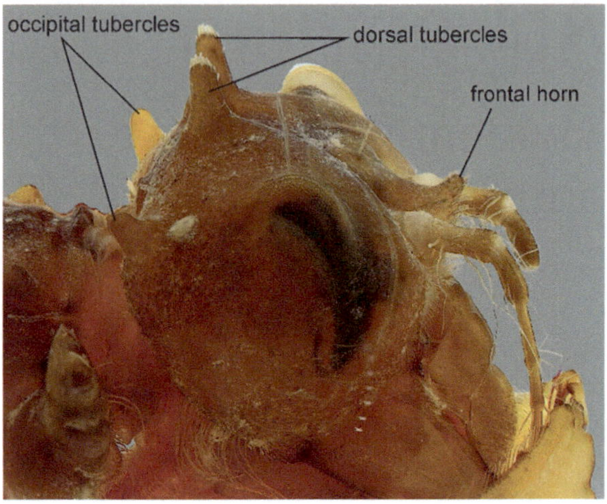

Fig. 3.5 Head of *Macromia arachnomima* (Macromiidae), laterodorsal view, showing the frontal horn and tubercles

5' Frons not reflexed as a horn, papilliform tubercles usually lacking; ligula usually prominent, its apical margin triangular or convex; sterna 3–9 variable 6

6. Prementum with a ventrobasal median sulcus (Fig. 3.6a); compound eyes small, smaller than cephalic lobes; male gonapophyses vestigial, female gonapophyses very reduced or lacking 7

6' Prementum without a ventrobasal median sulcus (Fig. 3.6b); compound eyes variable; male and female gonapophyses lacking . . . . LIBELLULIDAE

7(6) Anterior margin of frons, cephalic lobes, and/or dorsal rim of labial palp with scale-like setae intermingled with hair-like setae; sterna 3–8 divided into five sternites . . . . INCERTAE SEDIS

7' Anterior margin of frons, cephalic lobes, and/or dorsal rim of labial palp lacking scale-like setae; sterna 3– or 4–6 and sometimes sternum 8 divided into three sternites . . . . CORDULIIDAE

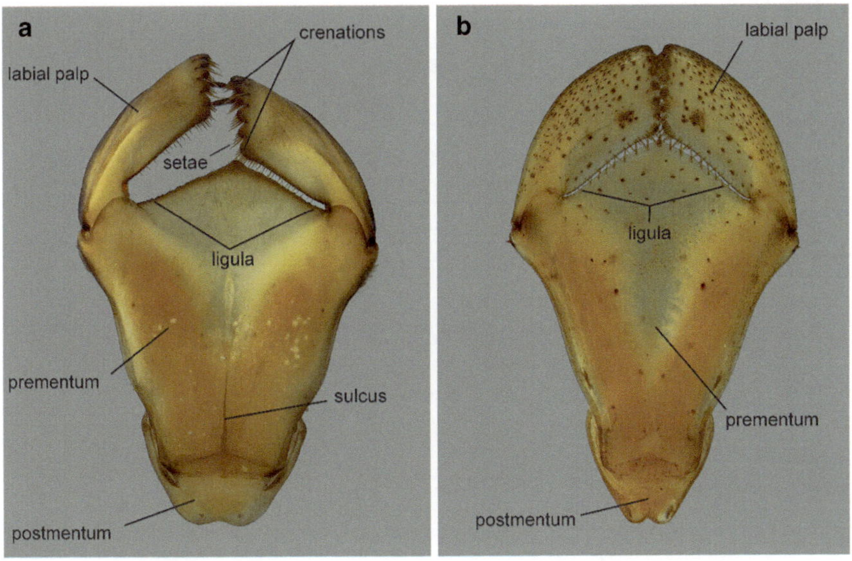

Fig. 3.6 Ventral view of prementa of *Hemicordulia* sp. (Corduliidae) (**a**) and *Hydrobasileus croceus* (Libellulidae) (**b**)

Aeshnidae

4

Contents

4.1 Key to the Genera of Aeshnidae 42
4.2 *Anaciaeschna* Selys, 1878 51
4.2.1 General Appearance 51
4.2.2 Description 51
4.2.3 Remarks 54
4.2.4 Distribution 54
4.3 *Anax* Leach, 1815 55
4.3.1 General Appearance 55
4.3.2 Description 55
4.3.3 Remarks 58
4.3.4 Distribution 58
4.4 *Cephalaeschna* Selys, 1883 59
4.4.1 General Appearance 60
4.4.2 Description 60
4.4.3 Remarks 63
4.4.4 Distribution 63
4.5 *Gynacantha* Rambur, 1842 64
4.5.1 General Appearance 64
4.5.2 Description 64
4.5.3 Remarks 67
4.5.4 Distribution 67
4.6 *Heliaeschna* Selys, 1882 68
4.6.1 General Appearance 68
4.6.2 Description 68
4.6.3 Remarks 69
4.6.4 Distribution 70
4.7 *Indaeschna* Fraser, 1926 71
4.7.1 General Appearance 71
4.7.2 Description 71
4.7.3 Remarks 72
4.7.4 Distribution 72
4.8 *Periaeschna* Martin, 1908 73
4.8.1 General Appearance 73

R. Novelo-Gutiérrez, R. W. Sites, *The Dragonfly Nymphs of Thailand (Odonata: Anisoptera)*, https://doi.org/10.1007/978-3-031-33712-3_4

4.8.2 Description. 73
4.8.3 Remarks. 76
4.8.4 Distribution. 76
4.9 *Petaliaeschna* Fraser, 1927. 77
4.9.1 General Appearance. 77
4.9.2 Description. 78
4.9.3 Remarks. 81
4.9.4 Distribution. 82
4.10 *Planaeschna* McLachlan, 1896. 83
4.10.1 General Appearance. 83
4.10.2 Description. 83
4.10.3 Remarks. 86
4.10.4 Distribution. 86
4.11 *Polycanthagyna* Fraser, 1933. 87
4.11.1 General Appearance. 87
4.11.2 Remarks. 90
4.11.3 Distribution. 90
4.12 *Sarasaeschna* Karube and Yeh, 2001. 91
4.12.1 General Appearance. 91
4.12.2 Description. 92
4.12.3 Remarks. 94
4.12.4 Distribution. 94
4.13 *Tetracanthagyna* Selys, 1883. 95
4.13.1 General Appearance. 95
4.13.2 Description. 96
4.13.3 Remarks. 99
4.13.4 Distribution. 99
References. 102

The Aeshnidae is an ancient family with moderate species richness. It is cosmopolitan and contains 498 species in 54 genera (Paulson et al. 2023). Together with the Austropetaliidae Carle and Luton (1994), they compose the superfamily Aeshnoidea (Blanke et al. 2013). Aeshnids are known commonly as darners and hawkers, are medium to large dragonflies, and are strong flyers and active hunters (Michalski 2012). The females possess a functional ovipositor and deposit their eggs endophytically (Tze-wai et al. 2011). They fly nearly continuously, hover intermittently, and hang vertically when perching (Paulson 2009).

The nymphs are active climbers and can be found among submerged roots, aquatic plants, and dead stems that have fallen into the water (Figs. 4.1 and 4.2). They are fierce predators, attacking even tadpoles and small fish. Most of the species inhabit lentic environments, including phytotelmata (Corbet 1999), although *Tetracanthagyna* is a notable stream-dwelling genus (Orr 2003). The aeshnid nymphs mostly lack hair-like setae and have the integument smooth, granulose, or tuberculate. The wing pads are always parallel, and the sides of the pterothorax slope steeply down. They rarely burrow into the substrate. Most of the Aeshnidae nymphs known from Thailand were described outside of the country. Recently, Makbun et al. (2022) described the nymph of *Anax aurantiacus* Makbun, Wongkamhaeng and Keetapithchayakul, 2022, based upon Thai material.

Fig. 4.1 Pond at Amphur Nuea Khlong, Tambon Nuea Khlong, Ban Paga Sai, Krabi Province

Fig. 4.2 Pond at Ao Nang, Amphur Mueang, Noppharat Thara Beach, Krabi Province

Presently, 13 genera containing 65 species of aeshnids have been recorded in Thailand (Plate 4.1), which is 12 more than the 23 species previously recorded by Hämäläinen and Pinratana (1999).

4.1 Key to the Genera of Aeshnidae

1 Abdomen with middorsal protuberances on S7–S10 or S8–S10, that on S9 largest (Figs. 2.12 and 4.80); head and thorax with multiple tubercles....... ***Tetracanthagyna***

1' Abdomen lacking middorsal protuberances; head and thorax usually without tubercles 2

2(1') Body setose with a shaggy appearance; compound eyes transverse, twice as wide as long, not bulging; occipital lobes with longitudinal rows of thick, stiff setae giving a striated appearance (Fig. 4.3); ventral carina of head strongly developed, ending anteriorly in a robust spine; movable hook shorter than labial palp (Fig. 4.4); lateral spine on S9 thick, three-dimensional (Figs. 4.70 and 4.73) .... ***Sarasaeschna***

Fig. 4.3 The head of a *Sarasaeschna* exuvia in dorsal view, showing occipital striae

2' Body mostly glabrous, granulose, or minutely scaly; compound eyes rounded, bulging; occipital lobes variable but never as described above; ventral carina of head moderately developed, not ending in a spine anteriorly; movable hook longer than labial palp; lateral spine on S9 two-dimensional 3

3(2') Caudal appendages shorter than dorsal length of S9 + S10, covered abundantly with short, peg-like tubercles (Fig. 4.5); propleural apophyses narrowly V-shaped (Fig. 4.6) .... ***Petaliaeschna***

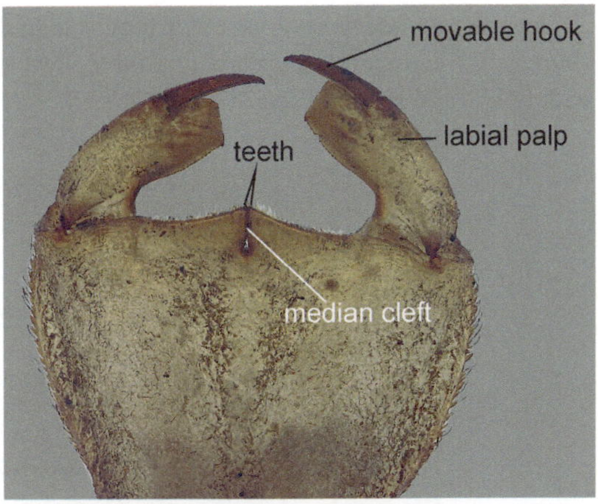

Fig. 4.4 Ligula and palps of *Sarasaeschna* in ventral view

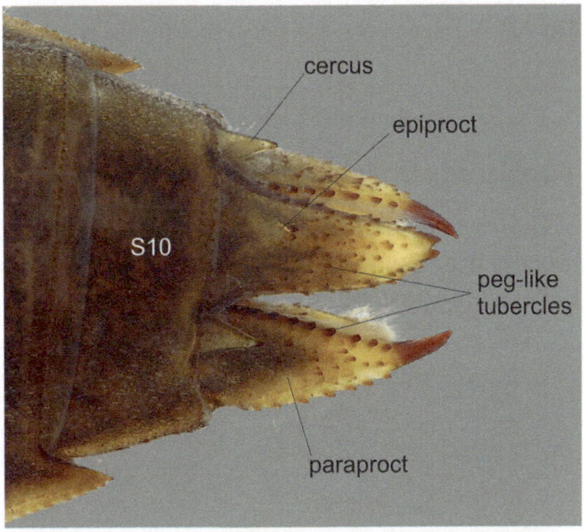

Fig. 4.5 Caudal appendages of *Petaliaeschna*

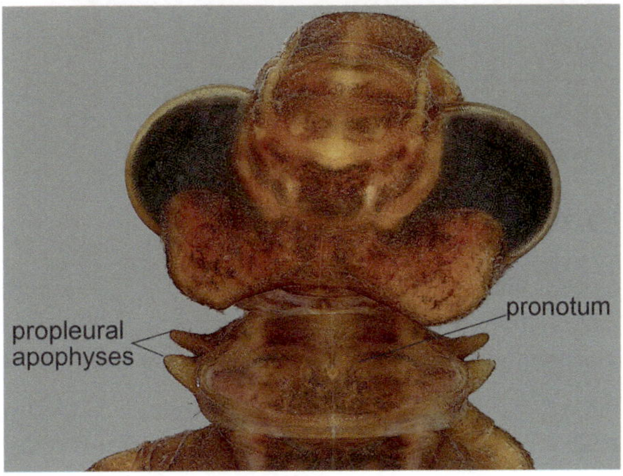

Fig. 4.6 Head and prothorax of *Petaliaeschna*

3' Caudal appendages as long as or longer than dorsal length of S9 + S10, lacking short, peg-like tubercles; epiproct variable, never cylindrical; propleural apophyses widely U- or V-shaped (Figs. 2.7 and 4.7) 4

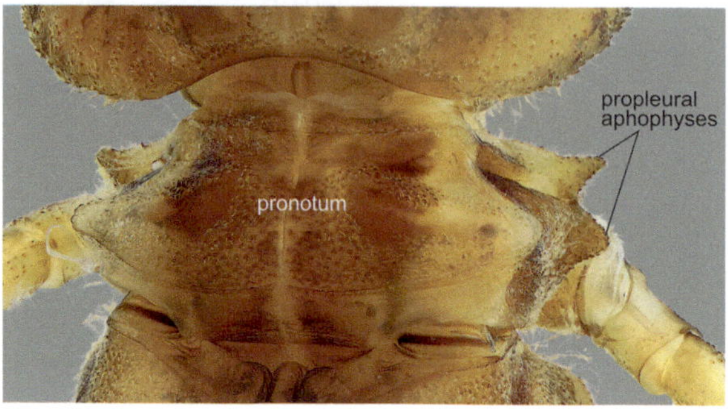

Fig. 4.7 Prothorax of *Gynacantha* showing the propleural apophyses

4(3') Labial palp with at least one long seta (Figs. 4.8 and 4.9) 5

4' Labial palp lacking long setae (Figs. 4.11, 4.12, 4.13, 4.15, and 4.18) 7

5(4) Movable hook with a row of long and short interspersed setae (Fig. 4.8a); ligula broadly triangular (Fig. 4.8a); cerci 0.75× length of epiproct; female gonapophyses not reaching posterior margin of S9 . . . . ***Anaciaeschna***

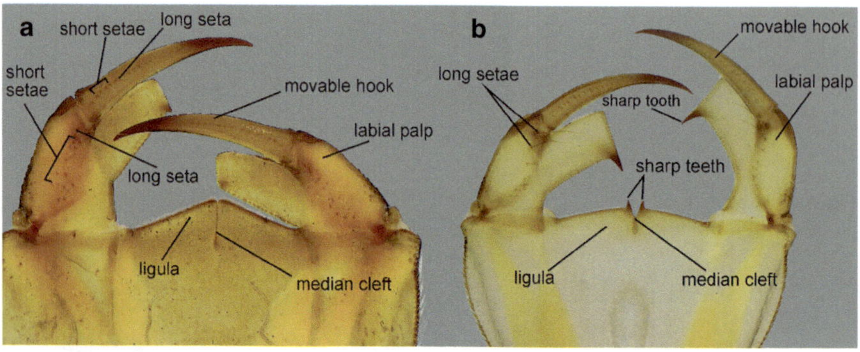

Fig. 4.8 Ligula and palps of *Anaciaeschna martini* (**a**) and *Gynacantha* sp. (**b**), dorsal view

5' Movable hook lacking long setae, with a row of short, fine setae only; ligula nearly straight or shallowly convex; cerci 0.90–0.95× length of epiproct; female gonapophyses reaching to at least the basal fourth of S10 6

6(5') Vertex lacking a tuft of setae; ligula nearly straight, median cleft V-shaped (Fig. 4.8b), with a tooth on each side; 10–12 stiff setae on labial palp, of which 2–5 are long (Fig. 4.8b); propleural apophyses widely V-shaped (Fig. 4.7); epiproct with a dorsal longitudinal keel (Fig. 4.10a) . . . . ***Gynacantha***

6' Vertex with a tuft of setae; ligula shallowly convex, median cleft closed, teeth absent; five robust, long setae on labial palp (Fig. 4.9); propleural apophyses widely U-shaped; epiproct lacking a dorsal keel . . . . ***Heliaeschna*** (in part)

7(4') Epiproct truncate (Fig. 4.10a) or widely U-shaped (Fig. 4.16b), not deflexed apically; lateral margins of occipital lobes not crenulate; proepisternal apophysis not curved anteriorly; body mostly bare or with minute scales or setae 8

7' Epiproct sharply pointed (Fig. 4.10b), slightly deflexed apically; lateral margins of occipital lobes crenulate (Fig. 2.3); proepisternal apophysis curved anteriorly (Fig. 2.3); body finely wrinkled-granulose . . . . ***Periaeschna***

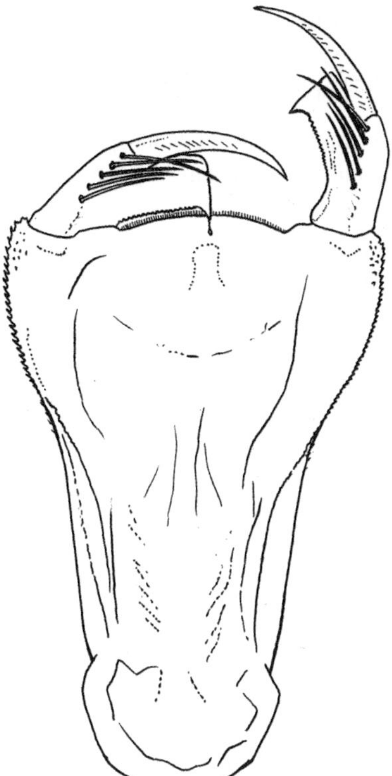

Fig. 4.9 Prementum of *Heliaeschna uninervulata* in dorsal view (after Orr and Ngiam 2011)

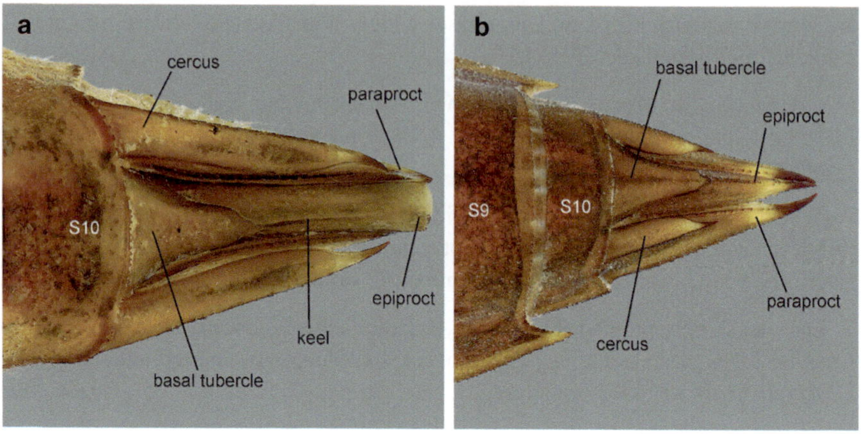

Fig. 4.10 Caudal appendages of *Gynacantha* (**a**) and *Periaeschna* (**b**), dorsal view

8(7) Apical lobe of labial palp squarely truncate (Figs. 4.11 and 4.12) 9

8' Apical lobe of labial palp not truncate but with apex sickle-shaped (Fig. 4.13), rounded or gradually tapering to a hooked tip (Fig. 4.15) 11

Fig. 4.11 Ligula and palps of *Polycanthagyna*

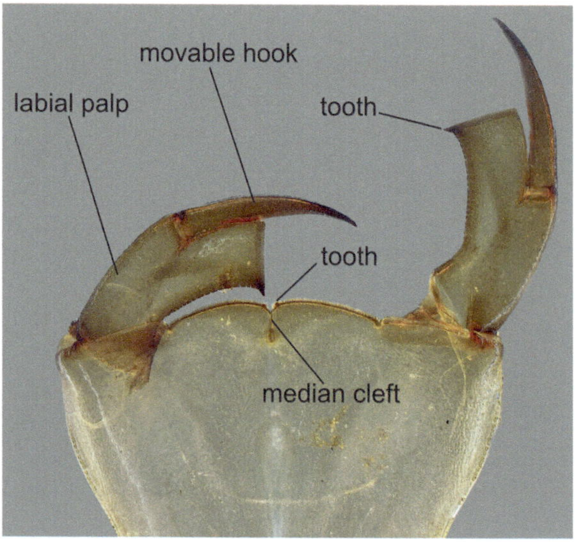

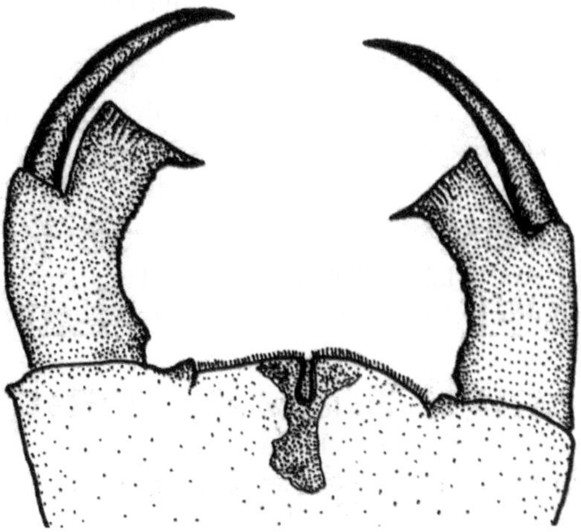

Fig. 4.12 Ligula and palps of in dorsal view *Indaeschna grubaueri*, dorsal view (after Orr 1994)

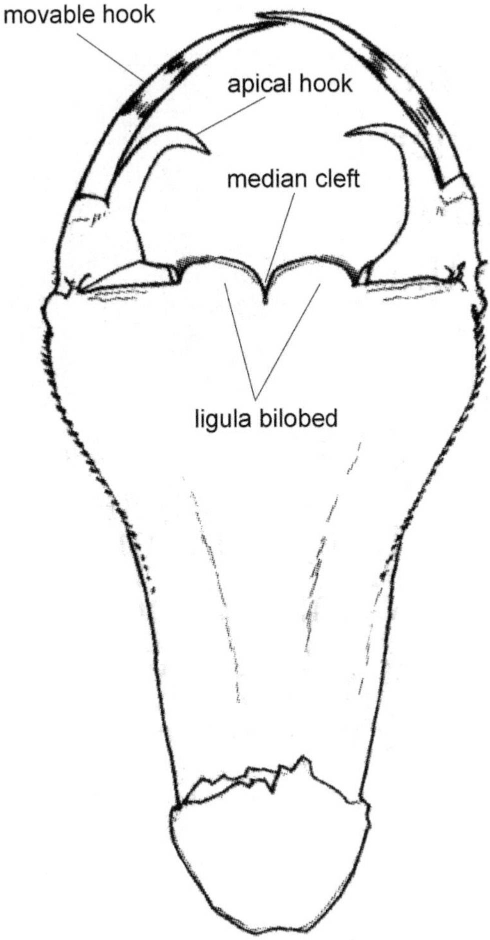

Fig. 4.13 Prementum of *Heliaeschna simplicia* in dorsal view (after Butler and Orr 2013)

9(8) Length/width ratio of prementum about 1:0.85 10

9' Length/width ratio of prementum 2:1 . . . . ***Anax*** (in part)

10(9) Apical lobe of labial palp with a short, sharply pointed tooth on mesodistal corner (Fig. 4.11); in ventral view, ligula with short, median, V-shaped cleft with a small, blunt tooth on each side (Fig. 4.11); epiproct with a dorsal keel (Fig. 4.64) . . . . ***Polycanthagyna***

10' Apical lobe of labial palp with a large, sharply pointed tooth on mesodistal corner (Fig. 4.12); median cleft of ligula nearly closed; epiproct lacking dorsal keel or carina . . . . ***Indaeschna***

11(8') Vertex with a tuft of setae; ligula bilobed, lobes widely rounded, with a deep V-shaped median cleft (Fig. 4.13); apical lobe of labial palp strongly sickle-shaped . . . . ***Heliaeschna*** (in part)

11' Vertex without a tuft of setae; ligula more or less triangular, with the median cleft closed (Fig. 4.15); apical lobe of labial palp not sickle-shaped 12

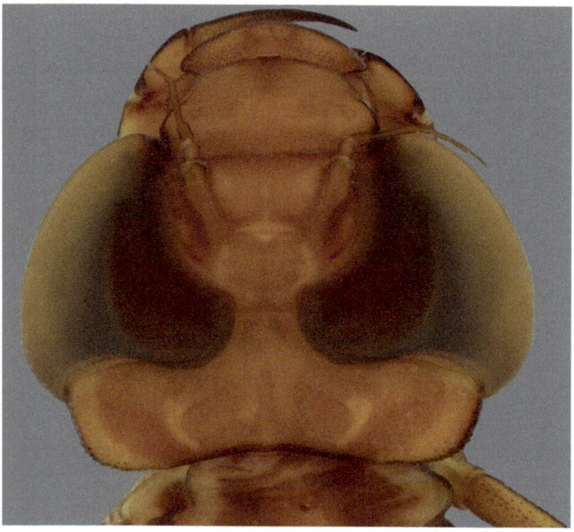

Fig. 4.14 Head of *Anax* in dorsal view

12(11') Compound eyes longer than wide (Fig. 4.14), scarcely protruding laterally; mandibles lacking *m* denticles; apical lobe of labial palp gradually tapering to a stout, apical hook (Fig. 4.15a); lateral spines on S7–S9; epiproct with a dorsal keel, apex truncate with two spines (Fig. 4.16a) . . . . ***Anax*** (in part)

12' Compound eyes wider than long, strongly bulging anterolaterally (Figs. 4.17 and 4.19); mandibles with 4–5 m denticles; apical lobe of labial palp mostly parallel-sided

(Fig. 4.15b) or gradually tapering (Fig. 4.18); lateral spines on S5 or S6–S9; epiproct without a dorsal keel, apex widely U-shaped (Fig. 4.16b) 13

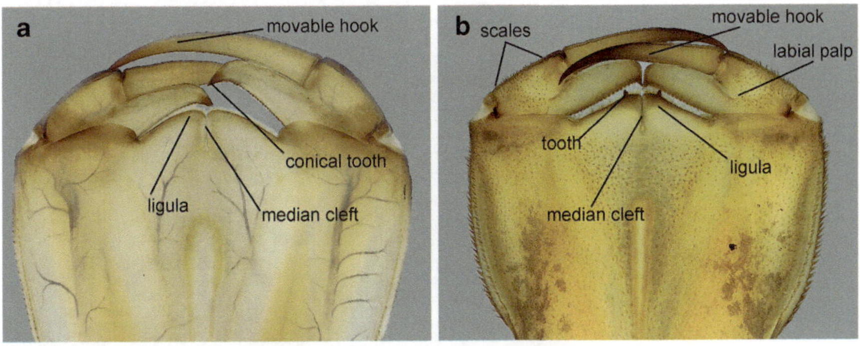

Fig. 4.15 Ligula and palps of *Anax* (**a**) and *Planaeschna* (**b**), ventral view

13(12') Head about twice as wide as thorax and abdomen; occiput lateral margins more or less straight and convergent posteriorly (Fig. 4.17); apical lobe of labial palp gradually tapering, ending in a short tooth (Fig. 4.18); femora and tibiae dark with thin yellow rings and multiple tubercles; abdomen slender, more or less parallel-sided (Figs. 4.32 and 4.35) . . . . ***Cephalaeschna***

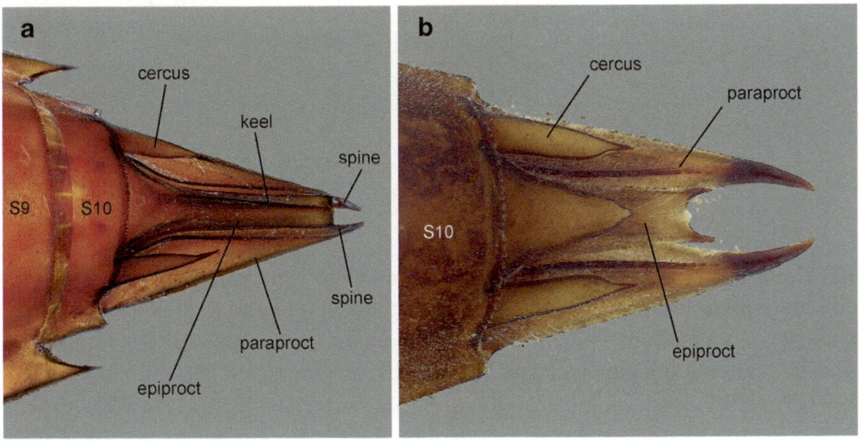

Fig. 4.16 Caudal appendages of *Anax* (**a**) and *Planaeschna* (**b**), dorsal view

Fig. 4.17 Head of *Cephalaeschna*, dorsal view

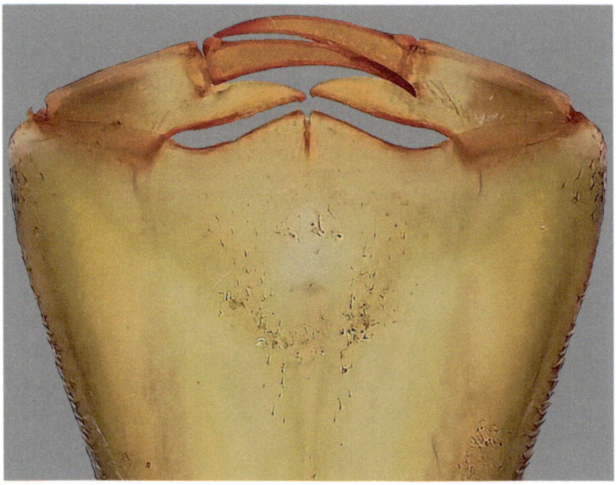

Fig. 4.18 Ligula and palps of *Cephalaeschna*, dorsal view

13' Head 0.30 and 0.15 wider than thorax and abdomen, respectively; occiput lateral margins straight and parallel (Fig. 4.19); apical lobe of labial palp mostly parallel-sided, with a stout, conical tooth on mesodistal corner (Fig. 4.15b); femora and tibia with alternating pale and dark wide bands, lacking tubercles; abdomen lanceolate (Fig. 4.58) . . . . ***Planaeschna***

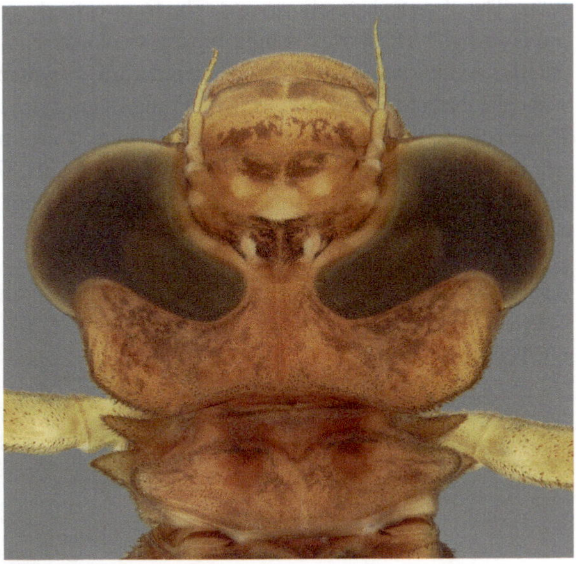

Fig. 4.19 Head of *Planaeschna*, dorsal view

Unknown nymph: *Oligoaeschna.*

4.2 *Anaciaeschna* Selys, 1878

Fig. 4.20 Dorsal habitus of the nymph of *Anaciaeschna martini*

4.2.1 General Appearance

The nymphs of *Anaciaeschna* are characterized by their medium to large size (30–37 mm), with the body slender and finely granulose, head flattened, abdomen lanceolate, and coloration light-brown (Fig. 4.20).

4.2.2 Description

Head. The head is wider than long (Fig. 4.21) and wider than the thorax and abdomen (Fig. 4.20), and the anterior part is mostly yellow. The labrum is granulose, 0.8 mm long, and the anterior margin is shallowly concave and fringed with short, delicate setae. The clypeus is finely striated. The frons is finely granulose and the anterior margin is straight. The vertex is light-brown, shallowly convex, and with three ocelli; the anterior ocellus is the largest. The antennae are 7-segmented; the scape is cup-like, the pedicel is barrel-shaped, and the flagellomeres are cylindrical, with antennomere 3 the longest. The compound eyes are large and strongly protrude anterolaterally (Fig. 4.21). The occipital lobes are gently rounded, not bulging, the sides converge posteriorly, and they have small spiniform setae. The posterior margin of the head is widely concave.

Fig. 4.21 The head in dorsal view, showing the compound eyes strongly protruding

Mandibles have a molar crest and the formula: $L\ 1234\ 0\ a(m^0)b\ /\ R\ 1234\ a(m^{1,2})b$, with $a > b$ in both mandibles. The maxilla's galeolacinia has seven moderately incurved teeth; the three dorsal teeth are subequal in length and robustness, the four ventral teeth are of different sizes, with the apical tooth the largest. The ventral pad of the hypopharynx is subrectangular and poorly sclerotized, the anterior margin is straight and has a few short, stiff setae, and the posterior margin is strongly convex, wide, and bare.

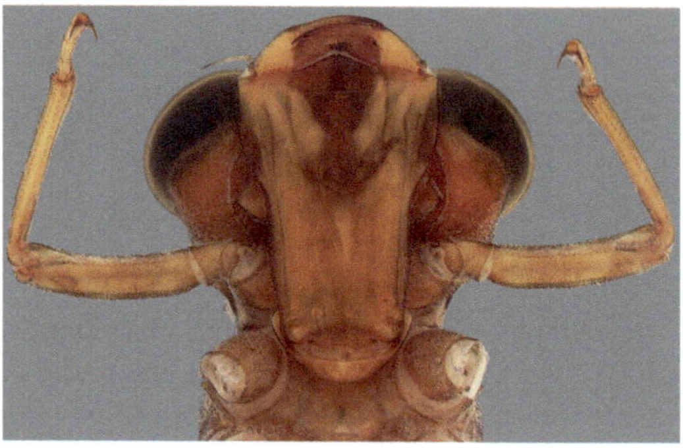

Fig. 4.22 The head and prementum in ventral view

Labium. The prementum (Fig. 4.22) is 0.25× longer than its widest part; the lateral margins are serrated, each serration has a small delicate seta, the apical half is strongly convex, and the basal half is narrower with the lateral margins sinuate. The ligula is triangular, well-developed, its apical margin has abundant piliform setae, the median cleft is deep, closed, and has a minute, blunt tooth on each side. The labial palp has a dorsomedial group of stiff setae of different sizes, with one or two long setae close to the articulation of the movable hook. The apical lobe of the labial palp is squarely truncate, with or without a small conical tooth on the mesal corner, and the distal and mesal margins are minutely denticulate. The movable hook is as long as or slightly shorter than the palp, sharp, and moderately incurved, with a row of intermingled long and small setae along the middorsal line.

Thorax. The pronotum (Figs. 4.20 and 4.21) is wide, granulose, with a pair of large, smooth, transverse, drop-shaped markings; its anterior margin is shallowly convex, the lateral margins are narrowly rounded, and the posterior margin is undulate. The ventral margin of the propleuron is V-shaped with the posterior branch stouter. The legs are relatively short (Figs. 4.20 and 4.23) (i.e., when fully extended, the hind legs scarcely reach the posterior margin of S8); the femora are granulose and with minute spiniform setae; the mesal surface of the fore tibiae and the ventral surface of the basal fore tarsal segment have tridentate setae, whereas the remainder of the tibiae and tarsi have small, spiniform setae. The anterior and posterior wing sheaths are parallel and reach the posterior margin of S4 and basal half of S5, respectively (Fig. 4.20).

Fig. 4.23 Ventral habitus of the nymph of *Anaciaeschna martini*

Abdomen. The abdomen (Figs. 4.20, 4.23, and 4.24) is lanceolate, slender, slightly widened at the middle, and then tapers posteriorly. The ventral surface is more or less flat, whereas the dorsal surface is convex and without dorsal protuberances. The posterior margins of terga S1–S10 have small, blunt spines; the lateral margins of S2–S9 are serrated; S6–S9 have sharply pointed posterolateral spines with those on S8 the largest (Figs. 4.20 and 4.23). The sterna are finely granulose with abundant minute, spiniform setae. The posterior margins of sterna 1–8 are smooth, sternum 9 has minute spines on the posterolateral corners of the laterosternites, and sternum 10 has small spines on the posterior margin. Sterna 3–8 are each divided into five sternites and sternum 9 into three sternites. The female gonapophyses almost reach the posterior margin of sternum 9 (Fig. 4.24). The caudal appendages (Figs. 4.20, 4.23, and 4.24) are longer than S9 + S10. The epiproct has a longitudinal dorsal carina; its apical margin is truncate and a small spine is on each corner. The epiproct of the male has the basal lamina along the basal half. The paraprocts and cerci are sharply pointed. The paraprocts are pyramidal and the longest of the caudal appendages (Figs. 4.23 and 4.24), with the dorsal and ventral margins strongly serrated. The cerci are slender with long tips, and 0.75× the length of the epiproct.

Fig. 4.24 The abdomen and caudal appendages in lateral view

Measurements (in mm). Total length including caudal appendages 30–37; abdomen (ventral) 19.8–25.0; maximum width of head 7.4–8.5; length of hind femur (lateral) 5.0–6.5.

4.2.3 Remarks

Only *Anaciaeschna jaspidea* (Burmeister, 1839) and *A. martini* (Selys, 1897) have been recorded from Thailand (Hämäläinen and Pinratana 1999; Hämäläinen 2002). The nymphs of both species are already described, and they are found in small, richly vegetated ponds. They can be distinguished from other Thai aeshnids by the presence of long setae on the movable hook of the labial palp (Matsuki and Obana 1985).

4.2.4 Distribution

In Thailand, *Anaciaeschna* has been recorded from only the provinces of Chiang Mai, Khon Kaen (Hämäläinen and Pinratana 1999; Hämäläinen 2002), Chiang Rai (Katatani et al. 2004), and more recently from Phetchabun (Day et al. 2012) (Fig. 4.25).

Fig. 4.25 Distribution of *Anaciaeschna*

4.3 *Anax* Leach, 1815

Fig. 4.26 Dorsal habitus of the nymph of *Anax*

4.3.1 General Appearance

The nymphs of *Anax* are characterized by their very large size (41–56 mm), with the body mostly bare, robust and slender, head flattened, abdomen lanceolate, and coloration light-yellow-brown (Fig. 4.26).

4.3.2 Description

Head. The head is mostly smooth, wider than long, and wider than the thorax and base of the abdomen (Fig. 4.27).

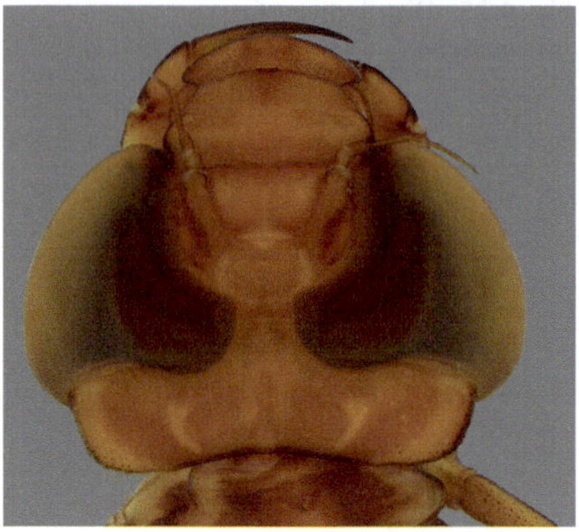

Fig. 4.27 The head in dorsal view, showing the large compound eyes

The labrum is 1.0 mm long, granulose in the basal third, and its ventral margin has a dense row of short, stiff setae. The clypeus is trapezoid. The antefrons is subequal in length to the clypeus, but the antefrons is wider; the anterior margin of the postfrons is rather straight and the vertex is flat. The antennae are 7-segmented and bare, with the scape and pedicel cylindrical, reddish-yellow, and wider than remaining antennomeres; the flagellomeres are yellow and antennomere 3 is the longest. The compound eyes are large, longer than wide, and protruding laterally (Fig. 4.27). The occipital lobes are rounded and have longitudinal rows of minute spinules. The posterior margin of the head is widely concave. Mandibles have the formula: *L* 1234 *y- a b k* / *R* 1234 *y* $a(m^{0})bb'$ *k*, and the left mandible has a molar crest but no *m* denticles. The maxilla's galeolacinia has seven moderately incurved teeth; the three dorsal teeth are subequal in length and robustness, and the four ventral teeth are of different sizes, with the apical tooth the largest. The ventral pad of the hypopharynx is trapezoidal, yellowish, softly sclerotized, and has long, abundant stiff setae on the anterior third.

Fig. 4.28 The prementum in ventral view

Labium. The prementum-postmentum articulation reaches the posterior margin of the mesocoxae or the middle of the metacoxae (Fig. 4.28). The prementum is flat, light-yellow, nearly twice longer than its widest part, with the lateral margins straight and divergent along the basal 0.60 and convex in the serrate apical 0.40; a longitudinal, bare, weak central sulcus is in the basal half of the ventral surface; the ligula is moderately prominent and convex or biconvex; its apical margin has abundant small, piliform setae; the median cleft is short and closed. The labial palp lacks raptorial setae, has the apical lobe securiform or narrow and tapering to a reddish, short, conical tooth; its mesal margin is straight or slightly concave and finely denticulate. The movable hook is yellow, sharp, moderately incurved, and longer than the palp.

Fig. 4.29 Ventral habitus of the nymph of *Anax*

Thorax. The thorax is slender, finely granulose, and minutely spiny laterally. The pronotum has the lateral margins rounded, the posterior margin widely convex, and all margins are finely granulose. The inferior margin of the propleuron is V-shaped, with each arm short, bluntly pointed, and spiny. The sides of the pterothorax slope steeply down. The legs are relatively short (Figs. 4.26 and 4.29) (i.e., when fully extended, the hind legs scarcely surpass the posterior margin of S7), yellowish, and have narrow, light-brown, subapical rings; the femora and tibiae margins are serrulate; the claws are simple and has a pulvilliform empodium. The wing sheaths are parallel and reach S4 (Fig. 4.26).

Abdomen. The abdomen is lanceolate, narrow basally, widened at the middle, and then tapers posteriorly (Figs. 4.26 and 4.29). The ventral surface is more or less flat, whereas the dorsal surface is convex and without dorsal protuberances (Fig. 4.30). S6–S9 or S7–S9 end in large, sharply pointed lateral spines (Figs. 4.26 and 4.29). Terga 2–10 are beset with minute, white, delicate setae, and the posterior margins are serrulated. The lateral margins of S2–S10 are serrulate. Sterna 1–10 are yellow (Fig. 4.29) and setae are as described for the terga. Sterna 1–8 are each divided into five sternites and sternum 9 into three sternites. The female gonapophyses reach the basal half of sternum 9 (Figs. 4.29 and 4.30), whereas the male gonapophyses are vestigial. The caudal appendages are long, longer than S9 + S10 (Fig. 4.30). The epiproct and paraprocts are pyramidal; the epiproct has a strong, longitudinal, dorsal keel beset with long, stiff setae, its apical margin is truncate, and it has a short spine at each corner (Fig. 4.13). The epiproct of the male has the basal lamina along the basal 0.25–0.33. The dorsal and ventral margins of the paraprocts are finely denticulate and the apices sharply pointed. The cerci are digitiform and sharply pointed (Fig. 4.30. The paraprocts are the longest of the caudal appendages, and the cerci are 0.40× the length of the epiproct (Fig. 4.30).

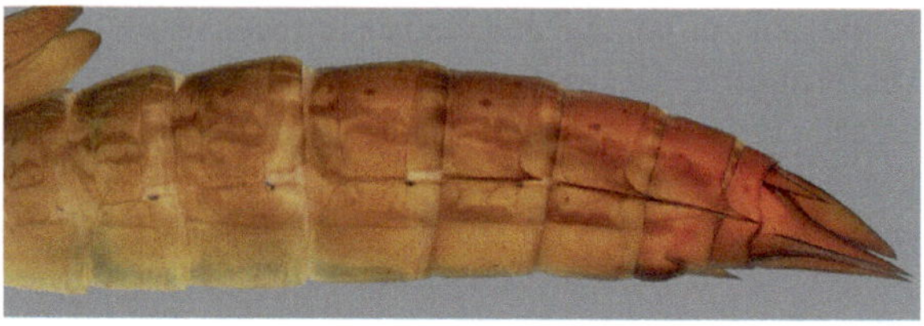

Fig. 4.30 The abdomen and caudal appendages in lateral view

Measurements (in mm). Total length including caudal appendages 41–56; abdomen (ventral) 26.5–33.3; maximum width of head 8.0–10.5; length of hind femur (lateral) 8.5–12.0.

4.3.3 Remarks

Six species of *Anax* have been recorded from Thailand (Hämäläinen and Pinratana 1999, 2000; Hämäläinen 2002; Pierce 2017; Makbun et al. 2022). The nymphs can be distinguished from other aeshnid nymphs inhabiting Thailand by their long and narrow prementum (10 mm length), and the presence of lateral spines usually on S7–S9 (Kumar 1973; Wan et al. 2011; Sasamoto et al. 2022) but *A. aurantiacus* Makbun, Wongkamhaeng and Keetapithchayakul, 2022, has spines on S6–S9. The nymphs of some species (e.g., *A. guttatus* (Burmeister, 1839) and *A. nigrofasciatus nigrolineatus* Fraser, 1935) are found in ponds with dense vegetation and temporary wetlands or marshes. In contrast, the nymph of *A. aurantiacus* inhabits pools in open areas of streams surrounded by trees, herbs, and shrubs (Makbun et al. 2022). They become top predators in fishless waterbodies (Paulson 2009) as fierce and agile hunters of tadpoles and other odonates (Orr 2003).

4.3.4 Distribution

In Thailand, *Anax* is widespread and has been recorded from the provinces of Bangkok, Chaiyaphum, Chanthaburi, Chiang Mai, Kalasin, Kanchanaburi, Loei, Nakhon Sawan, Nakhon Si Thammarat, Phrae, Phuket, Prachuap Khiri Khan, Ranong, Rayong, Sakon Nakhon, Songkhla, Ubon Ratchathani (Hämäläinen and Pinratana 1999; Hämäläinen 2002), Trang (Kitagawa and Katatani 2003), Chiang Rai (Katatani et al. 2004), Chon Buri (Kosterin and Vikhrev 2008), Petchabun, Phangnga, Surat Thani (Day et al. 2012), and more recently from Chumphon (Pierce 2017). ***New Province record***: Krabi (Fig. 4.31).

Fig. 4.31 Distribution of *Anax*

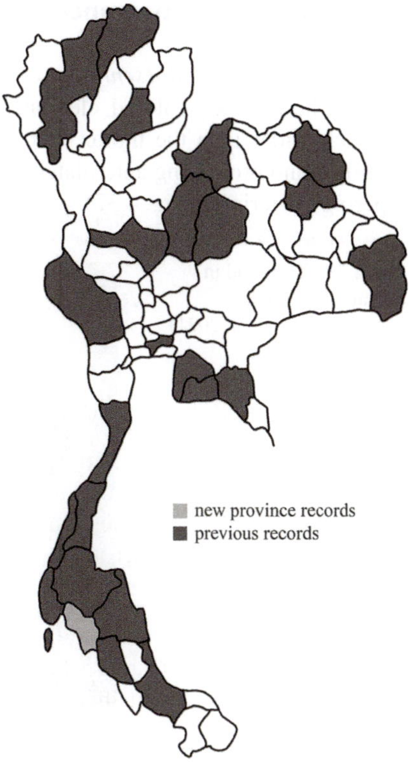

4.4 *Cephalaeschna* Selys, 1883

Fig. 4.32 Dorsal habitus of the nymph of *Cephalaeschna*. (Photo courtesy of Hu et al. 2021)

In the absence of specimens to examine, the description presented here has been adapted from Jiang and Zhang (2008) and Hu et al. (2021).

4.4.1 General Appearance

The nymphs of *Cephalaeschna* are characterized by their medium size (32–34 mm) with the body darkly colored, elongated, and covered with small tubercles. The head is much wider than the thorax and abdomen, and the compound eyes are large, rounded, and protruding anterolaterally. The abdomen is parallel-sided and slightly tapering posteriorly (Fig. 4.32).

Fig. 4.33 The head in dorsal view (photo courtesy of Hu et al. 2021)

4.4.2 Description

Head. The head is wider than long and approximately twice as wide as the thorax and abdomen, and with a complex, mottled color pattern (Fig. 4.33). The labrum is light-brown with black spots, the anterior margin is rounded, and has a dense row of long, light-gray setae. The clypeus is light-brown and slightly protruding in the middle. The vertex has three white ocelli. The antennae are very short, filiform, and 7-segmented; the scape and pedicel are cylindrical and wider than the remaining antennomeres; antennomere 3 is the longest and the seventh the shortest, and the tip of the antennae does not reach the anterior margin of the labrum. The compound eyes are large and strongly protrude anterolaterally (Fig. 4.33). The occiput is large and as long as the compound eyes, and each lobe has a comb-shaped protuberance. The sides of the occipital lobes are more or less straight and strongly convergent posteriorly. The posterolateral corners of the occiput are rounded, and the posterior margin of the head is widely concave.

Fig. 4.34 The head and prementum in ventral view. (Photo courtesy of Hu et al. 2021)

Mandibles have the formula: *L* 1234 0 $a(m^{1-5})b$ / *R* 1234 *y* $a(m^{1-5 \text{ or } 6})b$, with $a > b$ in both mandibles. The maxilla's galeolacinia is as described as for *Anax*. The ventral pad of the hypopharynx is short and has abundant golden setae on the anterior margin. Labium. The prementum-postmentum articulation reaches the anterior margin of the mesocoxae (Figs. 4.34 and 4.35). The prementum is brownish-yellow and 0.25x longer than its widest part (Fig. 4.34), the lateral margins are slightly convex along the apical half, strongly convergent posteriorly and with a row of spiniform setae, and the basal half is more or less sinuate and parallel-sided. The ligula is convex, moderately prominent, and its apical margin has abundant short setae, and a short, closed median cleft. The labial palp has a few short setae; the apical lobe of the labial palp is narrow and gradually tapering distally ending in a short, red, conical tooth; the mesal margin has a row of irregular, short teeth. The movable hook is almost as long as the palp, sharp, and moderately incurved.

Thorax. The pronotal disc is subtrapezoidal and wider at the middle; the lateral margins are rounded and strongly produced, and the posterior margin is more or less straight. The ventral margin of the propleuron is widely V-shaped. The pterothorax is granulose with abundant golden setae laterally. The mesospiracle is very large. The legs are short (i.e., when fully extended, the hind legs scarcely surpass the posterior margin of S7) (Fig. 4.35), dark, with narrow, yellow rings; the femora and tibiae are tuberculate with patches of spiniform setae; the pretarsal claws are simple. The anterior and posterior wing sheaths are parallel, dark-brown, and reach the basal half and surpass the posterior margin of S4, respectively (Fig. 4.32).

Fig. 4.35 Ventral habitus of the nymph of *Cephalaeschna*. (Photo courtesy of Hu et al. 2021)

Abdomen. The abdomen is cylindrical, enlarged, parallel-sided, and slightly tapers posteriorly (Figs. 4.32 and 4.35). The ventral surface is more or less flat, whereas the dorsal surface is convex; S5 or S6–S9 end in lateral spines, those on S9 are the largest. The terga are covered with small setae, the lateral surfaces have abundant, long, stiff, golden setae. The sterna are reddish-brown. The female gonapophyses are very long, reaching the basal 0.50–0.64 of the paraprocts. The caudal appendages are as long as S9 + S10 (Fig. 4.36). The epiproct is subrectangular, the lateral margins are straight and spiny, and the apical margin is widely U-shaped and ends in a sharply pointed spine on each side. The epiproct of the male has the basal lamina along the basal 0.60–0.83. The cerci are digitiform and end in small, sharply pointed spines, and their lateral margins are serrate. The paraprocts are pyramidal and the longest of the caudal appendages, and the cerci are the shortest, 0.50–0.94× the length of the epiproct.

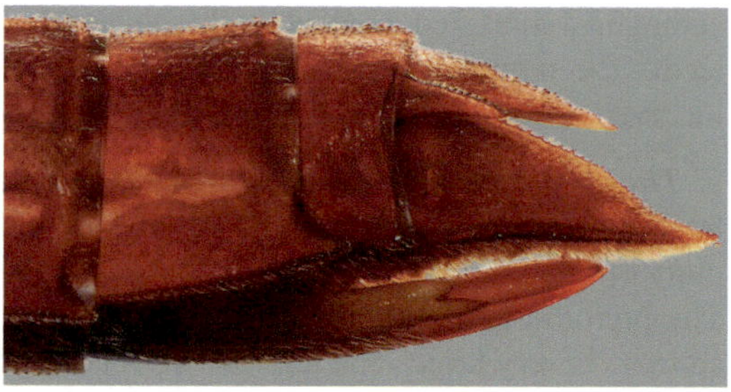

Fig. 4.36 The abdominal segments 8–10, caudal appendages, and female gonapophyses in lateral view. (Photo courtesy of Hu et al. 2021)

Measurements (in mm). Total length including caudal appendages 32.1–41.0; abdomen (ventral) 21–32; maximum width of head 7.0–8.2; length of hind femur (lateral) 5.4–5.8.

4.4.3 Remarks

Cephalaeschna appears to be closely related to the genus *Planaeschna*; the nymphs of both genera share several features in common, including the compound eyes wider than long and strongly bulging anterolaterally, mandibles with 4–5 m denticles, and epiproct apex U-shaped. Only one species of *Cephalaeschna* has been recorded in Thailand: *C. asahinai* Karube, 2011 (Buppachat et al. 2020). *Cephalaeschna* inhabits mountain forests; the nymph of *C. risi* Asahina, 1981, has been reported in Taiwan as semi-terrestrial with early instars found in mosses and final instars in moist leaf litter (Hu et al. 2021). Jiang and Zhang (2008) found exuviae of *C. patrorum* Needham, 1930, on shrubs, rocks, and weeds 0.5–1.5 m away from the water.

4.4.4 Distribution

In Thailand, *Cephalaeschna* has been recorded only from the province of Chiang Mai (Buppachat et al. 2020) (Fig. 4.37).

Fig. 4.37 Distribution of *Cephalaeschna*

4.5 *Gynacantha* Rambur, 1842

Fig. 4.38 Dorsal habitus of the nymph of *Gynacantha*

4.5.1 General Appearance

The nymphs of *Gynacantha* are characterized by their medium to large size (33–40 mm), with the body finely granulose and slender, head flattened, abdomen lanceolate, and coloration light-yellow-brown to dark-brown.

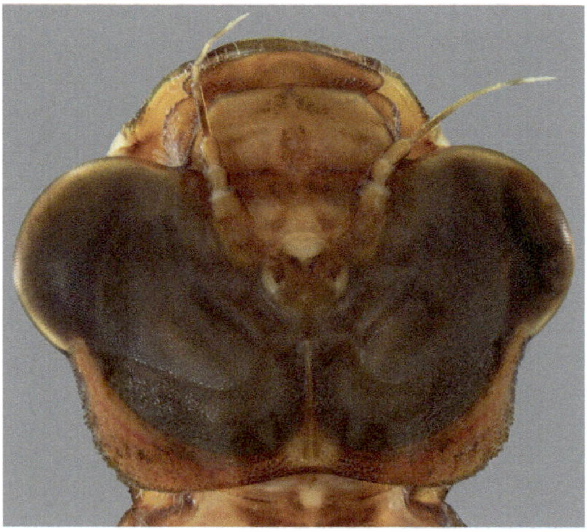

Fig. 4.39 The head in dorsal view, showing the protruding compound eyes

4.5.2 Description

Head. The head is wider than long and wider than the thorax and abdomen, and has a complex, mottled color pattern (Fig. 4.38). The labrum is 0.7 mm long, the anterior margin is rounded and more or less flattened ventrally, and has a dense row of setae. The clypeus and antefrons are the same length, but the antefrons is wider with the sides convex (Fig. 4.39); the anterior margin of the postfrons is almost straight. The vertex is domed. The antennae are 7-segmented; the scape and pedicel are cylindrical, wider than the remaining antennomeres, and have small setae; the flagellomeres are mostly bare, antennomere 3 is the longest, and the apical margin of antennomere 4 reaches the anterior margin of labrum. The compound eyes are large and strongly protrude anterolaterally (Fig. 4.39). The occiput is large and has small spines, the sides are gently rounded and not bulging, and converge caudally. The posterior margin of the head is widely concave. Mandibles have the formula: *L* 1234 *y- aa'*(m^0)*bb'kl R* 1234 *y a*(m^0)*bk*, with $a > b$ in both mandibles. The maxilla's galeolacinia is as described for *Anax*. The ventral pad of the hypopharynx is subpentagonal, yellowish, softly sclerotized in the basal half, and has a transverse U-shaped ridge on the posterior margin; the apical half is soft with regularly spaced, stiff setae.

Fig. 4.40 The prementum in ventral view

Labium. The prementum-postmentum articulation reaches the anterior margin of the metacoxae (Fig. 4.40). The prementum is light-reddish-yellow, much longer than its widest part, the anterolateral margins are strongly convex and with stout serrations along the apical 0.30; the basal 0.70 is strongly narrowed and has the lateral margins sinuate and finely serrate; a longitudinal, bare, feebly marked central sulcus is in the basal half of the ventral surface; the basal margin is strongly emarginate. The ligula is poorly developed and nearly straight, its apical margin has abundant piliform setae and two short or large, parallel or divergent, bluntly pointed or sharply pointed spines (Fig. 4.9); one spine is on each side of the short, V-shaped, median cleft (best seen in the ventral view). The labial palp (Fig. 4.9) has a dorsal row of 10–12 stiff setae of different sizes, the longest 2–5 setae are close to the articulation of the movable hook; the apical lobe of the labial palp is squarely truncate and has a large, sharply pointed tooth on the mesal corner; its mesal margin is minutely denticulate; the movable hook is bicolored, yellow in the basal 0.40–0.60, and dark-reddish-brown in the apical 0.40–0.60, as long as the palp, sharp and moderately incurved, and with a row of small, stout setae along the dorsum of the pale area.

Fig. 4.41 Ventral habitus of the nymph of *Gynacantha*

Thorax. The anterior margin of the pronotum is shallowly concave, the lateral margins are narrowly rounded, and the posterior margin is mostly convex, but concave at the middle. The pronotal disc is finely granulose. The ventral margin of the propleuron is V-shaped with each arm wider basally, bluntly pointed, and spiny (Fig. 4.7). The legs are relatively short (Figs. 4.38 and 4.41) (i.e., when fully extended, the hind legs scarcely reach the basal half of S7), uniformly colored or with alternating pale and dark, wide rings; the femora and tibiae margins are serrulate; the claws are simple and each has a pulvilliform empodium. The anterior and posterior wing sheaths are parallel and reach the posterior margin of S3 and the basal half of S4, respectively (Fig. 4.38).

Fig. 4.42 Lateral view of the abdomen, female gonapophyses, and caudal appendages

Abdomen. The abdomen is lanceolate, slender, slightly widened at the middle, tapers posteriorly, and has a complex, mottled color pattern (Figs. 4.38 and 4.41). The ventral surface is more or less flat (Fig. 4.41), whereas the dorsal surface is convex and without dorsal protuberances (Fig. 4.42). S5–S9 or S6–S9 end in sharply pointed lateral spines, which increase in size and robustness posteriorly (Figs. 4.38, 4.41, and 4.42). Terga 2–10 are beset with minute, white, delicate setae, and the posterior margins are serrulate. The lateral margins of S3–S10 are serrulate. Sterna 1–10 are mostly bare or as described for the terga, although the posterior margins are smooth. Sterna 3–8 are each divided into five sternites and sternum 9 into three sternites. The female gonapophyses usually reach the posterior margin of S10 (Figs. 4.41 and 4.42). The caudal appendages are as long as or longer than S9 + S10 (Figs. 4.11 and 4.42). The epiproct and paraprocts are pyramidal. The epiproct has a strong, longitudinal, dorsal keel beset with long stiff setae; its apical margin is truncate and a short spine is on each corner. The epiproct of the male has the basal lamina along the basal 0.40 (Fig. 4.11). The dorsal and ventral margins of the paraprocts are finely denticulate, and the apices are sharply pointed. The cerci are digitiform, very sharply pointed, and abruptly narrowed in the apical 0.20. The paraprocts are the longest of the caudal appendages, and the cerci are 0.90× the length of the epiproct.

Measurements (in mm). Total length including caudal appendages 33–40; abdomen (ventral) 21.8–24.8; maximum width of head 6.7–7.0; length of hind femur (lateral) 5.3–5.7.

4.5.3 Remarks

Nine species of *Gynacantha* have been recorded in Thailand (Hämäläinen and Pinratana 1999, 2000; Pierce and Makbun 2020). The nymphs are found in leafy-bottomed forest pools (Orr 2005).

4.5.4 Distribution

In Thailand, *Gynacantha* is widespread and has been recorded from the provinces of Bangkok, Chachoengsao, Chai Nat, Chaiyaphum, Chanthaburi, Chiang Mai, Chiang Rai, Chon Buri, Chumphon, Kanchanaburi, Khon Kaen, Loei, Mukdahan, Nakhon Nayok, Nakhon Ratchasima, Nakhon Si Thammarat, Phitsanulok, Phrae, Prachuap Khiri Khan, Ranong, Sa Kaew, Sakon Nakhon, Saraburi, Songkhla, Suphan Buri, Trang, Trat, Yala (Hämäläinen and Pinratana 1999, 2000; Hämäläinen 2002), Phayao (Ferro et al. 2009), and more recently from Ratchaburi, Samut Sakhon, and Surat Thani (Day et al. 2012). ***New province records***: Kalasin, Krabi, Phangnga, Phetchabun, Phuket, and Si Sa Ket (Fig. 4.43).

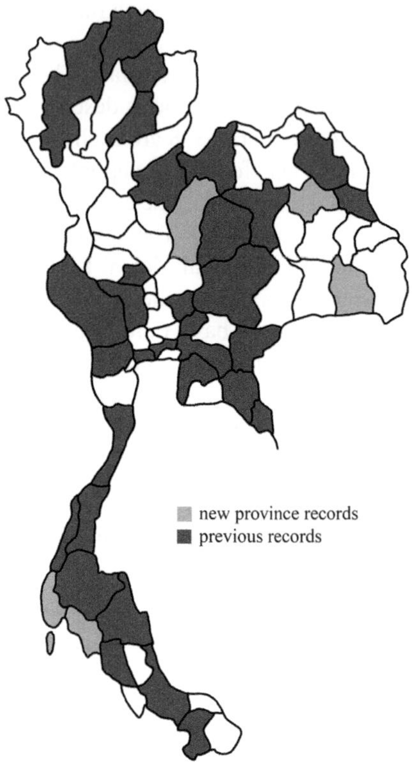

Fig. 4.43 Distribution of *Gynacantha*

4.6 *Heliaeschna* Selys, 1882

In the absence of specimens to examine, the description presented here has been adapted from Orr and Ngiam (2011) and Butler and Orr (2013).

4.6.1 General Appearance

The nymphs of *Heliaeschna* are characterized by their medium size (30 mm), with the body finely pilose and slender, head flattened, abdomen lanceolate, and coloration dark-brown with pale mottling.

4.6.2 Description

Head. The head is wider than long and wider than the thorax and abdomen, and it has a complex, mottled color pattern. The anterior margin of the frons is concave; the vertex is slightly swollen behind the ocelli and has a tuft of setae. The antennae are 7-segmented; the scape and pedicel are cylindrical and wider than the remainder antennomeres, and antennomere 3 is the longest. The compound eyes are large and strongly protrude anterolaterally. The occiput is large with small spines, evenly rounded, and not bulging. The posterior margin of the head is shallowly concave. Labium. The prementum is longer than its widest part (Fig. 4.10), the lateral margins are strongly convex along the apical 0.40 and have stout serrations intermingled with setae, and the basal 0.60 is narrower and has the lateral margins sinuate. The ligula is highly variable, from poorly developed and shallowly convex, with its apical margin with abundant piliform setae and median cleft very short (*H. uninervulata*), to well-developed with two large convex lobes separated by a wide and deep cleft (*H. simplicia*). The labial palp has a dorsal row of five very long and strong setae of which the anterior two are the longest, or lacking; the apical lobe of the labial palp can be squarely truncate with a large sharply pointed hook on the mesal corner, and the mesal margin is minutely denticulate (*H. uninervulata*) (Fig. 4.16), or the apical lobe is a long, sickle-shaped, thin curved hook, with a row of tiny denticles on the mesal margin (*H. simplicia*) Fig. 4.17). The movable hook is as long as the palp, sharp and moderately incurved, and has a row of small, stout setae along 0.66 of its length.

Thorax. The pronotal disc narrows laterally, ending in a triangular process directed anteriorly, and the posterior margin is evenly convex, but straight at the middle. The ventral margin of the propleuron is U- or V-shaped; each arm is wider at the base, bluntly pointed, and bears long setae and small spines. The legs are relatively short (i.e., when fully extended, the hind legs scarcely reach the posterior margin of S6) and have alternating pale and dark, wide rings; the femora and tibiae margins are serrulate; the claws are simple. The anterior and posterior wing sheaths are parallel and distinctly banded and reach the posterior margin of S3 and basal half of S4, respectively.

Abdomen. The abdomen is lanceolate, slender, slightly widened at the middle, tapers posteriorly, and has a complex, mottled color pattern. The ventral surface is more or less flat, whereas the dorsal surface is convex and without dorsal protuberances. S5–S9 end in sharply pointed lateral spines, which increase in size and robustness posteriorly. The female gonapophyses reach the basal 0.25 of S10. The caudal appendages are as long as S9 + S10. The epiproct lacks a longitudinal dorsal keel or carina, and its apical margin is truncate. The paraprocts are pyramidal, straight, and sharply pointed; their mesal and lateral margins are weakly serrate, and bear a few, long, fine setae. The cerci are very long, slightly hooked inwardly toward the tip, and are very sharply pointed. The paraprocts are the longest of the caudal appendages, and the cerci are 0.95× the length of the epiproct.

Measurements (in mm). Total length including caudal appendages 30; maximum width of head 7.3.

4.6.3 Remarks

Three species of *Heliaeschna* have been recorded from Thailand: *H. crassa* Krüger, 1899, *H. simplicia* (Karsch, 1891), and *H. uninervulata* Martin, 1909 (Hämäläinen and Pinratana 1999; Sribal et al. 2018). Of these, the nymphs of only

H. uninervulata and *H. simplicia* are known. They are found in secondary forest pools or tiny runnels in alluvial swamp forest with plenty of leaf litter, fallen twigs and branches, and with mud as the underlying substrate (Orr and Ngiam 2011; Butler and Orr 2013). The nymphs of *Heliaeschna* closely resemble those of *Gynacantha*. Some authors consider *Heliaeschna* to be polyphyletic because nymphal characters suggest this. In fact, genuine *Heliaeschna* may be only in Africa, so a new name would be needed for at least one Asian group (Orr, pers. com.), and probably the Oriental *Heliaeschna* represents at least three different genera (Orr and Ngiam 2011).

4.6.4 Distribution

In Thailand, *Heliaeschna* has been recorded from the provinces of Chiang Mai, Lampang, Nakhon Nayok, Nakhon Si Thammarat, Trang (Hämäläinen and Pinratana 1999), Chiang Rai (Katatani et al. 2004), Nan (Katatani et al. 2010), Chanthaburi (Kosterin et al. 2011), Phetchaburi (Day et al. 2012), and more recently from Narathiwat and Samut Sakhon (Sribal et al. 2018) (Fig. 4.44).

Fig. 4.44 Distribution of *Heliaeschna*

4.7 *Indaeschna* Fraser, 1926

In the absence of specimens to examine, the description presented here has been adapted from Orr (1994).

4.7.1 General Appearance

The nymphs of *Indaeschna* are characterized by their very large size (51–55 mm), body slender, head flattened, abdomen lanceolate, and coloration light-brown to nearly black.

4.7.2 Description

Head. The head is wider than long and wider than the thorax and abdomen. The antennae are long and 7-segmented. The compound eyes are large and strongly protrude anterolaterally. The occipital lobes are widely rounded, and the posterior margin of the head is widely concave. The prementum is long, longer than its widest part; the lateral margins are strongly convex along the apical 0.35, the basal 0.65 is narrower and with the lateral margins sinuate. The ligula is poorly developed and shallowly convex, its apical margin has abundant piliform setae, the median cleft is closed, and the vertical slit short. The labial palp (Fig. 4.16) is seruciform and lacks setae, its apical lobe is squarely truncate, the dorsal surface is shallowly concave, and a large, sharply pointed tooth is on the mesal corner; the mesal margin is shallowly concave, and the apical margin is blade-like and minutely crenulate. The movable hook is as long as the palp, sharp, and moderately incurved.

Thorax. The legs are relatively short (i.e., when fully extended, the hind legs scarcely reach the anterior margin of S6) and have alternating pale and dark wide rings. The anterior and posterior wing sheaths are parallel, transversely banded, and do not reach the posterior margin of S2 and basal half of S3, respectively.

Abdomen. The abdomen is lanceolate, slender, widened at the middle, tapers posteriorly, and the coloration is variably mottled. The ventral surface is more or less flat, whereas the dorsal surface is convex and lacks dorsal protuberances. S6–S9 end in sharply pointed lateral spines. The caudal appendages are longer than the dorsal length of S9 + S10. The epiproct and paraprocts are pyramidal. The epiproct has its apical margin truncate. The epiproct of the male has the basal lamina along the basal 0.40. The paraprocts are sharply pointed. The cerci are very sharply pointed and abruptly narrowed along the apical 0.25. The paraprocts are the longest of the caudal appendages, and the cerci are 0.35× (♀) and 0.50× (♂) the length of the epiproct.

Measurements (in mm). Total length including caudal appendages 51–55; maximum width of head 10.0; length of hind femur (lateral) 9.0.

4.7.3 Remarks

Only *Indaeschna grubaueri* (Förster, 1904) has been recorded in Thailand (Hämäläinen and Pinratana 1999). The nymph of *I. grubaueri* has been found living in small ponds, rock pools, large treeholes, and in water at the bottom of a disused boat (Orr 1994). Its development and behavior were studied by Orr (1994). Apparently, the nymph closely resembles that of *Polycanthagyna* in general morphology.

4.7.4 Distribution

In Thailand, *I. grubaueri* has been recorded from the following provinces: Nakhon Si Thammarat, Ranong, Trang, Yala (Hämäläinen and Pinratana 1999), and more recently from Surat Thani (Day et al. 2012) (Fig. 4.45).

Fig. 4.45 Distribution of *Indaeschna*

4.8 *Periaeschna* Martin, 1908

Fig. 4.46 Dorsal habitus of the nymph of *Periaeschna*

4.8.1 General Appearance

The nymphs of *Periaeschna* are characterized by their medium to large size (31.2–39.5 mm) with the body finely wrinkled-granulose and slender, head flattened, abdomen lanceolate, and coloration reddish-brown to brown and also brownish-green (Fig. 4.46).

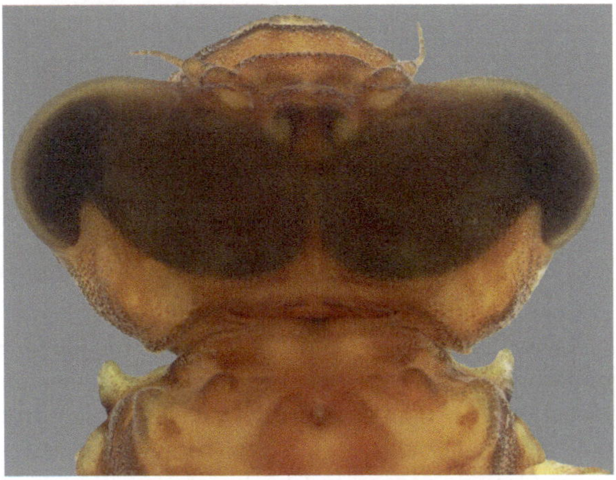

Fig. 4.47 The head in dorsal view

4.8.2 Description

Head. The head is wider than long and wider than the thorax and abdomen and has a complex, mottled color pattern (Figs. 4.46 and 4.47.). The labrum is 0.8 mm long, granulose, and the anterior margin is rounded and more or less flattened ventrally and has a dense row of small setae. The clypeus is bare and slightly wrinkled and the frons is granulose and longer. The anterior margin of the frons is straight. The vertex is domed. The antennae are 7-segmented; the scape and pedicel are cylindrical, wider than the remainder antennomeres, and have small setae; the flagellomeres are mostly bare, antennomere 3 is the longest, and the tip of the antennae does not reach the anterior margin of the labrum. The compound eyes are large and strongly protrude anterolaterally (Fig. 4.47). The occiput is large, and the margins of the occipital lobes are shallowly crenulate and granulose and have minute setae. The posterior margin of the head is widely concave. Mandibles have the formula: *L* 1234 *y*- $a(m^{1,2,3,\text{ or }4})b$/ *R* 1234 *y* $a(m^{1,2,3,4})b$, with $a > b$ in both mandibles. The maxilla's galeolacinia is as described for *Anax*. The ventral pad of the hypopharynx is subpentagonal, yellowish, and poorly sclerotized in the basal half; the apical half is soft and has long, stiff, regularly spaced setae, and the posterior margin is shallowly concave.

Fig. 4.48 The prementum in ventral view

Labium. The prementum-postmentum articulation reaches the posterior margin of the mesocoxae (Figs. 4.48 and 4.49). The prementum is light-reddish-yellow and longer than its widest part; the lateral margins are strongly convex along the apical 0.30 and with small serrations; the basal 0.70 is narrowed, and the lateral margins are moderately convex and finely serrulated; a longitudinal, bare, feebly marked central sulcus is in the basal half of the ventral surface; the basal margin is strongly emarginate. The ligula is well developed, convex, its apical margin has abundant piliform setae and two short, bluntly pointed teeth, one on each side of the widely open, V-shaped, short, median cleft; these teeth are distad of the vertical slit (best seen in ventral view). The labial palp lacks dorsal setae, its apical lobe is more or less spindle-shaped, squarely truncate and serrulated apically, it has a short conical tooth at the mesal corner, and the mesal margin has 10–12 small truncate teeth. The movable hook is bicolored, yellow in the basal 0.70 and reddish-brown in the apical 0.30, slightly shorter than the palp, sharp, moderately incurved, and without setae.

Fig. 4.49 Ventral habitus of the nymph of *Periaeschna*

Thorax. The anterior margin of the pronotum is biconvex, the lateral margins are squarely truncate and thickly ridged, and the posterior margin is mostly sinuate. The pronotal disc (Fig. 4.47) is granulose and has a median mound-like tubercle, and a large, transverse, subtriangular bare area is on each side of the midline. The ventral margin of the propleuron is V-shaped (Figs. 4.47 and 4.48), with the episternal arm digitiform and slightly curved anteriorly, and the epimeral arm conical, wider at the base, bluntly pointed, and gently curved posteriorly; both are covered with abundant white, minute, scale-like setae. The side of the pterothorax is finely granulose and has a submedian, mound-like tubercle on the mesepimeron. The legs are relatively short (Figs. 4.46 and 4.49) (i.e., when fully extended, the hind legs scarcely reach the basal half of S7), and have alternating pale and dark, wide rings on the femora and tibiae, which in turn have the margins serrulate; the pretarsal claws are simple and each has a pulvilliform empodium. The anterior and posterior wing sheaths are parallel and reach the basal half and posterior margin of S4, respectively.

Fig. 4.50 The abdomen and caudal appendages in lateral view

Abdomen. The abdomen is lanceolate, slender, slightly widened at the middle, tapers posteriorly, and has a complex, mottled color pattern (Fig. 4.46). The ventral surface is more or less flat, whereas the dorsal surface is convex and without dorsal protuberances (Fig. 4.50). S6–S9 end in sharply pointed lateral spines, which increase in size and robustness posteriorly (Figs. 4.46, 4.49, and 4.50). Terga 2–10 are finely granulose with minute, white setae, and the posterior margins are serrulated. The lateral margins of S3–S10 are serrulate. Sterna 1–10 are finely granulose; sterna 3–8 are each divided into five sternites (the anterolateral sternites are small and usually hidden by the preceding sternum) and sternum 9 into three sternites. The female gonapophyses scarcely reach the posterior margin of S10. The caudal appendages are longer than S9 + S10 and sharply pointed (Fig. 4.50). The epiproct and paraprocts are pyramidal, of the same length, and their margins are strongly serrated; the tip of the epiproct is slightly decurved and the tips of the paraprocts are slightly incurved and forceps-like. The epiproct of the male has the basal lamina along the basal half. The cerci are digitiform and 0.55× the length of the epiproct.

Measurements (in mm). Total length including caudal appendages 31.2–39.5; abdomen (ventral) 19.0–26.9; maximum width of head 7.2–7.7; length of hind femur (lateral) 5.8–6.0.

4.8.3 Remarks

Only *P. nocturnalis* Fraser, 1927, has been recorded in Thailand (Hämäläinen and Pinratana 1999). Specimens of this genus seem to be rare in collections. Nevertheless, the nymphs of six of the 11 species known to date have been described (Xu and Zhang 2014). The nymphs of *Periaeschna* are found in clear, shallow mountain streams under waterfalls (Kawashima and Sasamoto 2006), in forested streams in riffles, as well as in rice paddy field ditches (Zhang and Tong 2011). A nymph of *P. magdalena* Martin, 1909, was collected from the gravelly bottom of a small stream (Matsuki and Lien 1984); the nymphs of *P. yazhenae* Xu, 2012, were found "under fallen leaves in a small stone puddle adjacent to a large waterfall" (Xu 2012), and those of *P. zhangzhouensis* Xu, 2007, were found clinging to aquatic weeds (Xu and Zhang 2014). This genus is regarded as one of the most primitive groups in the family Aeshnidae (Kawashima and Sasamoto 2006).

4.8.4 Distribution

In Thailand, *Periaeschna* has been recorded only from Chiang Mai and Ranong provinces (Hämäläinen and Pinratana 1999). ***New province records***: Loei, Mae Hong Son, and Phitsanulok (Fig. 4.51).

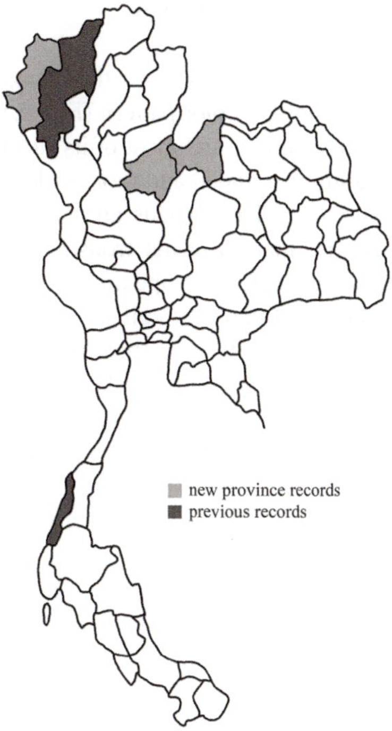

Fig. 4.51 Distribution of *Periaeschna*

4.9 *Petaliaeschna* Fraser, 1927

Fig. 4.52 Dorsal habitus of a young nymph of *Petaliaeschna*

4.9.1 General Appearance

The nymphs of *Petaliaeschna* are characterized by their medium size (31.8 mm), the body covered with minute scale-like setae, head flattened, abdomen lanceolate, coloration reddish-brown with a complex, mottled pattern, and the anal pyramid very short and tuberculated (Fig. 4.52).

4.9.2 Description

Head. The head is wider than long and much wider than the thorax, and has a complex, mottled color pattern (Fig. 4.53). The labrum is 0.8 mm long, spiny on the laterobasal areas, and most of its surface beset with minute scale-like setae; the anterior margin is rounded and more or less flattened ventrally and has a dense row of small, stiff setae. The clypeus and antefrons are scaly and both are the same length. The anterior margin of the frons is straight and the vertex is domed. The antennae are 7-segmented; the scape and pedicel are cylindrical, wider than the remaining antennomeres, and have small spiniform setae; the pedicel is the longest; the flagellomeres have small, stiff setae; the tips of the antennae reach the anterior margin of the labrum. The compound eyes are large and strongly protrude anterolaterally (Figs. 4.52 and 4.53). The occipital lobes are finely granulose and have longitudinal rows of minute, scale-like setae. The posterior margin of the head is widely concave (Fig. 4.53).

Fig. 4.53 The head and prothorax in dorsal view. Notice the deep and narrowly V-shaped margin of the propleuron

Mandibles have the formula: L 1234 $a(m^{1,2,3,4})bd/$ R 1234 y $a(m^{1,2,3,4,5,6})bd$, with $a = b$ in both mandibles. The maxilla's galeolacinia is as described for *Anax*. The ventral pad of the hypopharynx is subpentagonal, yellowish, and softly sclerotized in the basal half, and has a transverse U-shaped ridge close to the posterior margin; the apical half is soft and has long, stiff, regularly spaced setae.

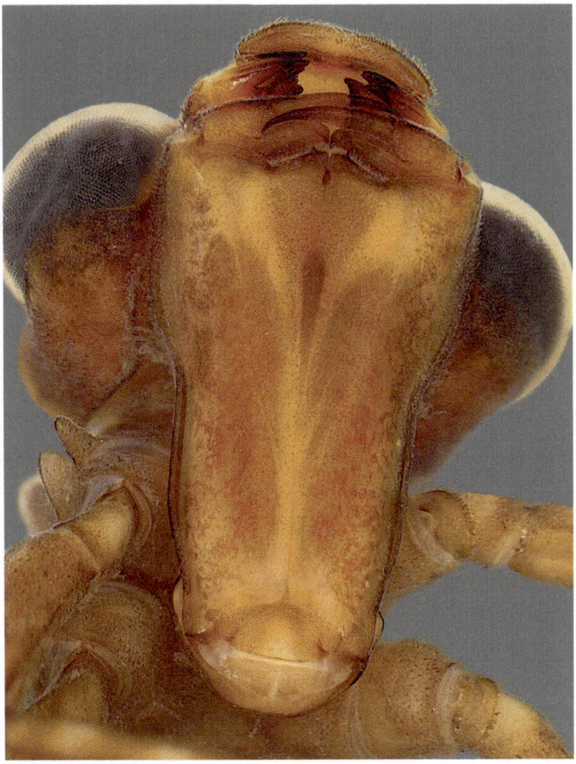

Fig. 4.54 The prementum in ventral view

Labium. The prementum-postmentum articulation reaches the posterior margin of the mesocoxae (Figs. 4.54 and 4.55). The prementum is yellowish-brown, longer than its widest part, the lateral margins are biconvex and strongly convex along the apical 0.40 with small, stout serrations (Fig. 4.54); the basal 0.60 is narrowed and moderately convex and ridged; the basal margin is strongly emarginate (Fig. 4.54). The ventral surface of the prementum has abundant minute, scale-like setae, and a longitudinal, bare, feebly marked central sulcus is in the basal half. The ligula is well developed, convex, and the apical margin has abundant piliform setae and two short, bluntly pointed teeth, one on each side of the closed, short, median cleft; these teeth are distad of the vertical slit (best seen in ventral view). The basal half of the labial palp has abundant small, scale-like setae on the ventral and lateral surfaces, and the dorsal surface has short hair-like setae; the apical lobe of the labial palp is more or less fusiform, the distal margin is nearly straight, and a short, conical tooth is at the rounded mesal corner; the mesal margin has 16–18 small truncate teeth. The movable hook is bicolored, yellowish-brown along the basal 0.70 and reddish-brown along the apical 0.30, as long as the palp, sharp, moderately incurved, and has a row of small, sparse setae on the dorsal surface.

Fig. 4.55 Ventral habitus of a young nymph of *Petaliaeschna*

Thorax. The anterior margin of the pronotum is shallowly biconvex, the lateral margins are rounded and feebly serrulated, and the posterior margin is widely convex. The pronotal disc is mostly bare and wide, and a submedian, gentle swelling beset with very minute, scale-like setae is on each side of the midline. The ventral margin of the propleuron is deep and narrowly V-shaped (Fig. 4.53), with the episternal arm digitiform and the epimeral arm conical and wider at the base (Figs. 4.54 and 4.55), and both are bluntly pointed and are beset with minute, scale-like setae. The pterothorax has minute spines and scale-like setae, and a submedian, mound-like tubercle is on the mesepimeron. The legs are relatively short (i.e., when fully extended, the hind legs scarcely surpass the posterior margin of S5) (Figs. 4.52 and 4.55), and have alternating pale and dark, wide bands on the femora and tibiae; the femora and tibiae have spiniform and scale-like setae on the margins; the claws are simple and each has a pulvilliform empodium. The anterior and posterior wing sheaths are parallel and reach the basal half and posterior margin of S3, respectively.

Abdomen. The abdomen is lanceolate, slender, slightly widened at the middle, gradually tapers posteriorly (Figs. 4.52 and 4.55), and has a complex, mottled color pattern. The ventral surface is more or less flat, whereas the dorsal surface is convex and without dorsal protuberances. S6–S9 end in triangular, sharply pointed lateral spines, which increase in size and robustness posteriorly (Fig. 4.55). Terga 2–10 are very finely granulose, covered with minute scale-like setae, and the posterior margins are serrulated. The lateral margins of S4–S5 are serrulate and those of S6–S9 moderately serrate. Sterna 1–8 have abundant minute, reddish setae, and sterna 9–10 have abundant long, stiff setae. Sterna 3 and 9 are each divided into three sternites and sterna 4–8 into five sternites. The female gonapophyses reach the basal half of S10 and have abundant long, stiff setae on the ventral margins.

Fig. 4.56 The caudal appendages in dorsolateral view

The caudal appendages are yellow and shorter than the dorsal length of S9 + S10 (Fig. 4.52). The epiproct, in dorsal view, appears more or less cylindrical, the lateral margins gradually converge to a shortly bifurcated apex with two short, bluntly pointed spines; it is slightly shorter than the paraprocts (Fig. 4.56). The paraprocts are pyramidal, the longest of the caudal appendages, and have the tips incurved and forceps-like (Fig. 4.55). The epiproct and paraprocts have short, roundly pointed, reddish-brown tubercles on the margins and surfaces (Fig. 4.56). The cerci are triangular, the shortest of the caudal appendages, mostly bare, and bluntly pointed (Fig. 4.56).

Measurements (in mm). (Taken from Kawashima and Karube 2006): Total length including caudal appendages 31.8; maximum width of head 6.7; length of hind femur (dorsal) 4.3.

4.9.3 Remarks

Only *Petaliaeschna flavipes* Karube, 1999, has been recorded from Thailand (Hämäläinen and Pinratana 1999 as *Petaliaeschna* sp., Yeh 1999 as *P. pinratanai*). The nymph of *P. flavipes* was described by Kawashima and Karube (2006) based on a female last instar exuvia; these authors reported that the exuvia "…was collected on a stem grass that grows along the bank at the source of a small river on the hill…." We collected a young nymph, probably F-4 or F-3, in a small mountain creek at 1100 m elevation. The nymph of *Petaliaeschna* shares most of its features with the nymph of *Planaeschna*. For this reason, we think both genera are the closest relatives of each other, although other authors, based exclusively on adults, have proposed close relationships with *Cephalaeschna*, *Gynacanthaeschna*, *Limnetron*, and *Periaeschna* (Asahina 1981, cited by Kawashima and Karube 2006, von

Ellenrieder 2002). Kawashima and Karube (2006) pointed out the differences of the nymph of *Petaliaeschna* from those of *Cephalaeschna* and *Periaeschna*.

4.9.4 Distribution

In Thailand, *Petaliaeschna* has been recorded only from Chiang Mai Province (Hämäläinen and Pinratana 1999; Yeh 1999) (Fig. 4.57).

Fig. 4.57 Distribution of *Petaliaeschna*

4.10 *Planaeschna* McLachlan, 1896

Fig. 4.58 Dorsal habitus of the nymph of *Planaeschna*

4.10.1 General Appearance

The nymphs of *Planaeschna* are characterized by their large size (34.5–35.5 mm) with the body covered abundantly by minute, scale-like setae, the head flattened, abdomen lanceolate, and coloration yellowish-brown to dark-brown with a complex, mottled pattern (Fig. 4.58).

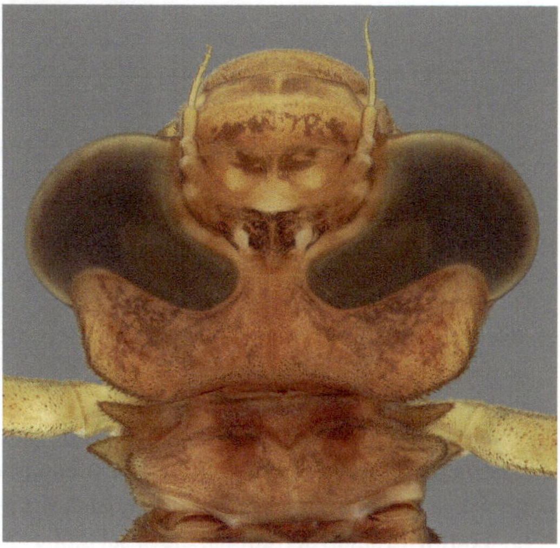

Fig. 4.59 The head in dorsal view

4.10.2 Description

Head. The head is wider than long (Fig. 4.59), much wider than the thorax and abdomen, and has a complex, mottled color pattern. The labrum is 0.9 mm long, spiny on the laterobasal areas, most of its surface has minute scale-like setae, the anterior margin is rounded and more or less flattened ventrally, and has a dense row of small, stiff setae. The clypeus is scaly, the frons is spiny-scaly, and both are the same length. The anterior margin of the frons is nearly straight and the vertex is domed. The antennae are 7-segmented; the scape and pedicel are cylindrical, wider than the remainder antennomeres, and have small spiniform setae; the flagellomeres have small, stiff setae and some long bristles; antennomere 3 is the longest and the distal margin of antennomere 6 reaches the anterior margin of the labrum. The compound eyes are large and strongly protrude anterolaterally (Fig. 4.59). The occipital lobes have a small rounded tubercle on the posterolateral margins and abundant minute, scale-like setae arranged in longitudinal rows. The posterior margin of the head is shallowly concave. Mandibles have the formula: *L* 1234 *y*- *a*($m^{1,2,3,4}$)*bb'dk*/*R* 1234 *y a*($m^{1,2,3,4,5}$)*bb'dk*, with $a > b$ in both mandibles. The maxilla's galeolacinia is as described for *Anax*. The ventral pad of the hypopharynx is subpentagonal, yellowish, poorly sclerotized in the basal half, and has a U-shaped ridge close to the posterior margin; the apical half is soft and has long, stiff, regularly spaced setae.

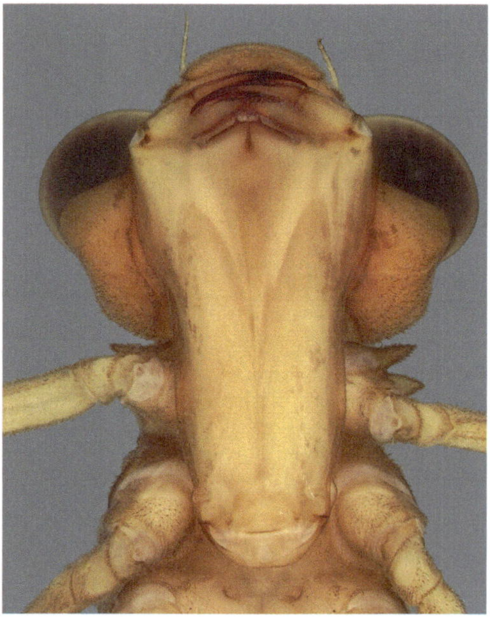

Fig. 4.60 The prementum in ventral view

Labium. The prementum-postmentum articulation reaches the posterior margin of the mesocoxae (Figs. 4.60 and 4.61). The prementum is yellow, longer than its widest part, the lateral margins are biconvex, and strongly convex along the apical 0.40 (Fig. 4.60) with small serrations; the basal 0.60 is narrowed and the margins are scaly; the basal margin is strongly emarginate. The ventral surface of the prementum has abundant minute, scale-like setae; a longitudinal, feebly marked, central sulcus is in the basal half. The ligula is well developed, convex, and its apical margin has abundant piliform setae and two short, blunt teeth (Fig. 4.60), one on each side of the closed, short, median cleft; these teeth are distad of the vertical slit (best seen in ventral view). The basal half of the labial palp has abundant small, scale-like setae on the ventral and lateral surfaces; a row of short, hair-like setae is on the dorsal surface; the apical lobe of the labial palp is more or less fusiform, the distal margin is straight and the lateral corner rounded; the mesal margin has 20–22 small, truncate teeth, and ends in a short, conical tooth at the distal corner. The movable hook is bicolored, yellow along the basal 0.60 and reddish-brown along the apical 0.40, longer than the palp, sharp, moderately incurved, and has a sparse row of small setae on the dorsal surface.

Fig. 4.61 Ventral habitus of the nymph of *Planaeschna*

Thorax. The anterior margin of the pronotum is biconvex, the lateral margins are rounded, and the posterior margin is strongly convex. The pronotal disc is scaly, wide, and has a submedian, gentle tumescence and a large, transverse, subtriangular bare area on each side of the midline. The ventral margin of the propleuron is deeply V-shaped (Figs. 4.59 and 4.60), with the episternal arm digitiform and slightly curved anteriorly, and the epimeral arm conical, wider at the base, bluntly pointed, and gently curved posteriorly; both are covered with abundant grayish, minute, scale-like setae. The side of the pterothorax is scaly, and a submedian, mound-like tubercle is on the mesepimeron. The legs are relatively short (i.e., when fully extended, the hind legs scarcely surpass the posterior margin of S6), and have alternating pale and dark, wide bands on the femora and tibiae, which have the margins abundantly scaly; the pretarsal claws are simple and each has a pulvilliform empodium. The anterior and posterior wing sheaths are parallel and reach the basal half and posterior margin of S3, respectively.

Fig. 4.62 The abdomen and caudal appendages in lateral view

Abdomen. The abdomen is lanceolate, slender, slightly widened at the middle, tapers posteriorly, and has a complex, mottled color pattern (Figs. 4.58 and 4.61). The ventral surface is more or less flat, whereas the dorsal surface is convex and without dorsal protuberances. S6–S9 end in triangular, sharply pointed lateral spines, which increase in size and robustness posteriorly. Terga 2–10 are covered with minute, grayish, scale-like setae, and the posterior margins are serrulated. The lateral margins of S4–S6 are serrulate and S7–S9 moderately serrate. Sterna 1–10 have abundant minute, grayish, scale-like setae intermingled with small, reddish setae. Sterna 3–8 are each divided into five sternites (the anterolateral sternites are small and usually hidden by the preceding sternum) and sternum 9 into three sternites. The female gonapophyses do not reach the posterior margin of S10. The caudal appendages are as long as or slightly longer than the dorsal length of S9 + S10, and sharply pointed. The epiproct is rectangular, the lateral margins are straight and spiny, and the apical margin is widely U-shaped and ends in a sharply pointed spine on each side (Fig. 4.58) and a small spine in the middle. The paraprocts are pyramidal, the longest of the caudal appendages (Fig. 4.62), and have the margins strongly serrated and the tips incurved and forceps-like (Fig. 4.61). The epiproct of the male has the basal lamina along the basal 0.60. The epiproct is 0.70× the length of the paraprocts. The cerci are digitiform and the shortest of the caudal appendages, 0.66× the length of the epiproct.

Measurements (in mm). Total length including caudal appendages 34.5–35.5; abdomen (ventral) 24.2–25.5; maximum width of head 7.3–7.5; length of hind femur (lateral) 5.5– 5.8.

4.10.3 Remarks

Three species of *Planaeschna* have been recorded from Thailand (Hämäläinen and Pinratana 1999, 2000). The nymphs have been found "clinging to the underside of a floating leaf in a forest brook" (Xu and Zhang 2011). The nymph of *Planaeschna* shares most of its features with those of *Petaliaeschna*.

4.10.4 Distribution

In Thailand, *Planaeschna* has been recorded from only the provinces of Chiang Mai, Loei (Hämäläinen and Pinratana 1999, 2000; Hämäläinen 2002), and Chiang Rai (Katatani et al. 2004). ***New province record***: Phitsanulok (Fig. 4.63).

Fig. 4.63 Distribution of *Planaeschna*

4.11 *Polycanthagyna* Fraser, 1933

Fig. 4.64 Dorsal habitus of the nymph of *Polycanthagyna erythromelas*

4.11.1 General Appearance

The nymphs of *Polycanthagyna* are characterized by their very large size (46.0–47.5 mm), the body slender with abundant minute, white setae, which give it a finely granulose appearance, head flattened, abdomen lanceolate, and coloration light-brown (Fig. 4.64).

Fig. 4.65 The head and pronotum in dorsal view

Head. The head is wider than long and wider than the thorax and abdomen (Fig. 4.65). The labrum is 1.3 mm long, mostly granulose, its anterior margin is rounded and more or less flattened ventrally, and it has a dense row of setae. The clypeus is striated. The antefrons is granulose and the same length than clypeus. The anterior margin of the postfrons is almost straight and the vertex is domed. The antennae are 7-segmented and mostly bare; the scape and pedicel are cylindrical and wider than the remaining antennomeres; antennomere 3 is the longest and the apical margin of antennomere 5 reaches the anterior margin of the labrum. The compound eyes are large and strongly protrude anterolaterally (Figs. 4.64 and 4.65). The occipital lobes are widely rounded and have abundant minute, spiniform setae; the posterior margin of the head is widely concave. The ventral carinae of the head are poorly developed and have a row of close-packed spiniform setae. Mandibles have the formula: *L* 1234 *y- a(m*^{0}*)bk*/ *R* 1234 *y a(m*$^{1+2+3+4+5}$*)bk*, with $a > b$ in both mandibles. The maxilla's galeolacinia is as described for *Anax*. The ventral pad of the hypopharynx is subpentagonal, yellowish, and poorly sclerotized in the basal half, a U-shaped ridge is close to the posterior margin; the apical half is soft and has stiff, regularly spaced setae.

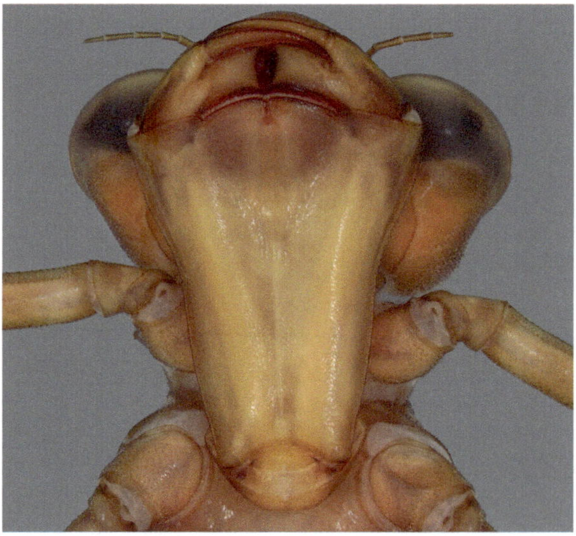

Fig. 4.66 The prementum in ventral view

Labium. The prementum-postmentum articulation almost reaches the posterior margin of the mesocoxae (Fig. 4.66). The prementum is light-yellow, longer than its widest part, and the lateral margins are strongly convex along the apical 0.30 and have stout serrations; the basal 0.70 is narrower and the lateral margins are sinuate and serrulate; the basal margin is strongly emarginate; there is no central sulcus in the basal half of the ventral surface. The ligula is poorly developed, convex (Fig. 4.66), its apical margin has abundant piliform setae and two small, bluntly pointed teeth, one on each side of the short, V-shaped, median cleft, and the vertical slit is short (best seen in ventral view). The labial palp is moderately and finely granulose on the ventral and lateral surfaces, and small setae are irregularly arranged on the dorsal surface. The apical lobe of the labial palp is squarely truncate (Fig. 4.66), its dorsal surface is shallowly concave, its mesal margin has 18–20 small, rectangular teeth, it ends in a large, sharply pointed tooth, and the apical margin is blade-like and minutely crenulate. The movable hook is reddish-yellow in the basal 0.80 and reddish in the apical 0.20, sharp, moderately incurved, shorter than the palp, and has a row of small, stout setae along most of its dorsal surface.

Fig. 4.67 Ventral habitus of the nymph of *Polycanthagyna erythromelas*

Thorax. The anterior margin of the prothorax is biconvex; the lateral margins of the pronotal disc are rounded, and the posterior margin is mostly convex and undulate. The ventral margin of the propleuron is widely V-shaped (Fig. 4.65), and each arm is wider basally, bluntly pointed, and with abundant spiniform setae. The legs are relatively short (Figs. 4.64 and 4.67) (i.e., when fully extended, the hind legs scarcely reach the posterior margin of S8), and have alternating narrow, pale and wide, dark rings; the femora and tibiae margins have spiniform setae; the pretarsal claws are simple and each has a pulvilliform empodium. The anterior and posterior wing sheaths are parallel and reach the posterior margin of S3 and basal half of S4, respectively (Fig. 4.64).

Abdomen. The abdomen is lanceolate, slender, widened at the middle, tapers posteriorly (Figs. 4.64 and 4.67), and the coloration is more or less uniformly variegated. The ventral surface is more or less flat, whereas the dorsal surface is convex and without dorsal protuberances. S6–S9 end in sharply pointed lateral spines, which increase in size and robustness posteriorly (Fig. 4.67). Terga 2–10 are beset with minute, white, delicate setae, and the posterior margins are serrulated. The lateral margins of S1–S10 are serrulate. Sterna 1–10 are as described for the terga, but the posterior margins are smooth. Sterna 3–8 are each divided into five sternites and sternum 9 into three sternites. The female gonapophyses reach the basal 0.70 of S10. The caudal appendages (Fig. 4.68) are longer than the dorsal length of S9 + S10. The epiproct and paraprocts are pyramidal. The epiproct has a strong, longitudinal, dorsal carina beset with long, stiff setae; its apical margin is truncate and a short spine is at each corner. The epiproct of the male has the basal lamina along the basal 0.40 (Fig. 4.64). The dorsal and ventral margins of the paraprocts are finely denticulate and the apices sharply pointed. The cerci are digitiform, very sharply pointed, and abruptly narrowed along the apical 0.20. The paraprocts are the longest of the caudal appendages, and the cerci are the shortest, 0.50–0.60× the length of the epiproct (Fig. 4.68).

Fig. 4.68 The abdomen and caudal appendages in lateral view

Measurements (in mm). Total length including caudal appendages 46.0–47.5; abdomen (ventral) 34.0–37.8; maximum width of head 9.3–10.0; length of hind femur (lateral) 9.0–9.2.

4.11.2 Remarks

Only two species of *Polycanthagyna* have been recorded from Thailand, *P. erythromelas* (McLachlan, 1896) (Hämäläinen and Pinratana 1999) and *P. ornithocephala* (McLachlan, 1896) (Hämäläinen 2002). The nymphs of *Polycanthagyna* are found in leaf litter accumulated at the muddy bottoms of pools of small waterbodies in the forest (Wong et al. 2012). They closely resemble the nymphs of *Gynacantha* in general morphology, but can be easily distinguished by the absence of long, raptorial setae on the labial palp, which are present in *Gynacantha.*

4.11.3 Distribution

In Thailand, *Polycanthagyna* has been recorded from the provinces of Chanthaburi, Chiang Mai, Kanchanaburi, Nakhon Nayok, Ranong (Hämäläinen and Pinratana 1999; Hämäläinen 2002), and Chiang Rai (Katatani et al. 2004). ***New province records***: Nakhon Ratchasima, Phitsanulok, and Tak (Fig. 4.69).

Fig. 4.69 Distribution of *Polycanthagyna*

4.12 *Sarasaeschna* Karube and Yeh, 2001

Fig. 4.70 Dorsal view of the exuvia of *Sarasaeschna pryeri*

4.12.1 General Appearance

The nymphs of *Sarasaeschna* are characterized by their medium to large size (28–34 mm) with the body slender and beset with long, thick, stiff setae giving it a shaggy appearance, head globose, eyes small, abdomen fusiform, and coloration yellowish-brown to reddish-brown (Fig. 4.70).

Fig. 4.71 The head in dorsal view (antennae broken)

4.12.2 Description

Head. The head is wider than long (Fig. 4.71) and slightly wider than the thorax and abdomen. The labrum is 1.0 mm long and has abundant small, spiniform setae, its anterior margin has a dense row of short, stiff setae, and its ventral surface is flat and bears a tuft of long setae. The clypeus is mostly bare and gently rugose, and the frons has abundant stiff setae. The anterior margin of the frons is nearly straight; the vertex is roughened and has a raised area with three ocelli. The antennae are very long and 7-segmented; the scape and pedicel are cylindrical and wider than the remaining antennomeres, and the antennomere 3 is the longest. The compound eyes are small, poorly developed, and slightly protrude laterally (Figs. 4.70 and 4.71). The occipital lobes are large, bulbous, with six dense rows of stiff setae arranged longitudinally on each side of the midline (Fig. 4.71). The posterior margin of the head is widely concave. A prominent, longitudinal, setose keel along the ventral surface of the head ends anteriorly in a large tusk-like spine. Mandibles have the formula: *L* 1234 *y*- $a(m^{1,2,3,4,5})bdk$ / *R* 1234 *y* $a(m^{1,2,3,4,5})bd$, $a > b$ in both mandibles. The maxilla's galeolacinia is as described for *Anax*. The ventral pad of the hypopharynx is subquadrate, the anterior and lateral margins are beset with long, stiff setae directed anteriorly, the central area is bare, and the posterior margin is concave.

Fig. 4.72 The prementum in ventral view

Labium. The prementum-postmentum articulation reaches the anterior margin of the mesocoxae (Figs. 4.72 and 4.73). The prementum is 0.20× longer than its widest part and densely covered with small, stiff setae on the ventral surface and lateral margins; the lateral margins are convex along the apical half (Fig. 4.72) and have short, robust spines, whereas the basal half is narrower and the sides converge basally; there is no central sulcus on the ventral surface; the basal margin is widely concave. The ligula is moderately developed, triangular (Fig. 4.72), its apical margin has abundant short, piliform setae and a minute tooth on each side of the median cleft; the median cleft is short, closed, and teardrop-shaped at the proximal end. The apical lobe of the labial palp is squarely truncate, the distal margin is straight and minutely serrulated, the lateral corner is rounded, the mesal corner has a minute tooth, and the mesal margin is serrulate. The movable hook is short, half the length of the palp, sharp, and moderately incurved.

Fig. 4.73 Ventral view of the exuvia of *Sarasaeschna pryeri*

Thorax. The thorax is abundantly covered by small, stiff setae. The pronotal disc is wide, its anterior margin is straight, the sides are produced laterally as conical, blunt, slightly raised tubercles, and the posterior margin is sinuate. The ventral margin of the proepisternum is more or less straight, and the ventral margin of the proepimeron is produced as a short, blunt, conical tubercle. The legs are setose and relatively short (i.e., when fully extended, the hind legs scarcely surpass the posterior margin of S6) (Figs. 4.70 and 4.73); the fore tibia and fore tarsus lack tridentate setae. The anterior and posterior wing sheaths are parallel and reach the anterior margin and basal half of S3, respectively.

Fig. 4.74 The abdomen and caudal appendages in lateral view

Abdomen. The abdomen is fusiform, slender, slightly widened at the middle, tapers posteriorly (Figs. 4.70 and 4.73), and is strongly setose with thick, long, stiff setae. The ventral surface is more or less flat, whereas the dorsal surface is convex and without dorsal protuberances. A middorsal longitudinal ridge of long, stiff setae is on S1–S8; the lateral margins of S6–S9 end in triangular, sharply pointed spines, which are small on S6, increasing in size and robustness posteriorly, and very broad on S9 (Fig. 4.73). Sterna 3 and 9 are each divided into three sternites and sterna 4–8 into five sternites. The female gonapophyses are strongly developed, bulky, and reach the posterior margin of S10 (Fig. 4.73). The caudal appendages are setose and shorter than the dorsal length of S9 + S10 (Fig. 4.74). The epiproct is cylindrical and truncate apically (Fig. 4.74); the epiproct of the male has the basal lamina along the basal 0.50 and is bilobed apically. The paraprocts are pyramidal, slightly longer than the epiproct, and have the margins strongly serrate with the tips incurved and forceps-like. The cerci are digitiform, sharply pointed, serrate at the sides, and the tips are elongate, incurved, and forceps-like; the cerci are 0.90× the length of the epiproct.

Measurements (in mm). Total length including caudal appendages 28–34; abdomen 16.0–22.5; maximum width of head 6.0–6.5; length of hind femur (lateral) 5.9–6.1.

4.12.3 Remarks

The genus *Sarasaeschna* was erected by Karube and Yeh (2001) to include those species formerly belonging to the genus *Oligoaeschna*, which are distributed from the eastern Himalayas to Japan (Karube and Yeh 2001). Two species of *Sarasaeschna* have been recorded from Thailand [as *Oligoaeschna*] (Hämäläinen and Pinratana 1999, 2000): *S. minuta* (Asahina, 1986) belonging to the *pryeri* group, and *S. pramoti* Yeh, 2000, belonging to the *niisatoi* group. The nymphs of *Sarasaeschna* have been found in stagnant waters such as "small bog pools and sphagnum beds supplied with seepage water … and from the surface mud and debris accumulated in a small pool …" (Lieftinck 1968). Apparently, the nymph of *S. pryeri* (Martin, 1909) is semi-terrestrial in Japan. To date, only the nymphs of *S. pryeri*, *S. kunigamiensis* (Ishida, 1972), and *S. niisatoi* (Karube, 1998) have been described (Asahina 1958; Matsuki 1993; Kawashima 2002).

4.12.4 Distribution

In Thailand, *Sarasaeschna* has been recorded only from the provinces of Chiang Mai and Loei (Hämäläinen and Pinratana 1999; Hämäläinen 2002) (Fig. 4.75).

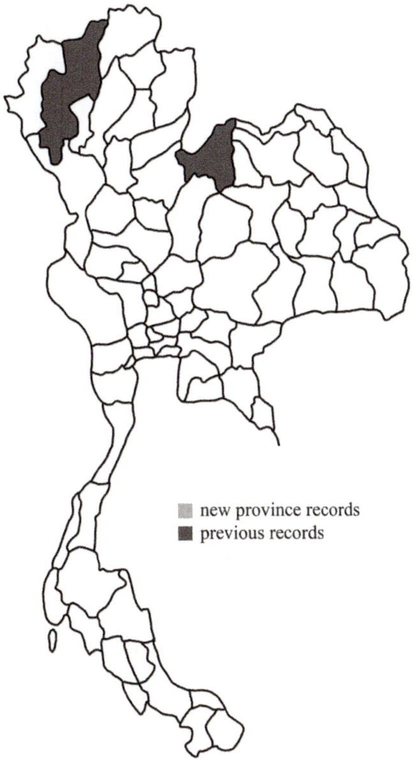

Fig. 4.75 Distribution of *Sarasaeschna*

4.13 *Tetracanthagyna* Selys, 1883

Fig. 4.76 Dorsal habitus of the nymph of *Tetracanthagyna*

4.13.1 General Appearance

The nymphs of *Tetracanthagyna* are characterized by their very large size (50–54 mm) with the body robust, enlarged, and beset with minute, scale-like and white setae, and with multiple tubercles on the head and thorax. The abdomen slightly tapers posteriorly, has dorsal protuberances on the last segments, and coloration is reddish-brown (Fig. 4.76).

Fig. 4.77 The head in dorsal view

4.13.2 Description

Head. The head is wider than long and wider than the thorax and abdomen, densely scaly, and with a complex, mottled color pattern (Fig. 4.77). The labrum is 1.1 mm long, the anterior margin is rounded and more or less flattened ventrally, and has a dense row of stiff setae. The clypeus is scaly and same length of antefrons, although the antefrons is wider and scalier than the clypeus. The anterior margin of frons is straight; the vertex is domed. The antennae are 7-segmented; the scape and pedicel are cylindrical and noticeably wider than the remainder antennomeres, and both have sparse and scale-like setae; the flagellomeres are bare, antennomere 3 is the longest, and the tip of the antennae does not reach the anterior margin of the labrum. The compound eyes are large and strongly protrude laterally (Fig. 4.77). The occiput is large and has four to six tubercles of different sizes as follows: one small and low tubercle on each side of the midline; one or two rounded tubercles on each lateral margin, the posteriormost is larger (Fig. 4.77). The sides of the occipital lobes are strongly sloping down to form broad flanges which have bare areas encircled by minute, scale-like setae, giving a honeycomb appearance (Fig. 4.77). The posterolateral corners of the occiput are more or less angulated, and the posterior margin of the head is widely concave. The ventral carinae of the head have large blunt, conical tubercles just below the posterior margin of each compound eye (Fig. 4.78).

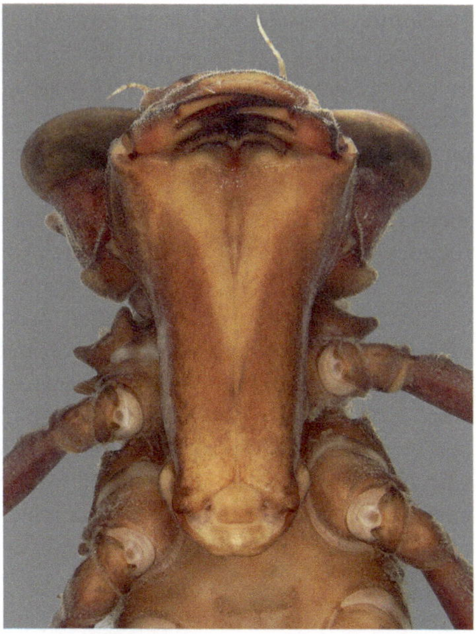

Fig. 4.78 The head and prementum in ventral view

Mandibles have the formula: *L* 1234 *abd-k–*/*R* 1234 *y abd-k-*, with *a* > *b* in both mandibles; in the left mandible a clear-cut intervening portion of the crest separates *a* and *b*; in the right mandible the entire crest is a high two-armed structure with *a* and *b* forming the pointed arms, and the margin between *a* and *b* is serrated. The maxilla's galeolacinia as described as for *Anax*. The ventral pad of the hypopharynx is poorly sclerotized, subpentagonal, yellowish, and has a transverse subbasal carina and abundant long, stiff setae in the anterior third, and a central concavity on the posterior margin. Labium. The prementum-postmentum articulation reaches the posterior margin of the mesocoxae (Figs. 4.78 and 4.79). The prementum is light-reddish-yellow and much longer than its widest part (Fig. 4.78), the lateral margins are strongly convex along the apical half and with stout serrations, and the basal half is more or less sinuate and has small scale-like setae on the margins; a longitudinal, central sulcus is in the basal 0.60, and minute, white, scale-like setae on the ventral surface; the basal margin is wide and deeply emarginated. The ligula is moderately prominent and biconvex, its apical margin has abundant piliform setae and two small, blunt spines, one on each side of the short, V-shaped, median cleft (best seen in ventral view). The labial palp has abundant scale-like setae on the lateral and ventral surfaces; the apical lobe of the labial palp is narrow and parallel-sided; the distal margin is nearly straight, with the lateral corner rounded and the mesal corner ending in a large, stout, red, conical hook; the mesal margin has a row of 18–20 short, square teeth. The movable hook is bicolored, yellow along the basal 0.60 and dark-reddish-brown along the apical 0.40, as long as the palp, sharp, and moderately incurved.

Thorax. The anterior margin of the prothorax is biconvex. The pronotal disc is subtrapezoidal and wider anteriorly; the anterolateral corners are rounded and moderately produced, the lateral margins are sinuate and strongly convergent posteriorly, and the posterior margin is more or less sinuate; a low, rounded tumescence is on each side of the midline. The ventral margin of the propleuron is V-shaped, and the proepisternal anterior margin is divided into two thick, conical, roundly pointed tubercles directed anteriorly; the proepimeral anterior margin is projected as a thick, roundly pointed tubercle gently curving posteriorly. The pterothorax has several low tubercles, mainly along the dorsal margin, and a conical, roundly pointed tubercle on the anterior corners of the ventral margins of the meso- and metathorax. The mesospiracle is very large and the metaspiracle is small. The legs are very short (i.e., when fully extended, the hind legs scarcely surpass the posterior margin of S6) (Fig. 4.79), reddish, with narrow, yellow rings; the femora and tibiae margins are beset with scale-like setae; the lateral surfaces of the femora have low tubercles and sparse, spiniform setae; the pretarsal claws are simple and each has a pulvilliform empodium. The anterior and posterior wing sheaths are parallel, yellow-brown, and reach the basal half and posterior margin of S3, respectively (Fig. 4.76).

Fig. 4.79 Ventral habitus of the nymph of *Tetracanthagyna*

Abdomen. The abdomen is cylindrical, enlarged, and slightly tapers posteriorly (Figs. 4.76 and 4.79). The ventral surface is more or less flat, whereas the dorsal surface is convex and has dorsal protuberances directed posteriorly on S7– or S8–S10 (Fig. 4.80), with that on S9 the largest; S5– or S6–S9 end in lateral spines. Terga 2–6 have stiff, short setae on the middle third; the posterior margins of S1–S5 have minute, scale-like setae, and those of S6–S10 have a row of dark-red, minute spines. The lateral margins of S1–S5 are scaly and of S6–S10 spiny. Sterna 1–7 are yellowish-brown, 8–10 are light-reddish-brown and covered with abundant small, delicate setae mainly on the posterior margin of 10. Sterna 3–8 are each divided into five sternites and sternum 9 into three sternites. The female gonapophyses reach the anterior margin of S10. The caudal appendages are scaly and shorter than S9 + S10 (Fig. 4.80). The epiproct and paraprocts are pyramidal and have small, blunt spines on the dorsal margin of the epiproct, and on the dorsal and ventral margins of the paraprocts; the ventral surfaces of the paraprocts have dense brushes of setae. The epiproct of the male has the basal lamina along the basal 0.65. The cerci are digitiform and end in small, bluntly pointed spines. The paraprocts are the longest of the caudal appendages and the cerci the shortest, 0.70× the length of the epiproct.

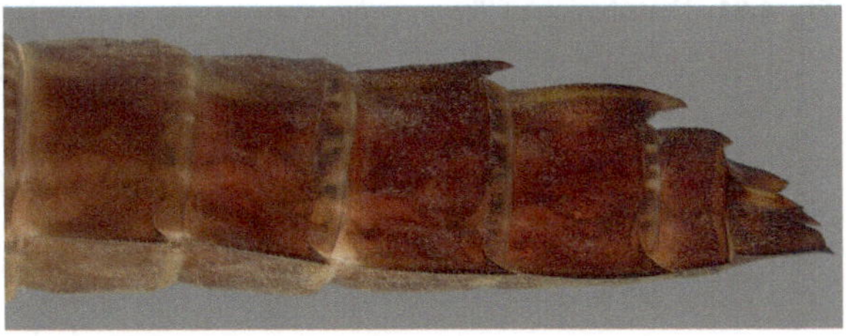

Fig. 4.80 The abdomen and caudal appendages in lateral view

Measurements (in mm). Total length including caudal appendages 50–62; abdomen (ventral) 35–45; maximum width of head 9.3–10.3; length of hind femur (lateral) 7.3–8.3.

4.13.3 Remarks

This genus appears to be closely related to the North American *Nasiaeschna pentacantha* (Rambur, 1842); the nymphs of both genera share several features in common, including the multituberculate body (on head and thorax), the shape of the head, and the lack of *m* denticles on the left mandible. Three species of *Tetracanthagyna* have been recorded in Thailand: *T. brunnea* McLachlan, 1898, *T. plagiata* (Waterhouse, 1877), and *T. waterhousei* McLachlan, 1898 (Hämäläinen and Pinratana 1999). They inhabit slow-flowing forest streams and can be found at streamsides in sand and mud substrate with debris (Orr et al. 2010). They feed upon shrimp, small fish, and other odonate nymphs (Orr 2001).

4.13.4 Distribution

In Thailand, *Tetracanthagyna* has been recorded from the provinces of Bangkok, Chanthaburi, Chiang Mai, Kanchanaburi, Phrae, Ranong, Trang, Yala (Hämäläinen and Pinratana 1999), Chiang Rai (Katatani et al. 2004), and more recently from Ratchaburi (Day et al. 2012). ***New province records:*** Loei, Mae Hong Son, Phitsanulok, Songkhla, and Uttaradit (Fig. 4.81).

Fig. 4.81 Distribution of *Tetracanthagyna*

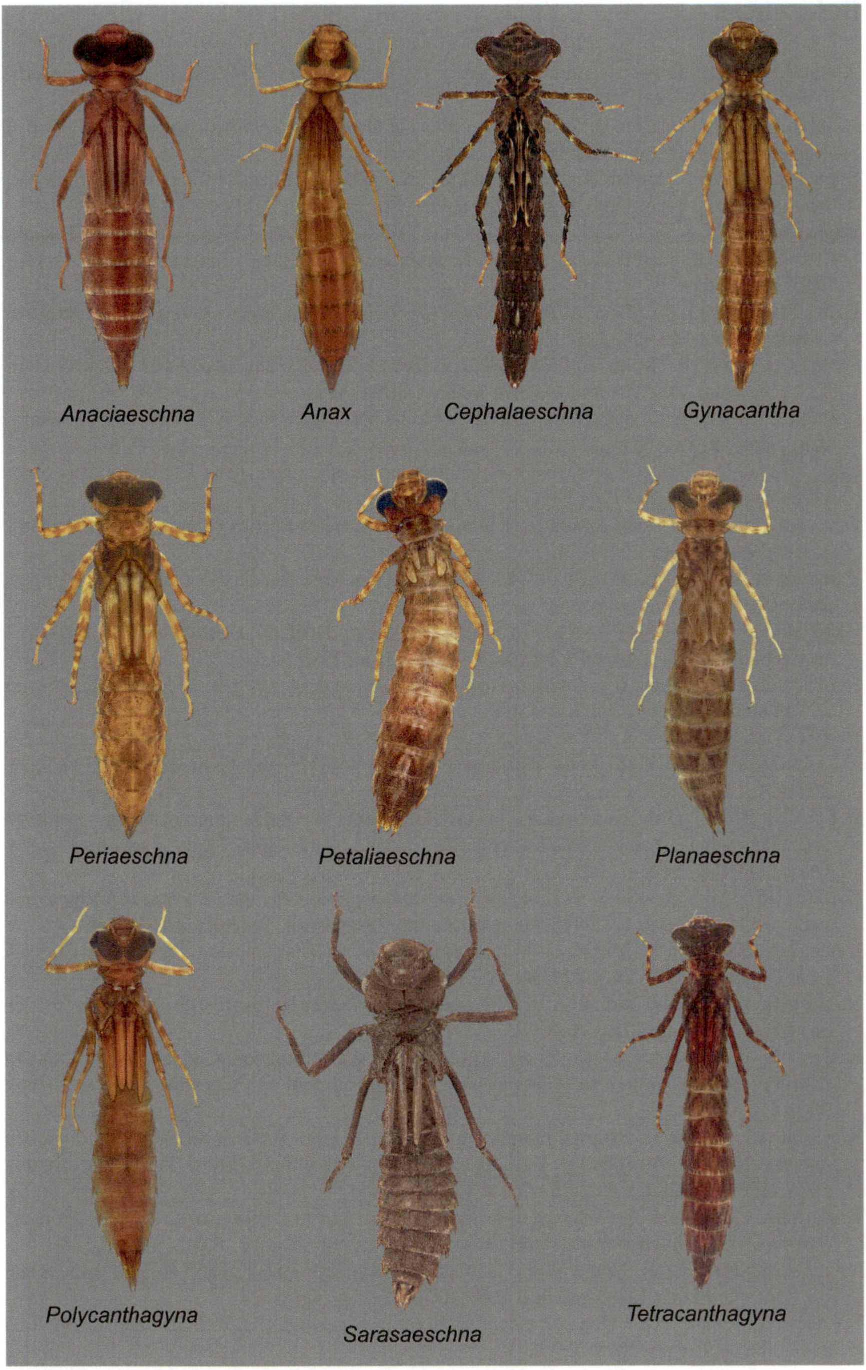

Plate 4.1 Dorsal habitus of Aeshnidae nymphs of Thailand. The nymphs of *Heliaeschna* and *Indaeschna* are not illustrated. The nymph of *Oligoaeschna* is unknown

References

Asahina S. On the discovery and description of the larval exuvia of *Oligoaeschna pryeri* Martin (Aeschnidae). Tombo. 1958;1(2/3):10–2.

Asahina S. A revision of the Himalayan dragonflies of the genus *Cephalaeschna* and its allies. Bull Natl Sci Mus Tokyo A. 1981;7:27–49.

Blanke AC, Greve R, Mokso F, et al. An updated phylogeny of Anisoptera including formal convergence analysis of morphological characters. Syst Entomol. 2013;38:474–90.

Buppachat N, Inkaew P, Thitiarchagul T, et al. Additions to the Odonata fauna of Thailand (Aeshnidae, Devadattidae, Euphaeidae, Gomphidae, Libellulidae, Platycnemididae). Agrion. 2020;20(3):184–91.

Butler SG, Orr AG. The larva of *Heliaeschna simplicia* Karsch, 1891 (Anisoptera: Aeshnidae). Odonatologica. 2013;42(2):151–6.

Carle FL, Louton JA. The larva of *Neopetalia punctata* and establishment of Austropetaliidae fam. nov. (Odonata). Proc Entomol Soc Washington. 1994;96(1):147–55.

Corbet PS. Dragonflies—behavior and ecology of Odonata. New York, NY: Comstock Publishing Associates; 1999.

Day L, Farrell D, Günther A, et al. New provincial records of Odonata from Thailand mostly based on photographs. Agrion. 2012;16(1):16–25.

Ferro ML, Sites RW, Vitheepradit A. Contributions to the faunistics of Odonata in Thailand. Insecta Mundi. 2009;104:1–24.

Hämäläinen M. The species list of Thai dragonflies increase steadily—an update. Malangpo. 2002;19:176–9.

Hämäläinen M, Pinratana BA. Atlas of the dragonflies of Thailand. Distribution maps by provinces. Bangkok: Brothers of Saint Gabriel in Thailand; 1999.

Hämäläinen M, Pinratana BA. Additions and corrections to dragonfly lists of five protected areas in Thailand. Malangpo. 2000;17:156–7.

Hu FS, Chan TW, Huang LC, Lee IL. Description of the final instar larva of *Cephalaeschna risi* Asahina, 1981 with notes on its semi-terrestrial lifestyle (Odonata: Aeshnidae). Int J Odonatol. 2021;24:139–48.

Jiang YH, Zhang HM. Descriptions of the full-grown Cephalaeschna patrorum Needham and Planaeschna shanxiensis Zhu and Zhang larvae from China (Anisoptera: Aeshnidae). Odonatologica. 2008;37(3):273–7.

Karube H, Yeh W-C. *Sarasaeschna* gen. nov., with descriptions of female *S. minuta* (Asahina) and male penile structures of *Linaeschna* (Anisoptera: Aeshnidae). Tombo. 2001;43:1–8.

Katatani N, Kitagawa K, Muraki A, Matuzawa T. Notes on Thai Odonata, Part 3. The Odonata from Chiang Rai. Aeschna. 2004;41:59–70.

Katatani N, Kitagawa K, Muraki A, et al. Notes on Thai Odonata, Part 5. The Odonata from Nan and Phrae. Aeschna. 2010;46:48–58.

Kawashima I. Description of the larva of the Aeshnid dragonfly *Sarasaeschna niisatoi* (Karube, 1998) (Aeshnidae: Gomphaeschninae) from northern Vietnam. Tombo (Tokyo). 2002;45(1–4):15–19.

Kawashima I, Karube H. External morphology of the last instar larva of probable *Petaliaeschna flavipes* Karube (Anisoptera, Aeshnidae, Brachytoninae) from Laos, Indochina. Tombo. 2006;48(1/4):7–11.

Kawashima I, Sasamoto A. Description of the last instar larva of *Periaeschna laidlawi* (Forster) (Anisoptera, Aeshnidae) from Malaysia, southwestern Asia. Tombo. 2006;48(1/4):12–7.

Kitagawa K, Katatani N. Notes on the Odonata of Thailand, Part 2. The Odonata from Trang Prefecture in March 1998. Aeschna. 2003;40:13–23.

Kosterin OE, Vikhrev NE. Odonatological field notes of two January trips to SE Thailand, in 2005 and 2006. Malangpo. 2008;22:222–36.

Kosterin OE, Vikhrev N, Zinchenko V, Korshunov A. More dragonfly records from Thailand. Agrion. 2011;15(1):12–6.

Kumar A. Description of the larvae of *Anax nigrofasciatus nigrolineatus* Fraser, 1935 and *A. parthenope parthenope* (Selys, 1839) from India, with a key to the known larvae of the Indian representatives of the genus *Anax* Leach, 1815 (Anisoptera: Aeshnidae). Odonatologica. 1973;2(2):83–90.

Lieftinck MA. A review of the genus *Oligoaeschna* Selys in southeast Asia. Tijdschr Ent. 1968;111(5):137–84.

Makbun N, Wongkamhaeng K, Keetapithchayakul TS. *Anax aurantiacus* sp. nov., a new dragonfly from mainland Southeast Asia (Odonata: Aeshnidae). Odonatologica. 2022;51(3/4):301–39.

Matsuki K. Description of the larval exuvia of *Oligoaeschna kunigamiensis* (Ishida) (Aeshnidae, Odonata). Gekkan-Mushi. 1993;263:15–7.

Matsuki K, Lien JC. Description of the larva of *Periaeschna magdalena* Martin from Taiwan (Anisoptera: Aeshnidae). Odonatologica. 1984;13(2):245–8.

Matsuki K, Obana S. Description of the larva of *Anaciaeschna jaspidea* (Burmeister) from Japan and Taiwan (Odonata: Aeshnidae). Gekkan-Mushi. 1985;175:29–31.

Michalski J. A manual for the identification of the dragonflies and damselflies of New Guinea, Maluku, and the Solomon Islands. Morristown, NJ: Kanduanum Books; 2012.

Orr AG. Life histories and ecology of Odonata breeding in phytotelmata in Bornean rainforest. Odonatologica. 1994;23(4):365–77.

Orr AG. An annotated checklist of the Odonata of Brunei with ecological notes and descriptions of hitherto unknown males and larvae. Int J Odonatol. 2001;4:167–220.

Orr AG. A guide to the dragonflies of Borneo. Their identification and biology. Kota Kinabalu, Sabah: Natural History Publications (Borneo); 2003.

Orr AG. Dragonflies of peninsular Malaysia and Singapore. A pocket guide. Kota Kinabalu, Sabah: Natural History Publications (Borneo); 2005.

Orr AG, Ngiam RWJ. A description of the larva of *Heliaeschna uninervulata* Martin (Odonata: Aeshnidae) from Singapore, with notes on its relationships. Int J Odonatol. 2011;14(2):163–9.

Orr AG, Ngiam RWJ, Leong TM. The larva of *Tetracanthagyna plagiata*, with notes on its biology and comparisons with congeneric species (Odonata: Aeshnidae). Int J Odonatol. 2010;13(2):153–66.

Paulson D. Dragonflies and damselflies of the West. Princeton, NJ: Princeton University Press; 2009.

Paulson DR, Schorr M, Abbott, J et al. World odonata list. Odonata Central, University of Alabama. 2023. https://www.odonatacentral.org/app/#/wol/. Accessed 18 July 2023

Pierce AJ. The first record of *Anax ephippiger* for Thailand and Southeast Asia (Odonata: Aeshnidae). Notul Odonatol. 2017;8(9):365–8.

Pierce AJ, Makbun N. First record of *Gynacantha limbalis* Karsch, 1892 from Thailand (Odonata: Anisoptera: Aeshnidae). Songklanakarin J Sci Technol. 2020;42(5):965–6.

Sasamoto A, Kotabe A, Takuma Y, et al. Description of larva of *Anax ephippiger* (Burmeister, 1839) from Japan, including changes and developments in external morphology (Odonata: Aeshnidae). Tombo. 2022;65:28–37.

Sribal U, Paweenpermsuk Y, Thitiarchagul T, et al. New national records of Odonata from Thailand based mostly on photographs (Odonata: Argiolestidae, Philosinidae, Aeshnidae, Libellulidae). Agrion. 2018;22(1):30–6.

Tze-wai T, Ka-king L, Kwan BSP, et al. The dragonflies of Hong Kong. Hong Kong: Friends of the Country Parks, Cosmos Books Ltd.; 2011.

von Ellenrieder N. A phylogenetic analysis of the extant Aeshnidae (Odonata: Anisoptera). Syst Entomol. 2002;27:437–67.

Wan FX, Jiang YH, Wan J. Descriptions of *Anax immaculifrons* Rambur and *Tetratcanthagyna waterhousei* McLachlan exuviae from China (Anisoptera: Aeshnidae). Odonatologica. 2011;40(4):339–45.

Wong KC, Yeh WC, Chan TW. Description of the final stadium larvae of *Polycanthagyna ornithocephala* (McLachlan, 1896) from Taiwan, with a key to the known larvae of the genus (Odonata: Anisoptera: Aeshnidae). Zootaxa. 2012;3238:64–8.

Xu QH. *Periaeschna yazhenae* sp. nov. from Fujian, China (Odonata: Anisoptera: Aeshnidae). Zootaxa. 2012;3526:72–8.

Xu QH, Zhang HM. The last-stadium larva and systematic status of *Planaeschna suichangensis* Zhou & Wei, 1980 (Odonata: Anisoptera: Aeshnidae). Zootaxa, 2011;3049:64–8.

Xu Q, Zhang H. Description of the final stadium larva of *Periaeschna zhangzhouensis* Xu, with discussion of taxonomic characters of the larvae of the genus *Periaeschna* Martin (Odonata: Aeshnidae). Int J Odonatol. 2014;17(1):53–8.

Yeh WC. Description of *Petaliaeschna pinratanai* spec. nov. from northern Thailand (Anisoptera: Aeshnidae). Odonatologica. 1999;28(3):283–8.

Zhang HM, Tong XL. Descriptions of *Boyeria karubei* Yokoi and *Periaeschna f. flinti* Asahina larvae from China (Anisoptera: Aeshnidae). Odonatologica. 2011;40(1):57–65.

Gomphidae 5

Contents

5.1 Key to Genera of Gomphidae 109
5.2 *Acrogomphus* Laidlaw, 1925 117
  5.2.1 General Appearance 117
  5.2.2 Description 117
  5.2.3 Remarks 120
  5.2.4 Distribution 120
5.3 *Amphigomphus* Chao, 1954 121
  5.3.1 General Appearance 121
  5.3.2 Description 121
  5.3.3 Remarks 123
  5.3.4 Distribution 124
5.4 *Anisogomphus* Selys, 1858 125
  5.4.1 General Appearance 125
  5.4.2 Description 125
  5.4.3 Remarks 128
  5.4.4 Distribution 128
5.5 *Asiagomphus* Asahina, 1985 129
  5.5.1 General Appearance 130
  5.5.2 Description 130
  5.5.3 Remarks 133
  5.5.4 Distribution 133
5.6 *Burmagomphus* Williamson, 1907 134
  5.6.1 General Appearance 134
  5.6.2 Description 134
  5.6.3 Remarks 137
  5.6.4 Distribution 137
5.7 *Davidius* Selys, 1878 138
  5.7.1 General Appearance 138
  5.7.2 Description 138
  5.7.3 Remarks 141
  5.7.4 Distribution 141

R. Novelo-Gutiérrez, R. W. Sites, *The Dragonfly Nymphs of Thailand (Odonata: Anisoptera)*, https://doi.org/10.1007/978-3-031-33712-3_5

5.8 *Euthygomphus* Kosterin, 2016 ... 142
5.8.1 General Appearance ... 142
5.8.2 Description ... 143
5.8.3 Remarks ... 145
5.8.4 Distribution ... 146
5.9 *Gomphidia* Selys, 1854 ... 146
5.9.1 General Appearance ... 147
5.9.2 Description ... 147
5.9.3 Remarks ... 150
5.9.4 Distribution ... 150
5.10 *Gomphidictinus* Fraser, 1942 ... 151
5.10.1 General Appearance ... 151
5.10.2 Description ... 152
5.10.3 Remarks ... 155
5.10.4 Distribution ... 155
5.11 *Heliogomphus* Laidlaw, 1922 ... 156
5.11.1 General Appearance ... 156
5.11.2 Description ... 157
5.11.3 Remarks ... 160
5.11.4 Distribution ... 161
5.12 *Ictinogomphus* Cowley, 1934 ... 162
5.12.1 General Appearance ... 162
5.12.2 Description ... 162
5.12.3 Remarks ... 165
5.12.4 Distribution ... 166
5.13 *Lamelligomphus* Fraser, 1922 ... 167
5.13.1 General Appearance ... 167
5.13.2 Description ... 167
5.13.3 Remarks ... 170
5.13.4 Distribution ... 170
5.14 *Leptogomphus* Selys, 1878 ... 171
5.14.1 General Appearance ... 171
5.14.2 Description ... 171
5.14.3 Remarks ... 174
5.14.4 Distribution ... 174
5.15 *Macrogomphus* Selys, 1857 ... 175
5.15.1 General Appearance ... 175
5.15.2 Description ... 175
5.15.3 Remarks ... 178
5.15.4 Distribution ... 178
5.16 *Megalogomphus* Campion, 1923 ... 179
5.16.1 General Appearance ... 179
5.16.2 Description ... 180
5.16.3 Remarks ... 183
5.16.4 Distribution ... 183
5.17 *Microgomphus* Selys, 1857 ... 184
5.17.1 General Appearance ... 184
5.17.2 Description ... 184
5.17.3 Remarks ... 187
5.17.4 Distribution ... 187
5.18 *Nepogomphus* Fraser, 1934 ... 188
5.18.1 General Appearance ... 188
5.18.2 Description ... 189

5.18.3 Remarks 191
5.18.4 Distribution 191
5.19 *Nihonogomphus* Oguma, 1926 192
5.19.1 General Appearance 192
5.19.2 Description 193
5.19.3 Remarks 195
5.19.4 Distribution 195
5.20 *Nychogomphus* Carle, 1986 196
5.20.1 General Appearance 196
5.20.2 Description 197
5.20.3 Remarks 199
5.20.4 Distribution 199
5.21 *Onychogomphus* Selys, 1854 200
5.21.1 General Appearance 200
5.21.2 Description 201
5.21.3 Remarks 203
5.21.4 Distribution 203
5.22 *Orientogomphus* Chao & Xu, 1987 204
5.22.1 General Appearance 204
5.22.2 Description 205
5.22.3 Remarks 207
5.22.4 Distribution 207
5.23 *Paragomphus* Cowley, 1934 208
5.23.1 General Appearance 208
5.23.2 Description 208
5.23.3 Remarks 210
5.23.4 Distribution 211
5.24 *Phaenandrogomphus* Lieftinck, 1964 212
5.24.1 General Appearance 212
5.24.2 Description 212
5.24.3 Remarks 215
5.24.4 Distribution 215
5.25 *Sieboldius* Selys, 1854 216
5.25.1 General Appearance 216
5.25.2 Description 217
5.25.3 Remarks 219
5.25.4 Distribution 220
5.26 *Sinictinogomphus* Fraser, 1939 221
5.26.1 General Appearance 221
5.26.2 Description 221
5.26.3 Remarks 224
5.26.4 Distribution 224
5.27 *Stylogomphus* Fraser, 1922 225
5.27.1 General Appearance 225
5.27.2 Description 226
5.27.3 Remarks 228
5.27.4 Distribution 228
5.28 Genus Undetermined 229
5.28.1 General Appearance 229
5.28.2 Description 230
5.28.3 Remarks 232
5.28.4 Distribution 233
References 235

The Gomphidae is a large and ancient family with high species richness comparable to that of Libellulidae. It is cosmopolitan with 1,009 species in 103 genera (Paulson et al. 2023). The family is considered to have probably originated in Southeast Asia from where it spread worldwide (Wilson 2001). A large debate still persists in its subfamily-level classification (Carle 1986; Belle 1996). A large consensus considers Gomphidae to be monophyletic (e.g., Dumont et al. 2010), with some debate whether Gomphidae is sister to Libelluloidea or Petaluridae (Fleck et al. 2008; Carle et al. 2015; Bybee et al. 2021), but with a wide consensus that Gomphidae should be placed as unique member into the superfamily Gomphoidea (Tennessen 2019).

The adult and nymphal stages are very distinct from all other dragonflies; they have retained several apparently ancestral characteristics, including the widely separated compound eyes and unspecialized wing venation in the adults, 2-segmented fore- and middle tarsi in nymphs and adults, and 4-segmented antennae with antennomere 4 minute or vestigial in the nymphs (Needham et al. 2014). The nymphs exhibit cryptic coloration (Paulson 2009), and the adults have a pattern of black with yellow or green stripes. Gomphid adults are typically low-perchers, preferring to land on the ground, rocks, logs, and other horizontal structures (Michalski 2012). Nymphs of most species inhabit streams and riverine habitats (Fig. 5.1), although a few are pond dwellers.

Fig. 5.1 Gomphid typical habitat in Nopphitam, Khlong Yod Leung, Nakhon Si Thammarat Province

The nymphs burrow shallowly in the mud, silt, sand, or gravel substrate and frequently hide in litter and other detritus, whereas others can be found at stream edges among detritus in pools, riffles, and runs, and have fairly specific habitat requirements (Wilson 2001). The body of gomphid nymphs shows adaptive reductions to a concealed mode of life in or on the substrate, including the 4-segmented antennae, short and flat labium, and 2-segmented pro- and mesothoracic tarsi (Tennessen 2019). Gomphid nymphs feed mainly during the night, so they depend mostly on tactile stimulation, including from structures such as the abundant and long setae on the antennae and legs. Nymphal development lasts from 1 to 3 years in the majority of the species.

In total, 29 genera containing 61 species of gomphids have been recorded in Thailand (Plates 5.1 and 5.2).

5.1 Key to Genera of Gomphidae

1 Wing sheaths divergent (Figs. 5.12, 5.17, 5.41, and 5.83) ... 2

1' Wing sheaths parallel (Figs. 5.23, 5.29, 5.59, and 5.101) ... 14

2(1) Sternum 7 divided into five sternites (as in Fig. 2.15) ... ***Leptogomphus***

2' Sternum 7 divided into three sternites ... 3

3(2') Sternum 8 divided into three sternites ... 4

3' Sternum 8 divided into five sternites ... 10

4(3) Mesocoxae closer to each other than are pro- and metacoxae (Fig. 5.2a); tibiae lacking burrowing hooks (Figs. 5.41 and 5.138) ... 5

4' Distance between mesocoxae same as or longer as between pro- and/or metacoxae (Fig. 5.2b); pro- and mesotibiae with or without burrowing hooks ... 7

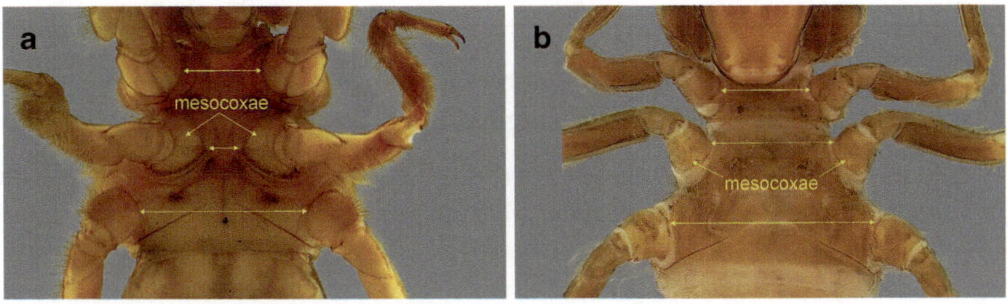

Fig. 5.2 Thoracic sternum of *Acrogomphus* (**a**) and *Davidius* (**b**)

5(4) Middorsal protuberances well-developed on S2–S9, spine-like (Fig. 5.15); lateral spines on S6–S9 (Figs. 5.12 and 5.14); pronotum large, bulky, shield-like, wider than head (Figs. 5.12 and 5.13) ... ***Acrogomphus***

5' Middorsal protuberances rudimentary on S2–S5, as short triangular projections on posterior margins of S6–S9 (Figs. 5.134 and 5.140); lateral spines on S4–S9 or S7–S9; pronotum not bulky nor shield-like, narrower than head (Figs. 5.131 and 5.136) ... 6

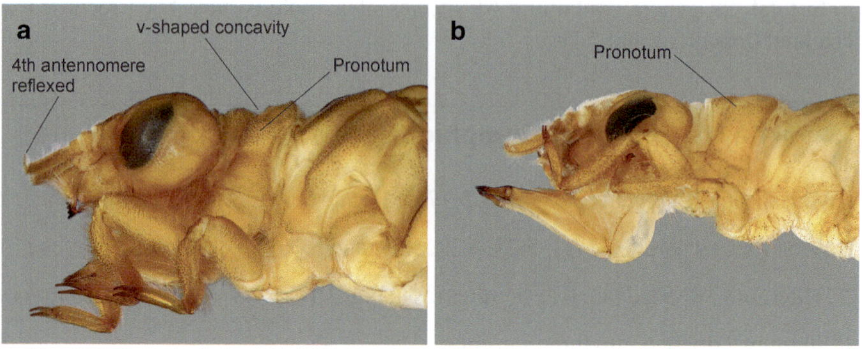

Fig. 5.3 Lateral view of head and prothorax of *Paragomphus capricornis* (**a**) and *Orientogomphus minor* (**b**)

6(5) Pronotum, in lateral view, with a wide v-shaped concavity between the anterior and posterior lobes (Fig. 5.3a); antennomere 4 finger-like, reflexed over the apex of antennomere 3 (Figs. 2.5 and 5.3a); lateral spines on S4–S9 (sometimes minute or lacking on S4) . . . . ***Paragomphus***

6’ Pronotum, in lateral view, with the anterior and posterior lobes continuous (Fig. 5.3b); antennomere 4 as a large blunt cone, directed upright (Fig. 5.132); lateral spines on S7–S9 . . . . ***Orientogomphus***

7(4’) Abdomen lacking middorsal protuberances (Figs. 5.41 and 5.45); ligula with 3–4 roundly pointed teeth on apical margin (Fig. 5.4a) . . . . ***Davidius***

7’ Middorsal protuberances on S2–S9 (Figs. 5.81, 5.99, and 5.111); ligula with more than 15 roundly pointed teeth on apical margin (Fig. 5.4b) 8

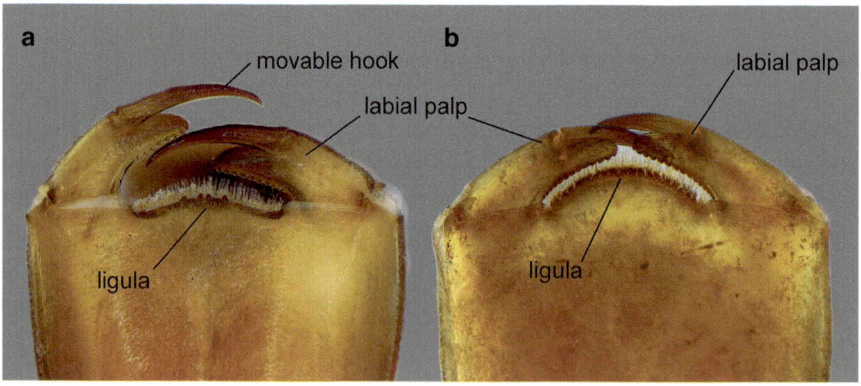

Fig. 5.4 Ligula and palps of *Davidius* (**a**) and *Nepogomphus* (**b**), ventral view

8(7') Apical margin of ligula straight (Fig. 5.5); apical lobe of labial palp sharply pointed and strongly incurved at tip, its mesal margin with 6–7 large, stout, sharp teeth (Fig. 5.5); middorsal protuberances on S7–S9 sharp, spine-like (Figs. 5.95 and 5.99) . . . . ***Megalogomphus***

8' Apical margin of ligula convex (as Fig. 5.4b); apical lobe of labial palp rounded, not strongly incurved at tip, its mesal margin with 11–15 small teeth (Fig. 5.4b); middorsal protuberances on S2–S9, roundly or bluntly pointed (Figs. 5.111 and 5.117) 9

9(8') Lateral spines on S2–S9 (Figs. 5.113 and 5.116); middorsal protuberances on S2–S9 low, increase in size but diminish in height posteriorly (Fig. 5.117); width/length ratio of ligula about 5:1 (Fig. 5.115) . . . . ***Nihonogomphus***

9' Lateral spines on S7–S9 (Figs. 5.107 and 5.110); middorsal protuberances on S2–S9 erect, well-developed and nearly of the same size and robustness on S2–S6, larger on S7–S9 (Fig. 5.111); width/length ratio of ligula about 3:1 (Fig. 5.4b) . . . . ***Nepogomphus***

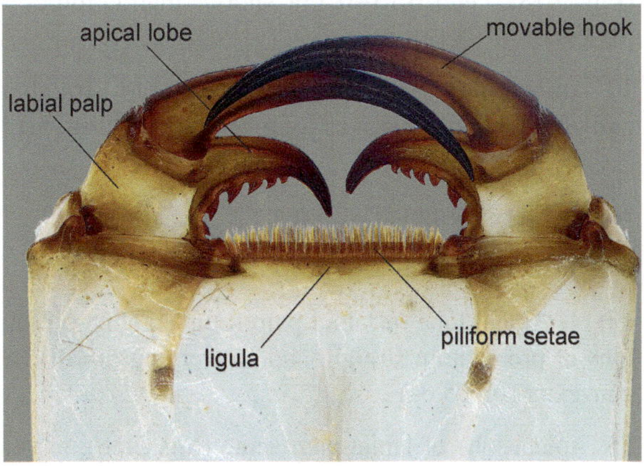

Fig. 5.5 Ligula and palps of *Megalogomphus* in dorsal view

10(3') Antennomere 3 largely hypertrophied, subrectangular, scoop-shaped (Figs. 5.126 and 5.127); clypeus, in dorsal view, mostly hidden by antefrons which is produced anteriorly as a triangle (Fig. 5.126); S2 with a blunt, conical, anteriorly directed middorsal protuberance, S3–S9 with very low, mound-like protuberances (Figs. 5.125 and 5.129); abdomen lacking lateral spines (Figs. 5.125 and 5.128) . . . . ***Onychogomphus***

10' Antennomere 3 cylindrical, spindle-shaped, claviform, widely ovoid or with the mesal margin strongly convex at middle; clypeus not hidden by antefrons; middorsal protuberances not as above; some of abdominal segments S6–S9 with lateral spines 11

11(10') Antennomere 3 cylindrical, parallel-sided (Figs. 5.17 and 5.18); 3 *m* denticles on left mandible; lateral margins of prementum nearly straight and parallel-sided in distal 0.4 (Fig. 5.19); burrowing hooks lacking; lateral surfaces of fore- and middle tibiae setose, but not tuberculate (Fig. 5.18); terga 2–10 with abundant minute setae; lateral spines on S7–S9 sharply pointed (Fig. 5.21) . . . . ***Amphigomphus***

11' Antennomere 3 variable, sides convex or straight and divergent apically; ≥ 6 *m* denticles on left mandible; lateral margins of prementum variable in distal half; with burrowing hooks; lateral surfaces of fore- and middle tibiae setose and tuberculate (as in Fig. 5.78); terga S2–S10 granulose; lateral spines on S6–S9, S7–S9, or S8–S9 variable 12

12(11') S8–S9 with small, posterolateral, bluntly pointed spines (Fig. 5.142); antennomere 3 fusiform or with the mesal margin strongly convex in the middle (Fig. 5.143); compound eyes strongly bulging anteriorly (Fig. 5.143) . . . . ***Phaenandrogomphus***

12' S6–S9 or S7–S9 with posterolateral, sharply pointed spines (Figs. 5.77, 5.80, and 5.119); antennomere 3 not as above (Figs. 5.78 and 5.120); compound eyes not bulging anteriorly (Figs. 5.78 and 5.120) 13

13(12') Antennomere 3 widely ovoid, with the lateral and mesal margins convex (Fig. 5.78); cerci 0.15–0.25× shorter than epiproct (Fig. 5.81); lateral spines on S6–S9 or S7–S9 . . . . ***Lamelligomphus***

13' Antennomere 3 claviform, with the sides straight and gradually diverging apically (Fig. 5.120); cerci as long as epiproct (Fig. 5.123); lateral spines on S7–S9 . . . . ***Nychogomphus***

14(1') Tarsal formula 2–2–2 15

14' Tarsal formula 2–2–3 18

15(14) Apical lobe of labial palp rounded (Fig. 5.6a); in ventral view, lateral margins of prementum straight and converging distally, with a strong lateral subbasal convexity (Fig. 5.6a) . . . . ***Gomphidia***

15' Apical lobe of labial palp acute (Fig. 5.6b); in ventral view, lateral margins of prementum sinuate or more or less straight in distal half (Fig. 5.6b) 16

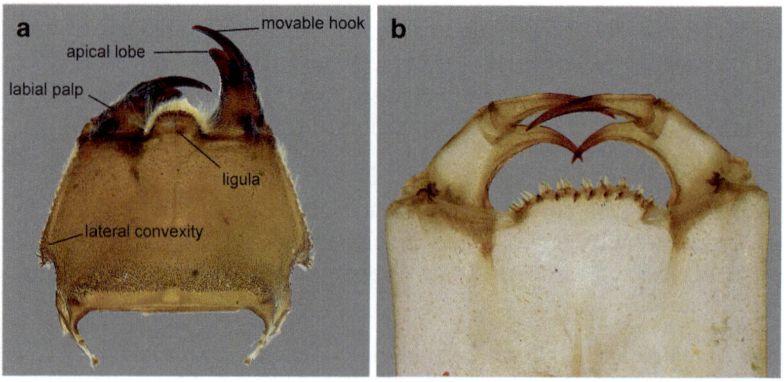

Fig. 5.6 Prementum of *Gomphidia* (**a**) (ventral) and *Sinictinogomphus* (**b**) (dorsal)

16(15) Apical lobe of labial palp strongly curved, its mesal margin sharp and smooth (Fig. 5.6b); abdomen distinctly longer than wide, tapering posteriorly (Figs. 5.154 and 5.156) . . . . ***Sinictinogomphus***

16' Apical lobe of labial palp not as above, its mesal margin with a row of small, squarely truncate teeth; abdomen about as wide as long, truncate posteriorly 17

17(16) Anterolateral corners and middle part of frons enlarged and armed with strong, sharp, recurved spines intermingled with elongate scale-like setae (Fig. 5.60); prementum covered with abundant scale-like setae on ventral surface (Fig. 5.61); ligula strongly convex; lateral spines of S3–S7 expanded laterally, those of S9 reaching tips of caudal appendages or beyond (Figs. 5.59 and 5.62) . . . . ***Gomphidictinus perakensis***

17' Anterolateral corners and middle part of frons not enlarged and without spines as described above (Fig. 5.72); prementum finely granulose on ventral surface (Fig. 5.73); ligula straight to shallowly convex; lateral spines of S3–S7 not expanded laterally, those of S9 not reaching tips of caudal appendages (Figs. 5.71 and 5.74) . . . . ***Ictinogomphus***

18(14') Body strongly flat and wide, leaf-like (Fig. 5.148); sterna 3–9 divided into three sternites . . . . ***Sieboldius***

18' Body more or less cylindrical, not leaf-like; sterna 3–9 variable 19

19(18') Sterna 3 and 4 each divided into five sternites 20

19' Sterna 3 and 4 each divided into three sternites 21

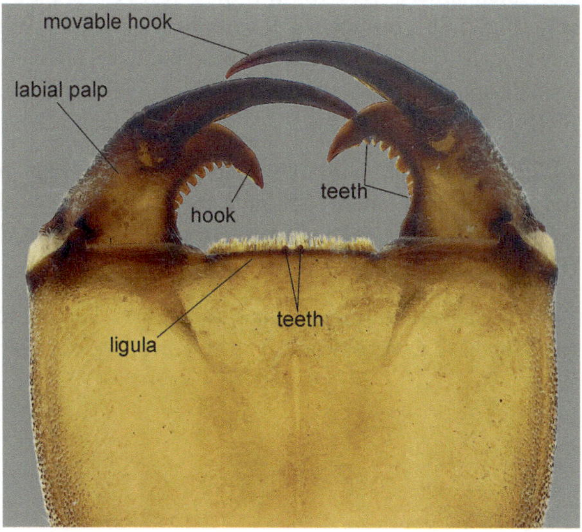

Fig. 5.7 Ligula and palps of undetermined genus in ventral view

20(19) Antennomere 3 hypertrophied, triangular, concave dorsally with thickened borders (Fig. 5.161); ligula nearly straight, with a row of 4 short, subconical teeth (Fig. 5.162); apical lobe mesal margin of labial palp more or less straight, ending in a short tooth; sterna 5–6 each divided into three sternites . . . . ***Stylogomphus***

20’ Antennomere 3 subrectangular; ligula straight with 2 short, subconical teeth (Fig. 5.7); apical lobe of labial palp strongly curved, ending in a stout hook (Fig. 5.7); sterna 5–6 each divided into five sternites . . . . **genus undetermined**

21(19’) Abdominal segment 9 greatly enlarged, forming together with S10 a long tube or “snorkel” (Figs. 5.89 and 5.93) . . . . ***Macrogomphus***

21’ Abdominal segment 9 of normal size, not forming with S10 a long tube 22

22(21’) Antennomere 3 hypertrophied, flattened, paddle-shaped (Fig. 5.66) or claviform (Fig. 5.102); ligula with a row of 17–24 small, quadrate teeth; prementum with a longitudinal, midventral sulcus; apical lobe of labial palp rounded 23

22’ Antennomere 3 more or less cylindrical; ligula with one small median tooth (Figs. 5.8–5.10) or teeth lacking; prementum lacking midventral sulcus; apical lobe of labial palp tapering to a strong hook or blunt tip 24

23(22) Antennomere 3 paddle-shaped, strongly flattened (Fig. 5.8a), its dorsal surface appearing as a sieve; central part of frons strongly projected anteriorly (Fig. 5.8a); a bilobed longitudinal carina below compound eye, covered with stiff setae (Fig. 5.67); middorsal protuberances on S3– or S4–S9; lateral spines on S7 short, bluntly pointed on S9 (Fig. 5.65) . . . . ***Heliogomphus***

23’ Antennomere 3 claviform (Fig. 5.8b); central part of frons not projected anteriorly (Fig. 5.8b); infraocular longitudinal carina not bilobed, rather simple and serrate (Fig. 5.103); with middorsal protuberances on S4– or S5–S9, sometimes only as a small, spine-like protuberance on S9; lateral spines on S7 large, sharply pointed on S9 (Fig. 5.101) . . . . ***Microgomphus***

24(22’) Labial palp with mesal margin toothed (Figs. 5.9 and 5.10); ligula, in ventral view, with a small or minute tooth at middle of distal margin (Figs. 5.9 and 5.10); burrowing hooks moderately to well-developed in fore- and middle tibiae (Figs. 5.23 and 5.29) 25

24’ Labial palp with mesal margin smooth, lacking teeth; ligula without median tooth on distal margin (Fig. 5.11); burrowing hooks vestigial 27

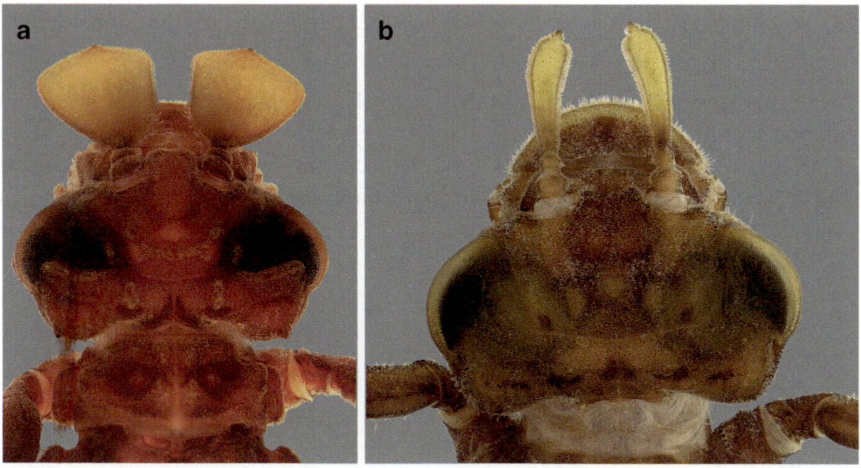

Fig. 5.8 Head of *Heliogomphus* (**a**) and *Microgomphus* (**b**), dorsal view

25(24) Labial palp ending in a stout hook, the mesal margin strongly curved (Fig. 5.9) with a row of 6–8 stout, short, quadrate teeth; lateral spines on S6–S9; epiproct and paraprocts of the same length, cerci shorter; terga 2–6 setose, 7–10 finely granulose, setose at sides; middorsal protuberances on S8–S9 small, low, spine-like (Fig. 5.33) . . . . ***Asiagomphus***

25' Labial palp gradually tapering to a bluntly pointed tip (Fig. 5.10a) or ending in a stout, conical tooth (Fig. 5.10b), with the mesal margin straight or slightly concave; lateral spines on S7–S9; other features variable 26

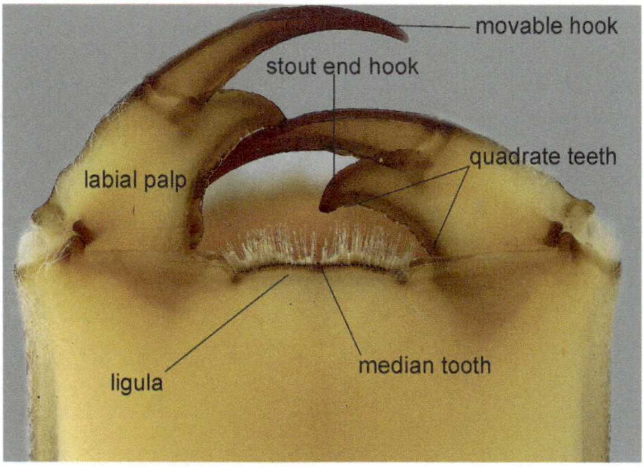

Fig. 5.9 Ligula and palps of *Asiagomphus* in ventral view

26(25') Ligula triangular (Fig. 5.10), not serrulate; labial palp ending in a stout, conical tooth, the mesal margin with serrate or quadrate teeth (Fig. 5.10a); terga 2–10 finely granulose; middorsal protuberances on S8–S9 short, low, triangular, serrate (Fig. 5.39); cerci, epiproct, and paraprocts of the same length .... ***Burmagomphus***

26' Ligula shallowly convex (Fig. 5.10), finely serrulate; labial palp gradually tapering to a blunt tip, the mesal margin irregularly toothed (Fig. 5.10); terga 2–10 finely setose; a middorsal protuberance only on S9 as a small spine (Fig. 5.27); cerci and epiproct equal in length, paraprocts slightly longer (Fig. 5.27) .... ***Anisogomphus***

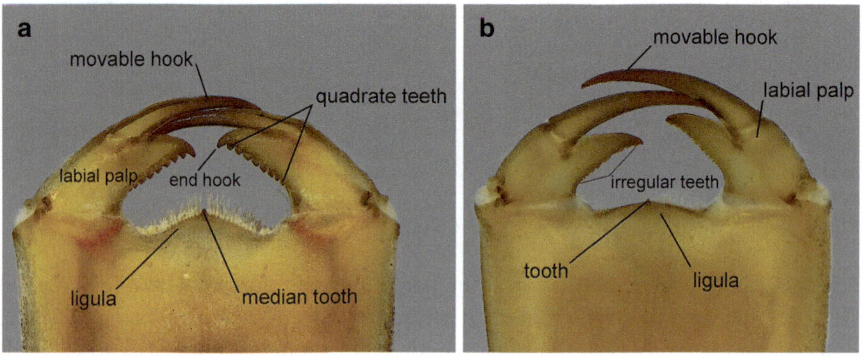

Fig. 5.10 Ligula and palps of *Burmagomphus* (**a**) and *Anisogomphus* (**b**), ventral view

27(24') Mesal margin of labial palp straight .... ***Merogomphus***

27' Mesal margin of labial palp concave (Fig. 5.11) .... ***Euthygomphus***

Nymphs unknown: *Asahinagomphus*, *Mattigomphus*.

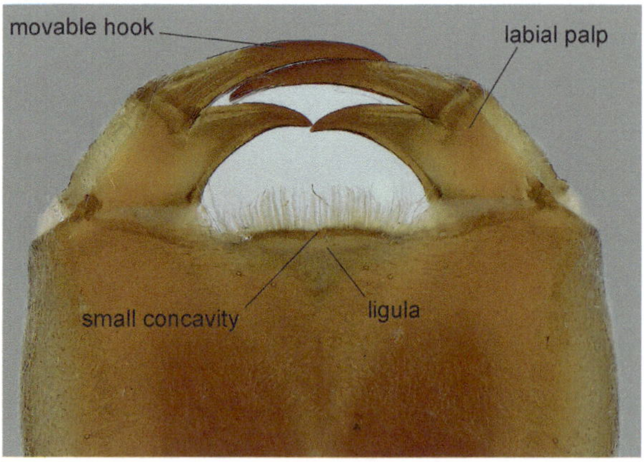

Fig. 5.11 Ligula and palps of *Euthygomphus* in ventral view

5.2 *Acrogomphus* Laidlaw, 1925

Fig. 5.12 Dorsal habitus of the nymph of *Acrogomphus*

5.2.1 General Appearance

The nymphs of *Acrogomphus* are characterized by their medium size (20.8 mm), with the body hairy, robust and enlarged, strongly tapering posteriorly, and coloration light-yellow-brown (Fig. 5.12).

5.2.2 Description

Head. The head is wider than long, but narrower than the thorax and abdomen (Fig. 5.13). The labrum is 0.6 mm long and flattened ventrally; it has very minute, spiniform setae and its anterior margin is densely setose. The clypeus is bare. The antefrons and postfrons are densely covered with robust, stiff setae. The anterior margin of the frons is almost straight. The vertex and occiput have small, spiniform setae. The posterior margin of the head is widely concave. The antennae are 4-segmented and densely covered with robust, reddish, stiff setae on the dorsal surface, and long, yellowish setae on the lateral margins; the scape is globose, the pedicel subglobose, antennomere 3 is cylindrical and the largest, with a tuft of long setae on the apical margin, and antennomere 4 is finger-like, reflexed over the apex of antennomere 3, and bears long setae. The compound eyes are moderately developed, and the occipital lobes are rounded and bulging. Mandibles have the formula: L 1234 0 a($m^{1,2,3,4,5}$)b/R 1234 y a($m^{1,2,3,4,5,6}$)b. The maxilla's galeolacinia has seven moderately incurved teeth; the three dorsal teeth are subequal in length and robustness, the four ventral teeth are of different sizes, with the apical tooth the largest. The ventral pad of the hypopharynx is subpentagonal, whitish, mostly soft and bare, and almost as long as wide.

Fig. 5.13 The head and pronotum in dorsal view

Labium: The prementum-postmentum articulation reaches the posterior margin of the fore coxae. The prementum is light-yellow-brown, longer than wide, and parallel-sided along the apical 0.40; it has a longitudinal, central sulcus in the basal half of the ventral surface with a row of long, stiff setae on each side; the ligula is moderately developed, its apical margin is convex, and it has a double row of piliform setae. The labial palp is yellowish-red, with the mesal margin slightly serrate. The movable hook is yellowish-red, shorter than the palp, sharp, and moderately incurved.

Thorax. The prothorax is wider than the head and setose ventrally. The pronotum is large, bulky, shield-like, subpentagonal (Figs. 5.12 and 5.13), and covered with small, spiniform setae, except on the large, dorsolateral, irregular, bare areas. The anterior lobe comprises two large, triangular sclerites separated by a middorsal suture; the anterior margin is straight, the lateral margins are rounded and continuous with the propleuron, and the posterior margin is mostly straight. In lateral view, the dorsal surface of the pronotum appears flat and continuous. The pterothorax is densely setose on the ventrolateral areas. The legs are short (i.e., when fully extended, the hind legs do not reach the posterior margin of S7) and strongly setose mainly on the fore- and middle legs; burrowing hooks are lacking; the middle coxae are closer to each other than are the fore- and hind coxae; the pretarsal claws are simple and chelate-like on the fore- and middle tarsi, but long and sharp on the hind tarsi, and each has a pulvilliform empodium. The wing sheaths are strongly divergent and mostly bare (Fig. 5.12), but with marginal setae; the anterior and posterior wing sheaths reach the posterior margin of S4 and basal half of S5, respectively.

Fig. 5.14 Ventral habitus of the nymph of *Acrogomphus*

Abdomen. The abdomen is lanceolate, widest at S4–S5, and suddenly tapers posteriorly; its ventral surface is more or less flat (Fig. 5.14), the dorsal surface is convex and has strong, well-developed, spine-like protuberances on S2–S9, which increase in size and height posteriorly (Fig. 5.15). S6–S9 end in strong lateral spines which increase in size and robustness posteriorly; those of S8 are the largest. Terga 2–10 are beset with minute, spiniform setae, and the posterior margins have a row of red, stiff setae; the lateral margins of S2–S8 are densely setose. Sterna 1–8 are creamy-pale and 2–10 are covered with abundant small, stiff setae+. Sterna 3–9 are each divided into three sternites (Fig. 5.14). The female gonapophyses are vestigial, whereas those of the male are lacking. The caudal appendages are long, but not longer than the dorsal length of S9+S10 (Fig. 5.15), and covered with minute, spiniform setae. The epiproct and paraprocts are of the same length and sharpness. The cerci are slightly shorter than the epiproct and paraprocts, and sharply pointed (Fig. 5.15).

Fig. 5.15 The abdomen and caudal appendages in lateral view

Measurements (in mm). Total length including caudal appendages 20.8; abdomen (ventral) 13.0; maximum width of head 4.4; length of hind femur (lateral) 3.1.

5.2.3 Remarks

The nymph of *Acrogomphus* differs from those of other gomphid genera inhabiting Thailand by the massive pronotum and that the middle coxae are positioned very close to each other. In this respect, it resembles the New World genus *Progomphus*. The nymphs inhabit sand banks of pristine forest streams (Orr 2005) and are considered to be rapid burrowers (Lieftinck 1941). By the nymphal morphology, *Acrogomphus* appears more similar to *Orientogomphus* and *Paragomphus*.

5.2.4 Distribution

Hämäläinen and Pinratana (1999) recorded *Acrogomphus minor* Laidlaw, 1931, from Phattalung Province; however, Wilson (2008) transferred this species to the genus *Orientogomphus*. A different and undetermined species, possibly *A. malayanus* Laidlaw, 1925, represents the first account of the genus *Acrogomphus* for Thailand. We have records from the provinces of Nakhon Si Thammarat, Satun, and Songkhla, all in the south (Fig. 5.16).

Fig. 5.16 Distribution of *Acrogomphus*

5.3 *Amphigomphus* Chao, 1954

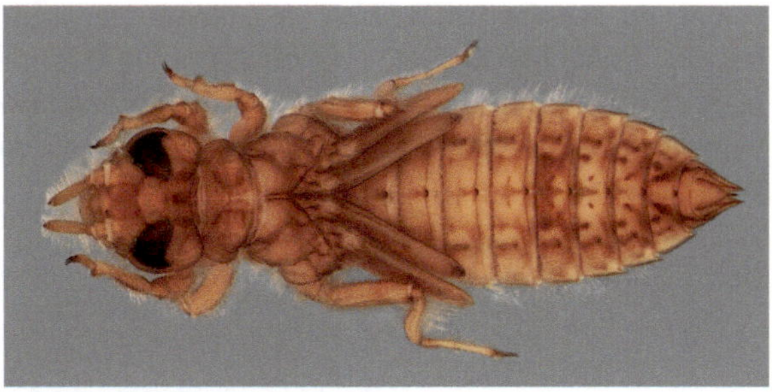

Fig. 5.17 Dorsal habitus of the nymph of *Amphigomphus somnuki* (after Novelo and Sites 2019, with permission of Magnolia Press)

5.3.1 General Appearance

The nymphs of *Amphigomphus* are characterized by their medium size (24–26 mm), with the body hairy at the sides, densely covered with minute dark setae, robust and oblong, and coloration yellow-brown (Fig. 5.17).

Fig. 5.18 The head in dorsal view

5.3.2 Description

Head. The head is wider than long (Fig. 5.18), but narrower than the thorax and abdomen. The labrum is 0.5 mm long and its anterior margin flattened ventrally and densely setose. The clypeus is bare; the antefrons and postfrons are covered with robust, stiff setae of different sizes; the vertex and occiput have small, stout setae. The antennae are densely covered with small, scale-like setae on the dorsal surface and have long, white setae on the lateral margins; the scape is globose, the pedicel subglobose, antennomere 3 is cylindrical and the largest, with a circular tuft of long setae on the apical margin, and antennomere 4 is a short, conical rudiment. The compound eyes are moderately developed. The occipital lobes are rounded and bulging. Mandibles have the formula: L 1234 0 a($m^{1,2,3}$)b/R 1234 y a($m^{1,2}$)b. The maxilla's galeolacinia is as described for *Acrogomphus*. The ventral pad of the hypopharynx is subpentagonal, whitish, mostly soft and bare, and has a U-shaped ventrobasal sclerite.

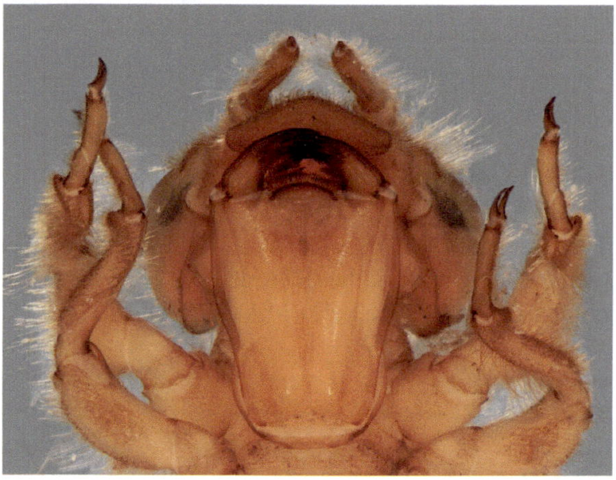

Fig. 5.19 The prementum in ventral view

Labium: The prementum-postmentum articulation reaches the posterior margin of the fore coxae. The prementum (Fig. 5.19) is light-yellow-brown, longer than wide, parallel-sided along the apical 0.40, and with a row of minute, scale-like setae on each side; a ventral sulcus is lacking. The ligula is poorly developed; its apical margin is convex, and it has a dorsal double row of piliform setae and ventral row of 27–28 small, truncate, reddish teeth. The labial palp is yellowish-red, with the apex rounded and the mesal margin beset with 13–15 small, truncate teeth. The movable hook is yellowish-red, shorter than the palp, sharp, and moderately incurved.

Thorax. The prothorax (Fig. 5.17) is narrower than the head and setose on the ventral margin. The pronotum is covered with minute, dark setae, except on the large, dorsolateral, bare, irregular areas; the anterior margin is sinuate, the lateral margins rounded, and the posterior margin widely convex. The pterothorax is yellow with the sutures dark, and densely setose on the ventrolateral areas. The legs are short (Figs. 5.17 and 5.20) (i.e., when fully extended, the hind legs do not reach the posterior margin of S7) and strongly setose mainly on the fore- and middle legs; burrowing hooks are lacking; the pretarsal claws are simple, chelate-like on the fore- and middle tarsi, long and sharp on hind tarsi, and each has a pulvilliform empodium. The anterior and posterior wing sheaths are strongly divergent and mostly bare, but with marginal setae, and reach the basal half and the posterior margin of S4, respectively (Fig. 5.17).

Fig. 5.20 Ventral habitus of the nymph of *Amphigomphus somnuki* (after Novelo and Sites 2019, with permission of Magnolia Press)

Abdomen. The abdomen is lanceolate, yellow, with a complex pattern of pale and dark spots and dots on S5–S10 (Fig. 5.17); the large dark spot on S6 and longitudinal, parallel stripes on S10 are predominant. The ventral surface is more or less flat, whereas the dorsal surface is convex and with short, dorsal protuberances on S2–S9 which diminish in size and height posteriorly (Fig. 5.21). S7–S9 end in lateral spines which diminish in size and robustness posteriorly. Terga 2–10 have abundant minute setae, and S2–S9 are beset with long, scattered, stiff setae including on the posterior margins; the lateral margins are setose on S2–S10 and serrate on S6–S9. Sterna 1–8 are creamy-pale, 9–10 light-brown (Fig. 5.20). Sterna 3–7 and 9 are each divided into three sternites and sternum 8 into five sternites. The female gonapophyses are vestigial and those of the male are lacking. The caudal appendages (Fig. 5.21) are longer than the length of S9+S10, sharply pointed, and densely covered with minute, spiniform setae. The epiproct and paraprocts are of the same length, whereas the cerci are slightly shorter.

Fig. 5.21 The abdomen and caudal appendages in lateral view (after Novelo and Sites 2019, with permission of Magnolia Press)

Measurements (in mm). Total length including caudal appendages 24.3–26.0; abdomen (ventral) 15.1–16.2; maximum width of head 5.0–5.2; length of hind femur (lateral) 3.8–4.0.

5.3.3 Remarks

Three species of *Amphigomphus* are known worldwide (Sasamoto and Kawashima 2007; Xu 2012a); of these, only *A. somnuki* Hämäläinen, 1996, is known from Thailand (Hämäläinen and Pinratana 1999). Sasamoto and Kawashima (2007) described the first nymph of the genus, *A. nakamurai* Karube, 2001, from Vietnam. A few years later, Xu (2012b) described the nymph of *A. hansoni* from China, and Novelo-Gutiérrez and Sites (2019) described the nymph of *A. somnuki*. The nymphs inhabit small streams buried deeply in sand (Xu 2012b). By the nymphal morphology, *Amphigomphus* appears more similar to *Phaenandrogomphus* and to a lesser extent to *Onychogomphus*.

5.3.4 Distribution

In Thailand, *Amphigomphus* has been recorded from the provinces of Chiang Mai (Hämäläinen and Pinratana 1999, 2000) and more recently from Kanchanaburi, Lampang, Loei, Mae Hong Son, Phayao, and Phitsanulok (Novelo-Gutiérrez and Sites 2019) (Fig. 5.22).

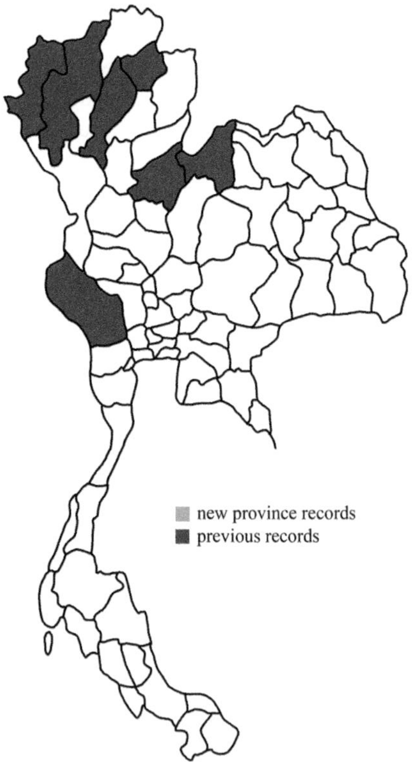

Fig. 5.22 Distribution of *Amphigomphus*

5.4 *Anisogomphus* Selys, 1858

Fig. 5.23 Dorsal habitus of the nymph of *Anisogomphus*

5.4.1 General Appearance

The nymphs of *Anisogomphus* are characterized by their large size (29.5–31.8 mm), with the body robust and flat (Fig. 5.23), abdomen parallel-sided with a low spine-like middorsal protuberance on S9 and large posterolateral spines on S7–S9, and coloration yellow-brown.

5.4.2 Description

Head. The head is wider than long and wider than the prothorax, but narrower than the pterothorax and abdomen (Figs. 5.23 and 5.24). The labrum is 0.5 mm long, mostly bare, with the anterior margin setose and flattened ventrally. The clypeus is bare; the frons, vertex, and occiput are covered with minute, reddish setae which are larger and stiff on the anterior margin of the postclypeus and cephalic lobes. The anterior margin of the frons is wide and shallowly concave. The occiput is short, and the occipital lobes have a low, rounded, posterolateral tubercle beset with short and long, stiff setae. The antennal scape is globose, the pedicel is annular, and both are covered with minute and long setae; antennomere 3 is subcylindrical, flattened, and slightly recurved (Fig. 5.24), its dorsal and ventral surfaces are finely granulose, and the lateral margins are beset with long, white, delicate setae; antennomere 4 is a hemispherical rudiment.

Fig. 5.24 The head in dorsal view

The compound eyes are moderately developed. Mandibles have the formula: L 1234 0 a($m^{1,2,3 \text{ or } 4}$)b/R 1234 y a($m^{1,2,3,4,5 \text{ or } 6}$)b, with a > b in both mandibles, the *m* denticles are sharp in the right mandible, but truncate in the left mandible. The maxilla's galeolacinia is as described for *Acrogomphus*. The ventral pad of the hypopharynx is subpentagonal, whitish, mostly soft, with a basal triangular sclerite on each side of the midline connected to each other by a basal, transverse, denticulate, sclerotized bridge; the posterior margin is sinuate and the anterior margin has a row of long, stiff setae. Labium: The prementum-postmentum articulation reaches the posterior margin of the fore coxae; the prementum is subquadrate (Fig. 5.25), shorter than wide, the lateral margins are straight, slightly divergent, and setose on the distal half, and the basal margin is nearly straight and ridged with a faint of ventromedian sulcus on the basal half; the ligula is poorly developed, shallowly convex, its anterior border is minutely serrulated, and usually has a small tooth in the middle (best seen in ventral view) (Fig. 5.25), a dorsal row of abundant long, piliform setae, and a ventral row of long setae which form a tuft on each side of the anterior margin. The apical lobe of the labial palp is stout and tapers to a bluntly pointed tip, the mesal margin is shallowly concave and has a row of 9–12 small, square teeth. The movable hook is yellow-brown in the basal half and reddish-brown in the apical half, sharp, moderately incurved, and as long as the palp.

Fig. 5.25 The prementum in ventral view

Thorax. The prothorax is narrower than the head and setose on the ventral margin of the propleuron (Fig. 5.23). The pronotum is covered with minute setae, except on the large, depressed, bare, irregular dorsolateral areas; the anterior margin is straight, the lateral margins are rounded, and the posterior margin is shallowly convex. The pterothorax is covered with minute setae, and setose on the ventrolateral areas. The hind legs are relatively short (Figs. 5.23 and 5.26) (i.e., when fully extended, do not reach the posterior margin of S9) and setose; the burrowing hooks are well-developed on the fore- and middle tibiae; the pretarsal claws are simple and each has a pulvilliform empodium. The wing sheaths are parallel (Fig. 5.23), their surfaces and margins have small setae, the margins also have delicate, long setae; the anterior and posterior wing sheaths reach the anterior margin and basal half of S4, respectively.

Fig. 5.26 Ventral habitus of the nymph of *Anisogomphus*

Abdomen. The abdomen is mostly parallel-sided, flattened, abruptly narrowing posteriorly, light-brown, and with a large, irregular, dark spot on each side of the midline on S4–S9; S9 is reddish-brown and S10 dark-brown (Fig. 5.23). Terga 2–10 are finely setose including on the posterior margins. Middorsal protuberances are lacking on S2–S8 (Fig. 5.27), S9 has a small, triangular, middorsal, posteriorly directed protuberance (Figs. 5.23 and 5.27). The lateral margins of S1–S9 are densely setose, usually spiny on the apical half of S7 and the entire margin of S8–S9; S7–S9 end in lateral, robust, sharply pointed spines which are divergent on S7–S8, parallel and the longest on S9 (as long as S10). Sterna 1–8 are light-brown and finely setose, and sterna 9–10 are yellowish-brown (Fig. 5.26). Sterna 3–6 and 9 are each divided into three sternites, sterna 7–8 into five sternites, and the posterior margin of sternum 10 is convex at the middle. The female gonapophyses are vestigial. The caudal appendages are robust, pyramidal, sharply pointed (Fig. 5.27), and as long as the dorsal length of S9. The paraprocts are slightly longer than the epiproct, and the cerci are as long as the epiproct. The male's epiproct is twice as long as its basal width, it has a pair of small, rounded tubercles at the basal 0.60, and the apical 0.40 is lanceolate.

Fig. 5.27 The abdomen and caudal appendages in lateral view

Measurements (in mm). Total length including caudal appendages 29.5–31.8; abdomen 18.0–18.8; maximum width of head 5.2–5.6; length of hind femur 6.0–6.5 (the larger dimensions pertain to exuviae).

5.4.3 Remarks

Anisogomphus inhabits from lowlands to middle elevations (170–1400 m asl). Two species of *Anisogomphus* have been recorded from Thailand: *A. yanagisawai* Sasamoto, 2015 (Sasamoto 2015) and *A. yiangsaki* Makbun, 2017 (Makbun 2017); the nymphs of both species are still unknown. The nymphs of *Anisogomphus* are found in swift, well-shaded streams (Wilson 1995) in the sandy, muddy bottom (Lieftinck et al. 1984). By the nymphal morphology, *Anisogomphus* appears more similar to *Asiagomphus* and *Burmagomphus*, and to a lesser extent to *Euthygomphus* and *Merogomphus*, and probably to the still unknown nymph of *Mattigomphus*.

5.4.4 Distribution

In Thailand, *Anisogomphus* has been recorded only from the provinces of Tak (Sasamoto 2015) and Sakon Nakhon (Makbun 2017). ***New province records*:** Mae Hong Son and Phitsanulok (Fig. 5.28).

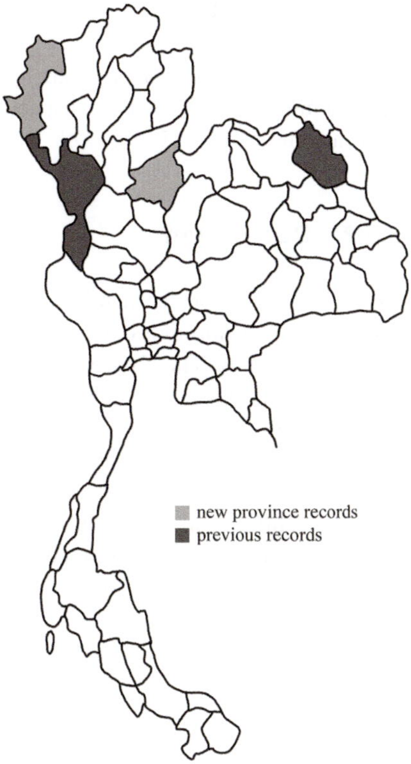

Fig. 5.28 Distribution of *Anisogomphus*

5.5 *Asiagomphus* Asahina, 1985

Fig. 5.29 Dorsal habitus of the nymph of *Asiagomphus*

5.5.1 General Appearance

The nymphs of *Asiagomphus* are characterized by their medium to large size (27–33 mm), with the body robust and flat, abdomen tapering posteriorly, and color pattern complex (Fig. 5.29).

5.5.2 Description

Head. The head is wider than long, but narrower than the pterothorax and abdomen (Fig. 5.29). The labrum is 0.5 mm long and bare, with the anterior margin setose and flattened ventrally. The clypeus is mostly bare. There is a row of long, stiff setae on the anterior margin of the antefrons, the postfrons, vertex, and occiput are mainly bare, but with sparse and very minute, reddish setae. The anterior margin of the postfrons is V-shaped. The antennal scape is globose, the pedicel subglobose, and both are covered with delicate setae; antennomere 3 is the largest, subcylindrical, flattened ventrally, slightly reflexed, its dorsal and ventral surfaces are finely granulose, and the lateral margins are beset with long, white, delicate setae; antennomere 4 is strongly reduced and mostly covered by delicate setae (Fig. 5.30).

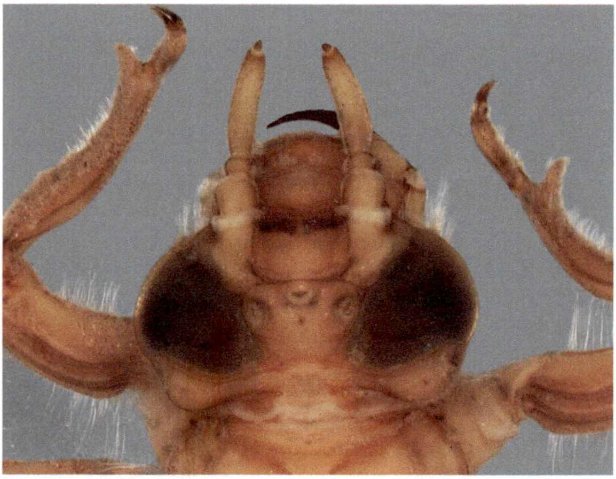

Fig. 5.30 The head in dorsal view

The compound eyes are large. The occipital lobes are small and rounded, slightly bulging, and have tufts of long, white setae (Fig. 5.30). Mandibles have the formula: L 1234 0 a($m^{1,2,3,4,5,6}$)b/R 1234 y a($m^{1,2,3,4}$)b, with a = b in both mandibles, the *m* denticles are sharp in the right mandible and truncate in the left mandible. The maxilla's galeolacinia is as described for *Acrogomphus*. The ventral pad of the hypopharynx is subpentagonal, whitish, mostly soft, with a basal triangular sclerite on each side of the midline, and the posterior margin is sinuate; the anterior margin has a row of long, stiff setae. Labium: The prementum-postmentum articulation reaches the posterior margin of the fore coxae. The prementum is subrectangular (Fig. 5.31), longer than wide, the lateral margins are straight, parallel and finely serrate on the distal half, and the basal margin is widely emarginate and ridged with a weakly defined ventromedian sulcus along the basal half; the ligula is poorly developed, shallowly biconvex (Fig. 5.31), its anterior margin has a dorsal row of abundant long, piliform setae, and a small median tooth. The apical lobe of the labial palp is strongly incurved, terminates in a sharp hook, and the mesal margin has a row of 6–7 short, robust, square teeth. The movable hook is as long as the palp, reddish, sharp, and moderately incurved.

Fig. 5.31 The prementum in ventral view

Thorax. The prothorax is narrower than the head and setose on the ventral margin of the propleuron (Fig. 5.29). The pronotum is covered with minute setae, except on the large, depressed, bare, irregular dorsolateral areas; the anterior margin is shallowly concave, the lateral margin rounded, and the posterior margin shallowly convex. The pterothorax is sparsely covered with minute setae but is setose on the ventrolateral surface. The hind legs are moderately long (i.e., when fully extended, they reach the posterior margin of S10) and setose; the burrowing hooks are well-developed on the fore- and middle tibiae (Figs. 5.29 and 5.30); the pretarsal claws are simple and each has a pulvilliform empodium. The anterior and posterior wing sheaths are parallel (Fig. 5.29), their surfaces mostly smooth, the margins have delicate, long setae, and slightly surpass the anterior margin of S4 and reach the basal half of S4, respectively.

Fig. 5.32 Ventral habitus of the nymph of *Asiagomphus*

Abdomen. The abdomen is wide, robust, parallel-sided on S3–S5, gradually tapers posteriorly, and has a complex color pattern (Figs. 5.29 and 5.32). The ventral surface is flat, whereas the dorsal surface is shallowly convex. Terga 2–6 are setose and 7–10 finely granulose. Middorsal protuberances on S2–S7 are lacking, but S8–S9 have small, spine-like, posteriorly directed, middorsal protuberances (Figs. 5.29 and 5.33). The lateral margins are densely setose on S1–S7 and spiny on the apical third of S8 and on the entire lateral margin of S9; S6–S9 end in robust, lateral, sharply pointed and slightly incurved spines, which are subequal in size and robustness, but those of S8 are slightly longer (Fig. 5.32). Sterna 1–4 are light-brown and finely setose, and sterna 5–10 are reddish-brown. Sterna 3–6 and 9 are each divided into three sternites and sterna 7–8 into five sternites. The female gonapophyses are rudimentary. The caudal appendages are short and robust, shorter than the dorsal length of terga 9+10 (Fig. 5.33). The epiproct and paraprocts are pyramidal and subequal in length. The cerci are triangular and the shortest of the caudal appendages. The male's epiproct has a pair of small, rounded tubercles at the basal 0.70.

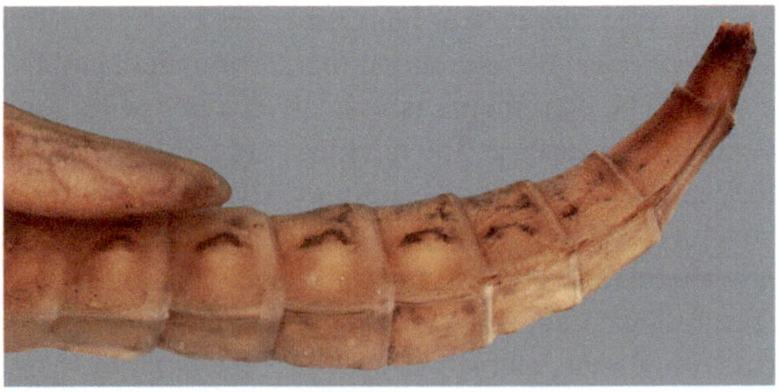

Fig. 5.33 The abdomen and caudal appendages in lateral view

Measurements (in mm). Total length including caudal appendages 27.7–33.0; abdomen 17.4–21.5; maximum width of head 5.7–6.1; length of hind femur 5.1–5.7.

5.5.3 Remarks

Only *Asiagomphus xanthenatus* (Williamson, 1907) has been recorded from Thailand (Hämäläinen and Pinratana 1999, 2000). The nymphs here described were inhabiting silt and mud near the banks of a bedrock stream. Some species have been recorded as "very common in stream pools and stream margins, in muddy areas with rooted weeds or leaf strewn mud substrates" (Wilson 1995). Egg development is direct, and the nymphal duration is usually 3 or 4 years in *A. pryeri* from Japan (Aoki 1999). By the nymphal morphology, *Asiagomphus* appears more similar to *Burmagomphus* and to a lesser extent to *Anisogomphus*.

5.5.4 Distribution

In Thailand, *Asiagomphus* has been recorded from the provinces of Chiang Mai, Kanchanaburi (Hämäläinen and Pinratana 1999, 2000), Chiang Rai (Katatani et al. 2004), and more recently from Ratchaburi (Day et al. 2012). ***New province records***: Kamphaeng Phet, Mae Hong Son, Nakhon Ratchasima, Nong Bua Lamphu, Phayao, Phrae, Phitsanulok, Songkhla, Udon Thani, and Yala (Fig. 5.34).

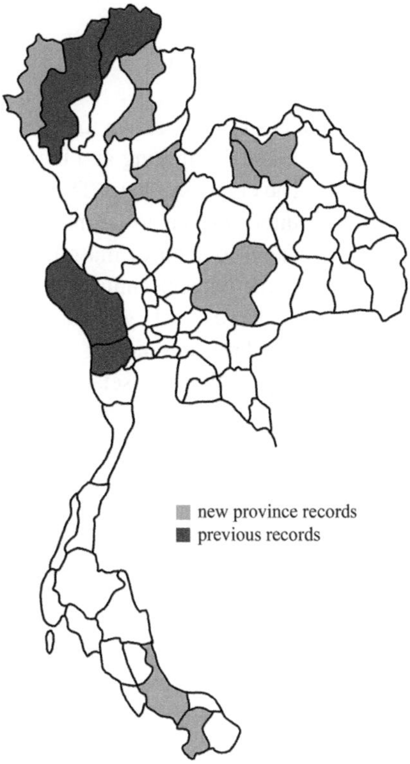

Fig. 5.34 Distribution of *Asiagomphus*

5.6 *Burmagomphus* Williamson, 1907

Fig. 5.35 Dorsal habitus of the nymph of *Burmagomphus*

5.6.1 General Appearance

The nymphs of *Burmagomphus* are characterized by their small to medium size (18–22 mm), with the body slender and somewhat flat, abdomen tapering posteriorly, and coloration yellow-brown to brown (Fig. 5.35).

5.6.2 Description

Head. The head is wider than long, but narrower than the pterothorax and abdomen. The labrum is 0.3 mm long and bare, with the anterior margin setose and flattened ventrally. The clypeus is bare. There is a row of long, stiff setae on the anterior margin of the antefrons, the postfrons, vertex, and occiput are covered with very minute, reddish setae. The anterior margin of the frons is V-shaped. The antennal scape is subglobose, the pedicel annular, and both are covered with small setae; antennomere 3 is the largest, subcylindrical, flattened ventrally, slightly reflexed, its dorsal and ventral surfaces are mostly smooth, and the lateral margins are beset with short and long, white, delicate setae; antennomere 4 is reduced and 0.18× as long as antennomere 3. The compound eyes are large (Fig. 5.36).

Fig. 5.36 The head in dorsal view

The occipital lobes are rounded, slightly bulging, and have small scale-like setae. Mandibles have the formula: L 1234 0 a($m^{1,2,3,4,5,6}$)b/R 1234 y a($m^{1,2,3}$)b, with a > b in both mandibles, and the *m* denticles are sharp in the right mandible and truncate in the left mandible. The maxilla's galeolacinia is as described for *Acrogomphus*. The ventral pad of the hypopharynx is subpentagonal, whitish, mostly soft, with a basal triangular sclerite on each side of the midline connected to each other by a basal, transverse, denticulate, sclerotized bridge; the posterior margin is sinuate and the anterior margin has a row of long, stiff setae. Labium: The prementum-postmentum articulation reaches the posterior margin of the fore coxae. The prementum is subrectangular (Fig. 5.37), longer than wide, the lateral margins are straight, slightly divergent, and finely serrate on the distal half, and the basal margin is widely emarginate and ridged with a weakly defined, ventromedian sulcus along the basal half; the ligula is poorly developed, angulate at the middle, its anterior margin has a dorsal row of abundant long, piliform setae and a well-developed median tooth (best seen in ventral view) (Fig. 5.37). The apical lobe of the labial palp tapers to a sharp hook, and the mesal margin is straight and has a row of 7–8 short, robust, square teeth. The movable hook is yellow-brown, shorter than the palp, sharp, and moderately incurved.

Fig. 5.37 The prementum in ventral view

Thorax. The prothorax is narrower than the head and setose on the ventral margin of the propleuron (Fig. 5.35). The pronotum is covered with minute setae, except on the large, depressed, bare, irregular dorsolateral areas; the anterior margin is straight, the lateral margin rounded, and the posterior margin shallowly convex. The pterothorax is covered with minute setae and the ventrolateral surface is setose. The hind legs are short (Figs. 5.35 and 5.38) (i.e., when fully extended, they scarcely reach the posterior margin of S7) and setose; the burrowing hooks are moderately developed on the fore- and middle tibiae (Fig. 5.37); the pretarsal claws are simple, and each has a pulvilliform empodium. The anterior and posterior wing sheaths are parallel, covered with minute setae, the margins have delicate, long setae, and nearly reach and surpass the anterior margin of S4, respectively.

Fig. 5.38 Ventral habitus of the nymph of *Burmagomphus*

Abdomen. The abdomen is slender, gradually tapers posteriorly, and has a complex color pattern (Fig. 5.35 and 5.38). The ventral surface is flat, whereas the dorsal surface is shallowly convex. Terga 2–10 are minutely setose including on the posterior margins. Terga 3–7 and the basal half of S8 have a longitudinal suture along the midline. Middorsal protuberances are lacking on S2–S6 or S2–S7, but S7–S9 or S8–S9 each has a short, low, serrate, triangular, dorsal, posteriorly directed protuberance (Fig. 5.39). The lateral margins are densely setose on S1−S7, spiny on the apical third of S7 and the entire lateral margins of S8−S9; S7−S9 end in a robust, lateral, sharply pointed spines, which are divergent and nearly of the same size and robustness on S7−S8, and those of S9 are parallel and the longest. Sterna 1−8 are light-brown to brown and finely setose, and sterna 9−10 are reddish-brown (Fig. 5.38). Sterna 3−6 and 9 are each divided into three sternites and sterna 7−8 into five sternites. The female gonapophyses are rudimentary. The caudal appendages are short (i.e., half of the middorsal length of S9) and robust (Fig. 5.39). The epiproct and paraprocts are pyramidal and bluntly pointed, the cerci are triangular and sharply pointed, and all are subequal in length. The male's epiproct has a pair of small, rounded tubercles at the basal 0.70.

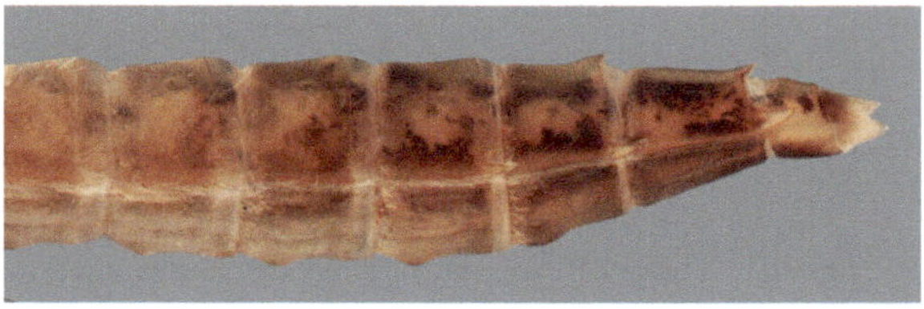

Fig. 5.39 The abdomen and caudal appendages in lateral view

Measurements (in mm). Total length including caudal appendages 18.0−21.5; abdomen 12.0−14.2; maximum width of head 3.5−4.0; length of hind femur 3.0−4.3.

5.6.3 Remarks

Eight species of *Burmagomphus* have been recorded from Thailand (Hämäläinen and Pinratana 1999, Kosterin et al. 2012, Makbun 2017a). The nymphs inhabit small, sluggish streams or faster, clear streams in forest or open country, and even occur in slightly brackish conditions (Orr 2005). However, the nymphs of *Burmagomphus* "are extremely inert, slow-moving creatures… capable only to burrow in silt or fine sand and avoid the tracts of streams with fast flowing water" (Lieftinck 1964). The nymphs of *B. williamsoni javicus* Schmidt, 1934, and *B. divaricatus* Lieftinck, 1964, were described by Lieftinck (1964) and Yang and Davies (1996), respectively. By the nymphal morphology, *Burmagomphus* appears more similar to *Anisogomphus* and *Asiagomphus*, and to a lesser extent to *Euthygomphus*.

5.6.4 Distribution

In Thailand, *Burmagomphus* has been recorded from the provinces of Chaiyaphum, Chanthaburi, Chiang Mai, Kanchanaburi, Loei, Nakhon Nayok, Nakhon Ratchasima, Phetchabun, Phrae, Saraburi, Tak, Trang, Songkhla (Hämäläinen and Pinratana 1999), Chiang Rai (Katatani et al. 2004), Phangnga, Prachuap Khiri Khan, Surat Thani (Ferro et al. 2009), Nan (Katatani et al. 2010), Phetchaburi, and Ratchaburi (Day et al. 2012). ***New province records***: Kalasin, Kamphaeng Phet, Lampang, Mae Hong Son, Mukdahan, Narathiwat, Phitsanulok, Ranong, Suphan Buri, Ubon Ratchathani, Udon Thani, and Uttaradit (Fig. 5.40).

Fig. 5.40 Distribution of *Burmagomphus*

5.7 *Davidius* Selys, 1878

Fig. 5.41 Dorsal habitus of the nymph of *Davidius*

5.7.1 General Appearance

The nymphs of *Davidius* are characterized by their small size (19 mm), with the body robust and flat, wing sheaths strongly divergent, abdomen lacking dorsal protuberances, and coloration yellow-brown (Fig. 5.41).

5.7.2 Description

Head. The head is wider than long (Fig. 4.42), but narrower than the pterothorax and abdomen. The labrum is 0.4 mm long, bare, with the anterior margin setose and flattened ventrally. The clypeus is bare; the antefrons is sparsely granulose with a row of small setae on the anterior margin, the postfrons, vertex, and occiput are finely granulose and with minute, sparse scale-like setae. The anterior margin of the frons is widely concave at the middle; the anterolateral portions are rounded, moderately produced frontally, and beset with long, stiff setae. The antennal scape is globose, the pedicel subglobose, and both are covered with delicate setae; antennomere 3 is the largest, flattened, and has the mesal margin straight and lateral margin convex (Fig. 5.42), its dorsal and ventral surfaces are finely granulose, and the lateral margins are beset with long, white, delicate setae; antennomere 4 is strongly reduced and mostly covered by delicate setae. The compound eyes are large (Fig. 5.42).

Fig. 5.42 The head in dorsal view

The occipital lobes appear angulated by the presence of a low, rounded tubercle directed laterally and slightly upturned (Fig. 5.42). Mandibles have the formula: L 1234 0 a($m^{1,2,3,4,5,6}$)b/R 1234 y a($m^{1,2,3}$)b, with a > b in both mandibles. The maxilla's galeolacinia is as described for *Acrogomphus*. The ventral pad of the hypopharynx is subpentagonal, yellowish, with the basal half lightly sclerotized and the apical half soft; the posterior margin is nearly straight with a small emargination at the middle, and the anterior margin has a row of long, stiff setae.

Fig. 5.43 The prementum in ventral view

Labium: The prementum-postmentum articulation reaches the posterior margin of the fore coxae. The prementum is subquadrate (Fig. 5.43), as long as wide, the lateral margins are convex along the apical 0.70 and serrulate, and the basal margin is widely emarginate and lacks a ventromedian sulcus. The ligula is poorly developed, shallowly convex, its anterior margin has a dorsal row of abundant long, piliform setae, and a ventral row of four short, red-brown, truncate teeth on the middle third (Fig. 5.43). The apical lobe of the labial palp is narrowly rounded at the tip, and the mesal margin is straight and has a row of 8–9 short, robust, square teeth (Fig. 5.43). The movable hook is reddish, as long as the palp, sharp, and moderately incurved.

Fig. 5.44 Ventral habitus of the nymph of *Davidius*

Thorax. The prothorax is narrower than the head and setose on the ventral margin of the propleuron (Fig. 5.41). The pronotum is covered with minute setae, except on the large, depressed, bare, irregular dorsolateral areas; the anterior margin is straight, the lateral margin rounded, and the posterior margin shallowly convex. The pterothorax is sparsely granulose and has long, stiff, reddish setae, and the ventrolateral surface is setose. The hind legs are moderately long (Figs. 5.41 and 5.44) (i.e., when fully extended, they reach the posterior margin of S9) and setose; burrowing hooks are lacking; the pretarsal claws are simple, and each has a pulvilliform empodium. The anterior and posterior wing sheaths are strongly divergent, their surfaces are mostly smooth, the margins have delicate, short setae, and both reach the basal half of S4.

Abdomen. The abdomen is wide and robust, parallel-sided at the middle, and has a complex color pattern. The ventral surface is flat and the dorsal surface is shallowly convex (Fig. 5.45). Terga 2–9 are finely granulose, sparsely covered with long, reddish setae, with spines on the posterior margins; S10 is sunken into S9. Dorsal protuberances are lacking. The lateral margins of S1–S5 are densely setose and those of S9 strongly convergent posteriorly. S7–S9 end in robust and sharply pointed lateral spines, which increase in size and robustness posteriorly; those of S9 are as long as S10. The sternum is reddish-brown and mostly bare. Sterna 3–9 are each divided into three sternites. The female gonapophyses are vestigial. The epiproct and paraprocts are pyramidal, robust, and longer than the dorsal length of S9+S10. The epiproct is bluntly pointed, and the male's epiproct has a pair of rounded dorsal tubercles at the basal 0.60. The paraprocts are the longest of the caudal appendages, sharply pointed, and with the tips divergent and decurved. The cerci are triangular, sharply pointed (Fig. 5.45), and the shortest of the caudal appendages.

Fig. 5.45 The abdomen and caudal appendages in lateral view

Measurements (in mm). Total length including caudal appendages 19.2; abdomen 11.5; maximum width of head 4.2; length of hind femur 3.9.

5.7.3 Remarks

Only two species of *Davidius* have been recorded from Thailand: *D. fruhstorferi* Martin, 1904, and an undetermined species (Hämäläinen and Pinratana 1999). Neither of these nymphs has been described. The nymphal description provided here was based on material collected in a mountain stream at the margins of a waterfall in Mae Hong Son Province (Fig. 5.46). In Japan, some species such as *D. moiwanus* (Okumura, 1935) inhabit narrow wetland streams (Kiyoshi & Sota 2006). The nymph is typical of sandy-gravel burrowers (Wilson, pers. com.). By the nymphal morphology, there is no apparent relationship of *Davidius* to any other genus inhabiting Thailand; however, it exhibits a close resemblance to the New World genus *Epigomphus*.

5.7.4 Distribution

In Thailand, *Davidius* has been recorded only from Chiang Mai Province (Hämäläinen and Pinratana 1999). ***New province record***: Mae Hong Son (Fig. 5.46b).

Fig. 5.46 (**a**) Stream at Namtok Maw Pang, Mae Hong Son Province. (**b**) Distribution of *Davidius*

5.8 *Euthygomphus* Kosterin, 2016

Fig. 5.47 Dorsal habitus of the nymph of *Euthygomphus*

5.8.1 General Appearance

The nymphs of *Euthygomphus* are characterized by their small to medium size (19–21 mm), with the body slender and flattened, legs with burrowing hooks vestigial, abdomen narrowed, with middorsal protuberances on S6–S9 or S7–S9, and posterolateral spines on S6–S9 (Fig. 5.47).

5.8.2 Description

Head. The head is wider than long (Fig. 5.48) and wider than the thorax and abdomen, and has tufts of delicate setae, except on some small areas around the antennae and on the occipital lobes. The labrum is 0.3 mm long, and its anterior margin is setose and ventrally flattened. The clypeus is glabrous, and the frons, vertex, and occiput are covered with minute, reddish setae, which are larger and stiff on the anterior margin of the antefrons and cephalic lobes. The posterolateral area of the frons just behind the base of the antenna is raised, thick, and set with a tuft of long, delicate setae; three large ocelli are on the vertex. The antennal scape is globose, the pedicel annular, and both are covered with minute and long setae; antennomere 3 is cylindrical and strongly incurved, the lateral margins are beset with long, white, delicate setae; antennomere 4 is a hemispherical rudiment.

Fig. 5.48 The head in dorsal view

The compound eyes are strongly developed. The occiput is short and the occipital lobes are rounded and bulging (Fig. 5.48). The posterior margin of the head is sinuate. Mandibles have the formula: L 1234 0 a($m^{1,2,3,4}$)b/R 1234 y a(m^0)b, with a = b in the right mandible and a > b in the left mandible. The maxilla's galeolacinia is as described for *Acrogomphus*. The ventral pad of the hypopharynx is subpentagonal, whitish, mostly soft and bare, and has a U-shaped sclerite. Labium: The prementum-postmentum articulation reaches the middle of the fore coxae. The prementum is subquadrate, wider than long (Fig. 5.49), the lateral margins are straight in the distal half, and the sides along the basal half are sinuate and divergent. The ligula is very poorly developed, narrow, straight on the anterior margin and with a slight median concavity furnished with long, white, piliform setae, and a tuft of long, bristle-like setae is close to the labial palp's articulation. The labial palp is stout, very broad at the base, abruptly narrowed and tapering to an acute tip, and the mesal margin is gently concave and smooth; the lateral surface is beset with long and white, bristle-like setae. The movable hook is robust, about as long as the palp, sharp, and moderately incurved.

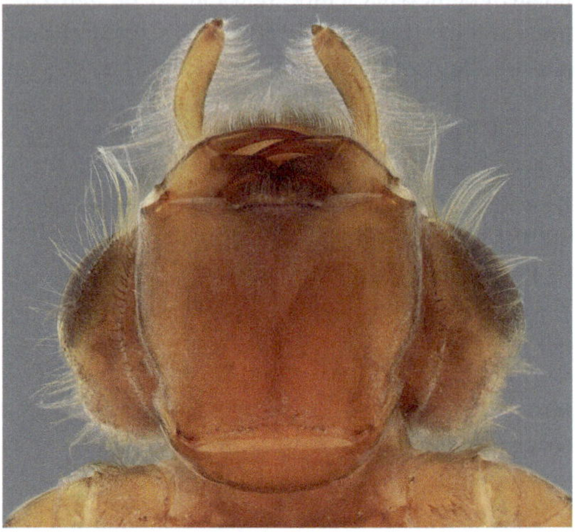

Fig. 5.49 The prementum in ventral view

Thorax. The prothorax is narrower than the head and setose on the ventral margin of the propleuron (Fig. 5.47). The pronotum is covered with short, reddish-brown setae, except on the large, depressed, smooth, irregular dorsolateral areas. The anterior margin is raised and thick, the lateral margin rounded, and the posterior margin shallowly convex. The pterothorax is covered with minute setae, which are long and white on the ventrolateral surface. The legs are covered with long, white setae (Figs. 5.47 and 5.50) except on the longitudinal, narrow, smooth stripes; the hind legs are relatively short (i.e., when fully extended, the tip of the tarsi scarcely reach the tips of the caudal appendages); the burrowing hooks are vestigial on the fore- and middle tibiae. The anterior and posterior wing sheaths are parallel, and their surfaces and margins have small setae and reach the anterior margin and the basal fourth of S4, respectively (Fig. 5.47).

Fig. 5.50 Ventral habitus of the nymph of *Euthygomphus*

Abdomen. The abdomen is flattened and tapers posteriorly, widest at S4–S5, and densely covered with minute, reddish-brown setae including on the posterior margins. Middorsal protuberances are lacking on S2–S4, vestigial on S5, but well-developed, triangular, raised, and directed posteriorly on S6–S9 or S7–S9 (Fig. 5.51), which increase in size posteriorly and have the posterior margins furnished with small spines. The lateral margins are setose, those of S6–S9 have a row of short spines that increase in length and robustness posteriorly and end in a large, postero-lateral, sharply pointed, posteriorly directed spine; the spines are moderately divergent and increase in length posteriorly, those on S9 are the longest, more or less fitting close against the lateral margins of S10, and reach the basal 0.60 of S10; S9 is shorter than its basal width, and S10 is short, wider than long, and set off from S9. The sternum is densely covered with minute setae. Sterna 1–6 are yellowish-brown and sterna 7–10 reddish-brown (Fig. 5.50). Sterna 3–6 and 9 are each divided into three sternites and sterna 7–8 into five sternites. The female gonapophyses are rudimentary and have the tips widely divergent. The caudal appendages are robust, triangular, obtusely pointed, subequal in length, and shorter than the dorsal length of S10 (Fig. 5.51). The epiproct is almost twice longer than its basal width. The cerci are as long as the epiproct, and the paraprocts are slightly longer (Fig. 5.51).

Fig. 5.51 The abdomen and caudal appendages in lateral view

Measurements (in mm). Total length including caudal appendages 19.3–20.5; abdomen 12.3–13.2; maximum width of head 3.7–4.2; length of hind femur 4.0–4.7.

5.8.3 Remarks

The genus *Euthygomphus* was recently erected by Kosterin (2016) to include several species previously belonging to *Anisogomphus*, *Leptogomphus*, and *Merogomphus*. To date, only one species of *Euthygomphus* has been recorded from Thailand: *E. yunnanensis* (Zhou and Wu, 1992) [as *Merogomphus parvus* (Krüger, 1899) by Hämäläinen and Pinratana 1999] (Kosterin 2016). The nymph of *E. parvus* is a sluggish, shallow-burrower that lives "in small clear streams with a sandy or somewhat gravelly bottom" (Lieftinck 1941).

5.8.4 Distribution

In Thailand, *Euthygomphus* has been recorded from the provinces of Chiang Mai, Kanchanaburi, Lampang, Mae Hong Son, Saraburi, Songkhla, Tak, Trang (Hämäläinen and Pinratana 1999), Chanthaburi (Kitagawa and Katatani 2002), Nakhon Si Thammarat (Ferro et al. 2009), Nan (Katatani et al. 2010), and more recently Nakhon Nayok, Nakhon Ratchasima, Phetchaburi, Ranong, and Surat Thani (Day et al. 2012). ***New province records:*** Phitsanulok and Yala (Fig. 5.52).

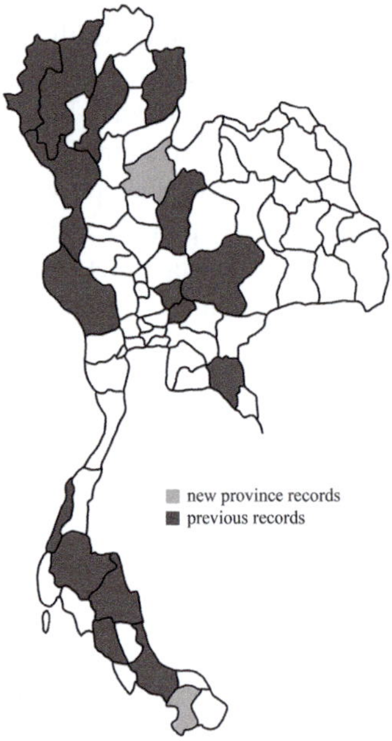

Fig. 5.52 Distribution of *Euthygomphus*

5.9 *Gomphidia* Selys, 1854

Fig. 5.53 Dorsal habitus of a young nymph of *Gomphidia*

5.9.1 General Appearance

The nymphs of *Gomphidia* are characterized by their medium size (25 mm), with the body very robust and appearing granulose due to the abundant small, scale-like setae, mouthparts setose, abdomen strongly convex dorsally and flattened ventrally, and coloration dark-brown (Fig. 5.53).

5.9.2 Description

Head. The head is wider than long (Fig. 5.54), and the labrum, frons, vertex, and occiput are densely covered with small scale-like setae, but the clypeus is bare. The anteromesal part of the frons is U-shaped, and the anterolateral and posterolateral corners protrude as blunt projections. The labrum is 0.9 mm long, and its anterior margin is densely covered with brushes of short, golden, fine setae, and flattened ventrally. The genae protrude ventrally as small tusks, and a blunt, conical projection is below the ventral margin of the compound eyes (Fig. 5.54). The antennal scape is subglobose and the pedicel cylindrical; antennomere 3 is the largest and distinctly longer than wide, cylindrical, covered with minute scale-like setae on the basal half and small, delicate white setae on the distal half; antennomere 4 is subconical, bare, and strongly reduced. The compound eyes are relatively large. The occipital lobes are large, not bulging, and the lateral margins are rounded and have a round dorsal ridge; a small, central, conical tubercle is on each side of the midline.

Fig. 5.54 The head in dorsal view

Mandibles have the formula: L 1234 0 a($m^{1,2,3,4}$)b/R 1234 y a($m^{1,2,3,4,5}$)b, with a > b in both mandibles and noticeably larger in the right mandible. The maxilla's galeolacinia is as described for *Acrogomphus*. The ventral pad of the hypopharynx is subcircular, light-yellow-brown, mostly sclerotized, its posterior margin is widely rounded, and the anterior margin has a row of long, stiff setae.

Fig. 5.55 The head and prementum in ventral view

Labium: The prementum-postmentum articulation reaches the midline of the middle coxae. The prementum is yellowish dorsally and reddish ventrally, subquadrate, wider than long, the lateral margins are serrate with the serrations stout, sharp, and incurved, and the sides are strongly convex at middle, straight basally, and convergent apically (Fig. 5.55). The ventral surface is strongly concave and without a median sulcus, although it has a subbasal, depressed, narrow, bare area along the basal margin. The ligula is small but well-developed anteriorly (Fig. 5.55), the apical margin is strongly convex and covered with three rows of setae and one row of teeth as follows: a ventral row of dense, white, delicate setae, a ventromesal row of 14 short, truncate, reddish teeth, a dorsomesal row of whitish piliform setae, and a dorsal row of dense, white, delicate setae. The labial palp is yellowish-brown, lateral margin setose, apical lobe rounded, and its mesal margin has a row of 18–19 small, squarely truncate teeth. The movable hook is dark-reddish-brown, shorter than the palp, sharp, and moderately incurved.

Fig. 5.56 Ventral habitus of the nymph of *Gomphidia*

Thorax. The prothorax is narrower than the head (Fig. 5.53). The ventral margin of the propleuron has small, blunt spines, the proepisternum is greatly reduced, and the proepimeron is produced laterally as a stout, conical, blunt projection; the anterior lobe of the pronotum is granulose, raised, and it has a mound-like protuberance at the middle; the posterior lobe is subrectangular with the lateral margins rounded and slightly reflexed, it has a mound-like protuberance at the middle of the posterior margin, and a large, elongate, bare area on each side of the midline. Pterothorax: The ventral margin of the mesepisternum is produced as a short, blunt, conical, spiny projection; the ventral margin of the mesepimeron is straight and ends in a small spine on the anteroventral corner; the ventral margin of the metepisternum has a stout, blunt, conical, spiny protuberance. The legs are relatively long and stocky (Figs. 5.53 and 5.56) (i.e., when fully extended, the hind legs reach the posterior margin of S10) and have abundant scale-like setae; the tibiae lack burrowing hooks; the tarsal formula is 2–2–2; the pretarsal claws are simple and each has a pulvilliform empodium. The anterior and posterior wing sheaths are parallel and reach the posterior margin of S5 and basal half of S6, respectively. The sternum is mostly bare; the mesosternum has a transverse, rounded, low ridge beset with long, stiff, white setae and a row of six small, blunt, reddish-brown spines on the middle third; a short, conical, scaly tubercle is close to the articulation of the middle coxae; the metasternum has a large, stout, rounded, scaly tubercle close to the articulation of the hind coxae.

Fig. 5.57 The abdomen and caudal appendages in lateral view

Abdomen. The abdomen is subcircular and widest at the posterior margin of S6. The ventral surface is flat, whereas the dorsal surface is strongly convex and reaches its highest point on S6 (Fig. 5.57); middorsal protuberances are present on S2–S10, which are vertical spines on S2–S3 (Fig. 5.53) and form a keel on S4–S10 (Fig. 5.57); they are fused on S4–S6 and those of S5–S9 end in a small, sharp spine (Figs. 5.53 and 5.56). Terga 2–10 are finely granulose, but the posterior margins have a row of dark, truncate spines. The lateral margins of S2–S9 are beset with a row of dark spines intermingled with light-brown, scale-like setae, which increase in size and robustness posteriorly, S3–S9 end in strong, sharp spines which are divergent on S3–S5, incurved on S6, convergent on S8, and directed posteriorly on S9; those on S7 are the largest; S10 is laterally concealed by the spines of S9. The sternum is reddish and finely granulose (Fig. 5.56). Sterna 3–6 and 9 are each divided into three sternites, sterna 7–8 into five sternites, and the median sutures converge apically on 2, are parallel on 3–6, and divergent apically on 7–9; the posterior margins are convex on sterna 2–4, straight on 5–6 and 9, sinuate on 1, 7, and 10, and concave on 8. The female gonapophyses are conical and rudimentary. The caudal appendages are short, pyramidal, and as long as the dorsolateral length of S10 (Fig. 5.57). The epiproct and cerci are dark-red, mostly bare, and shining; the paraprocts are light-brown dorsally, reddish ventrally, and densely covered with scale-like setae. The epiproct is obtusely pointed. The cerci are rounded apically but have a small spine at the tip and the tips are convergent. The paraprocts are obtusely pointed and the tips are straight; the paraprocts are the longest of the caudal appendages and the cerci are the shortest.

Measurements (in mm). Total length including caudal appendages 25; abdomen 16.5; maximum width of head 8.6; length of hind femur 6.2.

5.9.3 Remarks

The nymph of *Gomphidia* resembles those of *Gomphidictinus* and *Ictinogomphus*, all belonging to the subfamily Lindeniinae (Carle 1986). At close examination, *Gomphidia* is distinguished from *Ictinogomphus* by the truncate spines on the posterior margins of S2–S8 (fan-shaped or tridentate setae on the latter) and from *Gomphidictinus* by the anterolateral corners of the frons not enlarged (enlarged and with strong spines in the latter). Three species have been recorded from Thailand: *Gomphidia abbotti* Williamson, 1907, *G. kruegeri* Martin, 1904, and *G. maclachlani* Selys, 1873 (Hämäläinen and Pinratana 1999). The nymphs of *Gomphidia* inhabit leaf litter, sand, or mud in sluggish forest streams (Orr 2005).

5.9.4 Distribution

In Thailand, *Gomphidia* is widespread and has been recorded from the following provinces: Bangkok, Chaiyaphum, Chanthaburi, Chumphon, Kamphaeng Phet, Kanchanaburi, Krabi, Nakhon Ratchasima, Narathiwat, Phangnga, Phrae, Prachuap Khiri Khan, Ranong, Satun, Songkhla, Tak, Trang (Hämäläinen and Pinratana 1999), Chiang Rai (Katatani et al. 2004), Phuket (Kosterin and Vikhrev 2009), Nan (Katatani et al. 2010), and more recently Nakhon Si Tammarat and Ratchaburi (Day et al. 2012). ***New province records***: Nakhon Nayok, Phatthalung, Phayao, Phetchabun, Sa Kaew, and Surat Thani (Fig. 5.58).

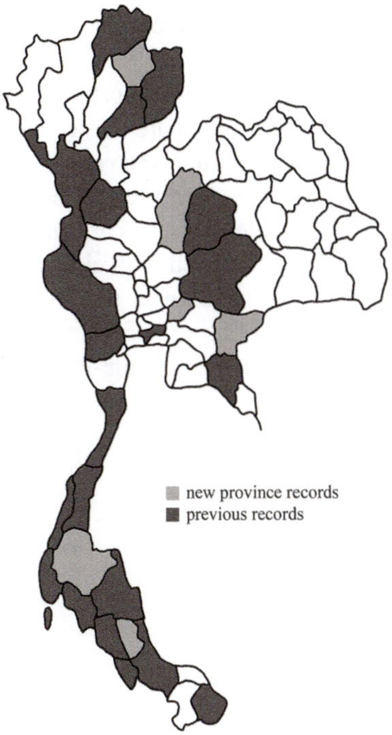

Fig. 5.58 Distribution of *Gomphidia*

5.10 *Gomphidictinus* Fraser, 1942

Fig. 5.59 Dorsal habitus of the nymph of *Gomphidictinus perakensis* showing strong spines on the abdomen

5.10.1 General Appearance

The nymphs of *Gomphidictinus* are characterized by their medium size (27 mm), with the body robust and appearing granulose and spiny, mouthparts setose, abdomen strongly convex dorsally and flattened ventrally, and coloration reddish-yellow to yellowish-brown with pale areas on the head and thorax (Fig. 5.59).

5.10.2 Description

Head. The head is almost twice wider than long (Fig. 5.60). The labrum, frons, vertex, and occiput are densely covered with small scale-like setae, but the clypeus is bare. The labrum is 0.6 mm long, and its anterior margin is flattened ventrally and densely covered with setae. The anteromesal part of the frons is narrowly U-shaped, the anterolateral corners are enlarged and armed with strong, sharp, dark, recurved spines intermingled with elongate, scale-like setae (Fig. 5.60). The genae protrude ventrally as small tusks (Fig. 5.61). Below the ventral margin of the compound eye is a mound-like projection which is covered with strong, sharp, dark, recurved spines and elongate, scale-like setae that are visible in dorsal view.

Fig. 5.60 The head in dorsal view, showing strong spines on the frons

The antennal scape is globose, the pedicel cylindrical, antennomere 3 is the largest and distinctly longer than wide, cylindrical, straight, covered with minute, scale-like seta and abundant delicate, white setae; antennomere 4 is strongly reduced, subconical, and bare. The compound eyes are relatively large. The occipital lobes are large (Fig. 5.60), pale, slightly bulging, with large bare areas, the lateral margins are rounded and have a rounded dorsal ridge covered with abundant scale-like setae and 3–4 stout, dark, sharp spines. Mandibles have the formula: L 1234 0 a($m^{1,2}$)b/R 1234 y a($m^{1,2}$)b, with a > b in both mandibles; the *m* denticles in the left mandible are almost the same size as tooth "b." The maxilla's galeolacinia is as described for *Acrogomphus*. The ventral pad of the hypopharynx is subcircular, light-yellow-brown, mostly sclerotized, with the posterior margin widely rounded, and the anterior margin lacking setae. Labium: The prementum-postmentum articulation almost reaches the posterior margin of the fore coxae. The prementum is subquadrate (Fig. 5.61) , slightly wider than long, and reddish-yellow ventrally; the lateral margins are serrate, with the serrations stout, sharp and incurved; the sides are sinuate and convergent basally; the ventral surface is covered with minute, reddish, scale-like setae and a large concavity is below the ligula; there is no median sulcus but a sub-basal, depressed, narrow, bare area is present along the basal margin.

Fig. 5.61 The head and prementum in ventral view

The ligula is moderately developed (Fig. 5.61), trapezoidal, and the apical margin is serrate with three rows of setae and one of teeth as follows: a ventral row of dense, white, delicate setae, a ventromesal row of 16 min, truncate, reddish teeth, a dorsomesal row of whitish, piliform setae, and a dorsal row of dense, white, delicate setae. The labial palp is yellow to yellowish-brown and setose on the lateral margin, the apical lobe is sharply pointed and moderately incurved, and the mesal margin has a row of 10–11 small, squarely truncate teeth. The movable hook is reddish-brown, shorter than palp, sharp, and moderately incurved.

Fig. 5.62 Ventral habitus of the nymph of *Gomphidictinus perakensis*

Thorax. The prothorax is narrower than the head (Fig. 5.59). The ventral margin of the propleuron has stout sharp spines, the proepisternum is greatly reduced, and the proepimeron is produced outward as a large, rounded, spiny projection. The pronotum is largely pale and its anterior lobe is scaly and raised; a large, elongate bare area is on each side of the midline, which has a central, oval, scaly area. The posterior lobe is subrectangular, the lateral margins are rounded, and the posterior margin is sinuate with a mound-like protuberance in the middle. The pterothorax is largely pale dorsally, and the ventral margin of the pleuron is spiny. The legs are relatively short (Figs. 5.59 and 5.62) (i.e., when fully extended, the hind legs hardly reach the posterior margin of S10) and have abundant scale-like setae; the femora are reddish with yellow markings; the tibiae are yellow with two wide, reddish rings, lack burrowing hooks, and have some spines on the apical margins; the tarsal formula is 2–2–2, and the tarsi are yellow with a wide, dark band on all 2nd tarsomeres; the pretarsal claws are simple and each has a pulvilliform empodium. The anterior and posterior wing sheaths are parallel and reach the basal half and posterior margin of S6, respectively (Fig. 5.59). The sternum is mostly bare; the mesosternum has a transverse, rounded, low ridge beset with long, stiff, light-brown setae, which are intermingled with elongate, scale-like setae; a low, rounded tubercle is close to the articulation of the middle and hind coxae, which are covered with elongate, scale-like setae.

Fig. 5.63 The abdomen and caudal appendages in lateral view

Abdomen. The abdomen is suboval and widest at S7 (Figs. 5.59 and 5.62). The ventral surface is flat, whereas the dorsal surface is strongly convex and reaches its highest point on S6 (Fig. 5.63); dorsal protuberances are on S2–S10, that on S2 is a vertical spine, those on S3–S5 are more or less hook-like, those on S6–S10 form a setose keel, and those on S5–S9 end in a spine. The lateral margins of S3–S9 have a row of dark spines intermingled with light-brown, scale-like setae, which increase in size and robustness posteriorly, ending in a strong, sharp, slightly incurved spine (Figs. 5.59 and 5.62) with those on S7 and S9 the largest; S10 and the caudal appendages are laterally concealed by the lateral spines of S9. Terga 2–10 are covered by small, spiniform and scale-like setae, except on the central, glabrous part of S2–S5; the posterior margins of S2–S8 have a row of spiniform setae and on S9–S10 scale-like setae; S3–S8 have a complex pattern of sub-lateral blotches bordered with spiniform setae. The sterna are covered with minute, tetradentate, scale-like setae and have a reddish-yellow to yellowish-brown coloration with a complex pattern of blotches. Sterna 3–6 and 9 are each divided into three sternites and sterna 7–8 into five sternites; the median sutures are convergent apically on 2, parallel on 3–6, and divergent apically on 7–9. The posterior margins of sterna 2–5 are straight, sinuate on 1, 6, and 9–10, and concave on 7–8. The female gonapophyses are vestigial and ovate. The caudal appendages are pyramidal, reddish, and longer than the dorsolateral length of S10 (Figs. 5.59, 5.62, and 5.63). The epiproct and cerci are mostly glabrous; the epiproct is obtusely pointed, and the cerci and paraprocts are sharply pointed. The paraprocts, in ventral view, are divergent and the longest of the caudal appendages, and the cerci are the shortest.

Measurements (in mm). Total length including caudal appendages 27–28; abdomen 16–17; maximum width of head 7.4–7.5; length of hind femur 7.0–7.2.

5.10.3 Remarks

This nymph is the most impressive of the Lindeniinae present in Thailand due to the presence of strong, sharply pointed spines in several parts of the body. It is quickly recognized by "its extremely well-developed posterolateral flanges of abdominal segments 3–9" (Ferro and Sites 2006), who also referred to the shape of the abdomen as shurikenate. See also comments under *Gomphidia*. This is a monotypic genus represented by *G. perakensis* (Laidlaw, 1902). The nymphs burrow deep into accumulated leaf litter (Orr 2005).

5.10.4 Distribution

In Thailand, *Gomphidictinus perakensis* has been recorded from the provinces of Chanthaburi, Chiang Mai, Chumphon, Kanchanaburi, Nakhon Nayok, Phetchabun, Phrae, Tak, Yala (Hämäläinen and Pinratana 1999), Chiang Rai (Katatani et al. 2004), Phangnga, Prachuap Khiri Khan, Surat Thani (Ferro and Sites 2006), and Phetchaburi (Day et al. 2012). ***New province records:*** Mae Hong Son, Nakhon Ratchasima, Nan, Phayao, and Ubon Ratchathani (Fig. 5.64).

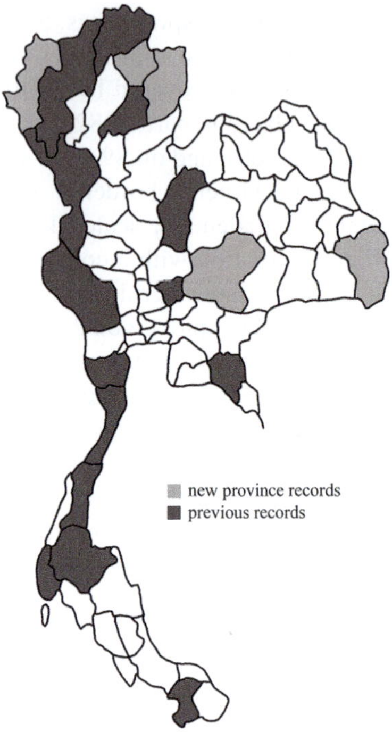

Fig. 5.64 Distribution of *Gomphidictinus perakensis*

5.11 *Heliogomphus* Laidlaw, 1922

Fig. 5.65 Dorsal habitus of the nymph of *Heliogomphus*

5.11.1 General Appearance

The nymphs of *Heliogomphus* are characterized by their small size (18.4–19.4 mm), with the body robust and slightly flattened, antennae hypertrophied, integument finely granulose and lacking long setae, and coloration reddish-yellow to dark-brown (Fig. 5.65).

5.11.2 Description

Head. The head is wider than long, wider than the prothorax but narrower than the pterothorax and abdomen, and densely granulose (Figs. 5.65 and 5.66). The labrum is 0.4 mm long, and its anterior margin is flattened ventrally and densely covered with claviform setae and short, stiff setae. The anterior margins of the clypeus and antefrons are slightly produced at the middle as a short, narrow convexity, and is slightly raised on the antefrons. The medial part of the postfrons is strongly projected anteriorly (Fig. 5.66), covering most of the antefrons; this projection is thick, strongly convex laterally, and widely concave apically. The antennal scape is globose and the pedicel ring-shaped; antennomere 3 is the largest (Fig. 5.66), yellow, paddle-shaped, and strongly flattened dorsoventrally, its lateral margin is strongly divergent and subapically convex, the mesal margin is straight, the apical margin is more or less straight with the vestigial, nipple-shaped antennomere 4 at the middle, the lateral margin has rows of short claviform setae, the dorsal and ventral surfaces are mostly finely granulose with some scattered claviform setae, and the dorsal surface has minute bare areas appearing as a sieve.

Fig. 5.66 The head and prothorax in dorsal view, showing the enlarged antennae

The compound eyes are moderately developed and strongly concave on the posterior margin (Fig. 5.66); a longitudinal carina is below each compound eye, which has two strong, ventrally directed lobes in the apical third; the anteriormost lobe is

the largest and covered with short, robust, stiff setae (Fig. 5.67). The occipital lobes are rounded, bulging, and sometimes have a small conical tubercle on the postero-lateral corner, and a low, submesal, dorsal tubercle on each lobe (Fig. 5.66). Mandibles have the formula: L 1234 0 a($m^{1,2,3,4,5,6,7}$)b/R 1234 y a($m^{1,2,3,4,5,6}$)b, with a > b in both mandibles and noticeably larger in the right mandible. The maxilla's galeolacinia is as described for *Acrogomphus*. The ventral pad of the hypopharynx is subpentagonal, yellowish-brown, and mostly sclerotized, the basocentral area is soft and whitish, the posterior margin is widely concave, and a row of short, transverse, delicate setae is on the apical third.

Fig. 5.67 The head, prementum, and prothorax in ventral view

Labium: The prementum-postmentum articulation reaches the posterior margin of the fore coxae. The prementum (Fig. 5.67) is yellow dorsally, reddish-yellow ventrally, subquadrate, slightly wider than long, the sides are straight and convergent basally, the ventral surface has a longitudinal medial sulcus along the basal 0.70 of its length, which widens distally, and the basal margin is widely concave. The ligula is poorly developed, its apical margin is convex, and it has a ventral row of about 24 short, subconical, reddish teeth (Fig. 5.67) and dorsal rows of whitish, piliform setae on the apical margin. The labial palp is yellow with the apical lobe widely rounded, and the mesal margin has a row of 12–13 small, squarely truncate teeth. The movable hook is reddish-brown, shorter than the palp, sharp, and moderately incurved.

Fig. 5.68 Ventral habitus of the nymph of *Heliogomphus*

Thorax. The prothorax is narrower than the head (Figs. 5.65 and 5.68). The ventral margin of the propleuron is thick, slightly reflexed, and has abundant short, claviform setae. The pronotum is mostly granulose, except on the large, depressed, bare, dorso-lateral areas; the anterior margin is almost straight and raised, the posterior margin is widely convex, and the lateral margins are rounded and slightly reflexed (Fig. 5.66). The pterothorax is granulose. The legs are long (Figs. 5.65 and 5.68) (i.e., when fully extended, the hind legs reach the posterior margin of S10), slender, granulose, and have abundant claviform setae; the legs are completely reddish-yellow or dark-brown with the tarsi pale; the tibiae lack burrowing hooks; the femur and tibia of the hind legs are the same length; the pretarsal claws are reddish-brown and each has a pulvilliform empodium. The wing sheaths are parallel; they reach the posterior margin of S4 and basal half of S5, respectively. The sternum is very finely granulose; the prosternum has a transverse, thick ridge set with a row of short, robust, stiff setae, and a short, blunt, conical tubercle densely covered with claviform setae is close to the articulation of the fore coxae; the meso- and metasterna have a low tubercle densely covered with claviform setae close to the articulation of the middle and hind coxae, respectively.

Fig. 5.69 The abdomen and caudal appendages in lateral view

Abdomen. The abdomen is large, parallel-sided, and widest at S5 (Figs. 5.65 and 5.68). The ventral surface is flat, whereas the dorsal surface is shallowly convex and has dorsal protuberances on S4–S9; the protuberances are small on S4, low and blunt on S5–S7, and roundly pointed and ridged on S8–S9 (Fig. 5.69); a pair of small brown spines are on the apices of protuberances on S7–S9. The lateral margins of S2–S10 have rows of short, claviform setae which are intermingled with short, spiniform setae on S8–S9; the margins of S7–S9 end in divergent posterolateral spines which increase in length and robustness posteriorly; S10 is very short and concealed laterally by the spines of S9; S9 is the longest segment (Figs. 5.65 and 5.68). Terga 2–10 are very finely granulose, S2–S5 are yellowish, and S6–S10 reddish with a complex, mottled color pattern. The sternum is finely granulose and reddish-yellow (Fig. 5.68). Sterna 3 and 7–8 are each divided into five sternites and sterna 4–6 and 9 into three sternites; the median sutures are convergent apically on 2, parallel on 3–8, and divergent apically on 9; the posterior margins are straight on sterna 1–7 and sinuate on 8–10. The caudal appendages are short, pyramidal, densely granulose, and barely reach half of the dorsal length of S9 (Fig. 5.69). The lateral margins of the cerci and paraprocts have minute, spiniform setae. The male's epiproct is as long as its basal width, it has a pair of small tubercles at 0.80 of its length, and its tip is yellow and blunt. The cerci are pointed with the tips divergent. The paraprocts are obtusely pointed and their tips are slightly divergent; the paraprocts are the longest of the caudal appendages, and the cerci are the shortest (Fig. 5.69).

Measurements (in mm). Total length including caudal appendages 18.4–19.4; abdomen 10.9–12.9; maximum width of head 4.4–4.7; length of hind femur 3.9–4.3.

5.11.3 Remarks

This nymph is easily recognizable by the paddle-shaped third antennal segment; other than this feature, it resembles at first glance the nymph of *Microgomphus chelifer* Selys. Ngiam et al. (2011) argued that the nymphs of *Heliogomphus kelantanensis* and *M. chelifer* can be separated only by the shape of the third antennomere. However, a detailed examination reveals other features that distinguish both genera (those of *Microgomphus* in parentheses): the medial part of the frons is strongly projected anteriorly (straight, not projected); the carina below the compound eye is bilobed (simple, not bilobed); the ventral pad of the hypopharynx is mostly sclerotized (soft); the hind femur and tibia are the same length (hind tibia shorter); the lateral keels on the abdomen are poorly developed and have small spines (strongly developed with large spines); the cerci are shorter than the epiproct (same length); sternum 3 is divided into five sternites (divided into three sternites). Two species of *Heliogomphus* have been recorded from Thailand: *H. selysi* Fraser, 1925, and

Heliogomphus sp. (Hämäläinen and Pinratana 1999). Of these, the nymph of only *H. selysi* has been described to date (Boonsoong and Chainthong 2014a). The nymph of *H. kelantanensis* inhabits rubble in swift water (Orr 2005) and *H. scorpio* Ris, 1912, known from Hong Kong and Vietnam, lives in fast-flowing streams with gravel/cobble substrates (Wilson 1995). The nymph of *H. selysi* has been found in pools with leaf litter in headwater streams (Boonsoong and Chainthong 2014a).

5.11.4 Distribution

In Thailand, *Heliogomphus* has been recorded from the provinces of Chiang Mai, Nakhon Nayok, Nakhon Ratchasima, Nakhon Si Thammarat, Tak, Trat (Hämäläinen and Pinratana 1999; Hämäläinen 2002), Chiang Rai (Katatani et al. 2004), Kanchanaburi (Kitagawa and Katatani 2005), Phrae (Katatani et al. 2010), Lampang (Kosterin et al. 2011), and more recently from Ratchaburi (Day et al. 2012; Boonsoong and Chainthong 2014a). ***New province records***: Phangnga, Phatthalung, Phayao, and Phetchaburi (Fig. 5.70).

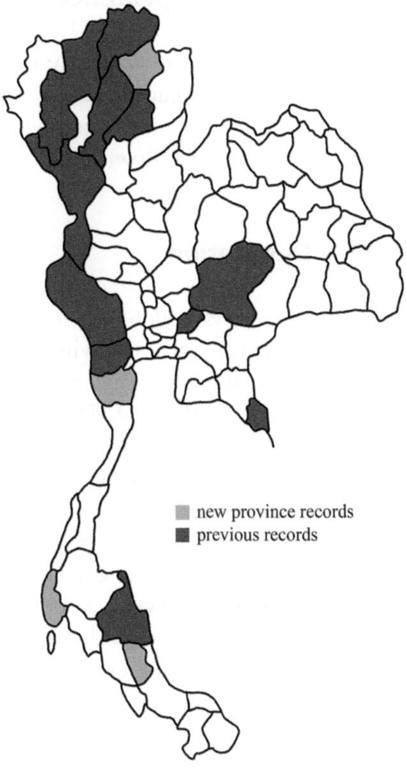

Fig. 5.70 Distribution of *Heliogomphus*

5.12 *Ictinogomphus* Cowley, 1934

Fig. 5.71 Dorsal habitus of the nymph of *Ictinogomphus*

5.12.1 General Appearance

The nymphs of *Ictinogomphus* are characterized by their medium size (24.0–26.5 mm), with the body sparsely setose and robust, appearance granulose due to abundant small, scale-like setae, abdomen strongly convex dorsally and flattened ventrally, and coloration yellowish-brown (Fig. 5.71).

5.12.2 Description

Head. The head is wider than long (Fig. 5.72). The labrum, frons, vertex, and occiput are densely covered with small scale-like setae, but the clypeus is bare. The labrum is 0.7 mm long, and its anterior margin is flattened ventrally and densely covered with setae. The anteromesal part of the postfrons is U-shaped, and the anterolateral corners are beset with long, stiff setae intermingled with small spines and long and short scale-like setae. The genae protrude ventrally as small, bluntly pointed tusks. The antennal scape is globose, the pedicel cylindrical, antennomere 3 is the largest and distinctly longer than wide, cylindrical, straight, covered with minute scale-like setae and delicate, white setae; antennomere 4 is strongly reduced, subconical, and bare. The compound eyes are relatively large and have a mound-like tubercle below the ventral margins, which are usually visible in dorsal view (Fig. 5.72).

Fig. 5.72 The head in dorsal view

The occipital lobes are large, pale, slightly bulging, with large bare areas, and the lateral margins are rounded and have a dorsal rounded ridge covered with abundant scale-like setae and 3–4 stout, dark, sharp spines. Mandibles have the formula: L 1234 0 a($m^{1,2,3,4,5}$)b/R 1234 y a($m^{1,2,3}$)b, with a > b in both mandibles; the *m* denticles in the left mandible are almost the same size as tooth "b." The maxilla's galeolacinia is as described for *Acrogomphus*. The ventral pad of the hypopharynx is subrectangular, its basal half is yellow-brown and sclerotized, the apical half is whitish and soft, the anterior margin is shallowly concave with a row of long, stiff setae, and the posterior margin is widely rounded. Labium: The prementum-postmentum articulation reaches the midline or the posterior margin of fore coxae. The prementum is yellow to reddish-yellow ventrally, quadrate or slightly wider than long (Fig. 5.73); the lateral margins are serrate or setose, the serrations bear tridentate scale-like setae, the sides are sinuate or straight apically and convergent basally, the ventral surface has a large concavity below the ligula, there is no median sulcus in the basal half, and a sub-basal, depressed, narrow, bare area is along the basal margin. The ligula is convex (Fig. 5.73), poorly developed, and the apical margin is armed with a row of 16 short, truncate, reddish teeth and a dorsomesal row of whitish, piliform setae. The labial palp is yellow to yellowish-brown and has the apical lobe sharply pointed, the tip slightly incurved, and the mesal margin has a row of 12–13 small, squarely truncate teeth. The movable hook is reddish-brown, as long as or slightly shorter than the palp, sharp, and moderately incurved.

Fig. 5.73 The head and prementum in ventral view

Thorax. The prothorax is narrower than the head (Fig. 5.71). The ventral margin of the propleuron is granulose with small, sharp spines; the proepisternum is greatly reduced, and the proepimeron is produced laterally as a large, rounded projection. The anterior lobe of the pronotum is scaly and raised, the posterior lobe is subrectangular with the lateral margins rounded, the posterior margin has a mound-like protuberance at the middle, and a large, elongate, bare area on each side of the midline, which has a central, oval, granulose zone. The ventral margins of the meso- and metapleura are more or less straight and setose. The legs are relatively long (Figs. 5.71 and 5.74) (i.e., when fully extended, the hind legs reach the posterior margin of S10) and have abundant scale-like setae and short or long, stiff setae. The femora are reddish or yellow with two reddish, wide rings; the fore- and middle tibiae lack burrowing hooks, but have strong spines at the apical margins. The tarsal formula is 2–2–2; the pretarsal claws are simple and each has a pulvilliform empodium. The anterior and posterior wing sheaths are parallel and reach the basal half and posterior margin of S6, respectively (Fig. 5.71). The sternum is mostly bare; the mesosternum has a transverse, rounded, low ridge beset with long, stiff, light-brown setae intermingled with elongate, scale-like, tridentate setae, and a small or large, low, rounded tubercle is close to the articulation of each hind coxa, and is covered with hair-like setae and scale-like, tridentate setae.

Fig. 5.74 Ventral habitus of the nymph of *Ictinogomphus*

Abdomen. The abdomen is oval in shape and widest at the posterior margin of S6 (Figs. 5.71 and 5.74). The ventral surface is flat, whereas the dorsal surface is convex and reaches its highest point at S7. Dorsal protuberances are present on S2–S9, which are vertical spines on S2–S5, hook-like and forming a setose keel on S6–S9, each ending in a short spine (Figs. 5.71 and 5.75). The lateral margins of S3–S9 or S4–S9 have a row of dark spines intermingled with light-brown setae, which increase in size and robustness posteriorly and end in a strong, sharp spine; the spines are slightly divergent on S3–S4, incurved on S5–S9, and those on S7 or S9 are the largest (Figs. 5.71, 5.74, and 5.75); S10 is concealed laterally by the spines of S9. Terga 2–10 are finely granulose, and the posterior margins have a row of minute, fan-shaped or tridentate setae.

Fig. 5.75 The abdomen and caudal appendages in lateral view

The sternum is reddish-yellow to yellowish-brown (Fig. 5.74), finely granulose, and with a complex pattern of blotches. Sterna 3–6 and 9 are each divided into three sternites and sterna 7–8 into five sternites; the median sutures are convergent apically on 2, parallel on 3–5, and divergent apically on 6–9; the posterior margins of sterna 2–5 are straight, sinuate on 1, 6, and 9, and concave on 7–8 and 10. The female gonapophyses are rudimentary and hemispherical. The caudal appendages are pyramidal, granulose, and setose, mostly yellowish-brown but yellow at the apex (Fig. 5.75), and twice longer than the dorsal length of S10. The epiproct and paraprocts are obtusely pointed and the same length. The cerci are sharply pointed and the shortest of the caudal appendages. The paraprocts are strongly divergent in ventral view (Fig. 5.74).

Measurements (in mm). Total length including caudal appendages 24.0–26.5; abdomen 15.3–16.9; maximum width of head 6.6–6.8; length of hind femur 6.2–6.6.

5.12.3 Remarks

At first glance, the nymphs of *Ictinogomphus* appear similar to those of *Gomphidictinus* and *Gomphidia* (see comments under *Gomphidia*). Two species have been recorded from Thailand: *I. decoratus* (Selys, 1854) and *I. rapax* (Rambur, 1842) (Hämäläinen and Pinratana 1999). The nymphs of *I. decoratus* cling to the undersides of lily pads (Orr 2005).

5.12.4 Distribution

In Thailand, *Ictinogomphus* has been recorded from the provinces of Bangkok, Buri Ram, Chaiyaphum, Chanthaburi, Chiang Mai, Chon Buri, Chumphon, Kalasin, Kamphaeng Phet, Kanchanaburi, Khon Kaen, Krabi, Lampang, Loei, Mae Hong Son, Maha Sarakhan, Nakhon Ratchasima, Nakhon Sawan, Nakhon Si Ayutthaya, Nakhon Si Thammarat, Nonthaburi, Phangnga, Phayao, Phetchabun, Phetchaburi, Phrae, Phuket, Prachuap Khiri Khan, Ratchaburi, Rayong, Sakhon Nakhon, Saraburi, Satun, Songkhla, Sa Kaew, Surat Thani, Tak, Trang, Yala (Hämäläinen and Pinratana 1999), Chiang Rai (Katatani et al. 2004), Nan (Katatani et al. 2010), Prachin Buri, Sukhothai, and Trat (Day et al. 2012). ***New province records***: Phatthalung and Phitsanulok (Fig. 5.76).

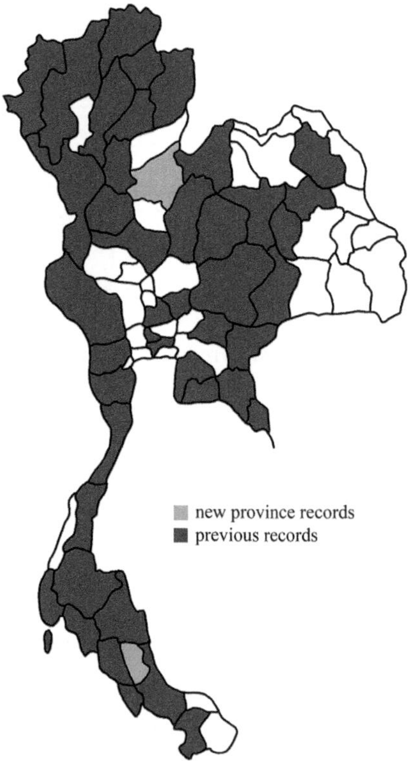

Fig. 5.76 Distribution of *Ictinogomphus*

5.13 *Lamelligomphus* Fraser, 1922

Fig. 5.77 Dorsal habitus of the nymph of *Lamelligomphus*

5.13.1 General Appearance

The nymphs of *Lamelligomphus* are characterized by their medium size (22–28 mm length), with the body robust and more or less flat, abdomen with dorsal protuberances and lateral spines, and coloration light-yellow-brown (Fig. 5.77).

5.13.2 Description

Head. The head is wider than long and wider than prothorax, but narrower than the pterothorax and abdomen (Fig. 5.77). The clypeus is bare, the frons, vertex, and occiput are densely covered with reddish-brown granuli and white, elongate, scale-like setae. The labrum is 0.7 mm long and bare, with the anterior margin setose and flattened ventrally. The frons is produced anterolaterally along the mesal margin of each compound eye, ending in a rounded projection which is beset with a tuft of long, stiff, reflexed setae (Fig. 5.78); the anterior margin of the mesal part of the postfrons is shallowly concave. The antennal scape is globose, the pedicel short and subglobose, and both are densely covered with scale-like setae or are finely granulose; antennomere 3 is claviform, strongly flattened, the largest antennomere (Fig. 5.78), and its dorsal and ventral surfaces are bare or finely granulose, except the ventrolateral margins which have a dense brush of long and stiff setae; antennomere 4 is a mere subconical rudiment mostly covered by claviform setae except at the extreme tip.

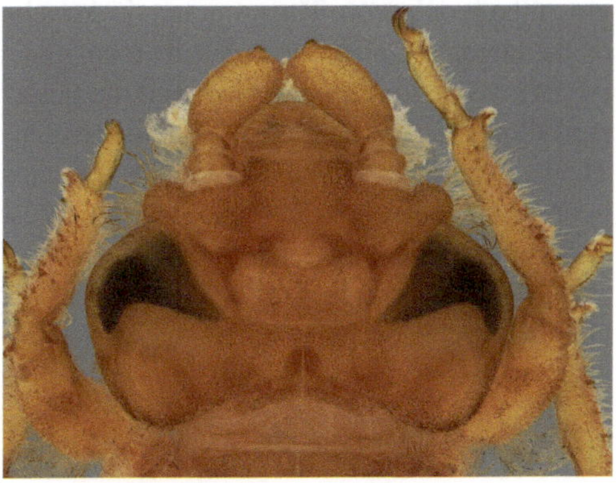

Fig. 5.78 The head and forelegs in dorsal view

The compound eyes are moderately developed. The occipital lobes are rounded and bulging (Fig. 5.78). The posterior margin of the head is widely concave. Mandibles have the formula: L 1234 0 a($m^{1-9 \text{ or } 1-10}$)b/R 1234 y a(m^{1-6})b with a > b in both mandibles and noticeably larger in the right mandible. The maxilla's galeo-lacinia is as described for *Acrogomphus*. The ventral pad of the hypopharynx is subpentagonal, whitish, mostly soft, the posterior margin is strongly concave and has a low, blunt tubercle on each side of the concavity, and the anterior margin has a row of long, stiff setae. Labium: The prementum-postmentum articulation reaches the middle of prosternum (Fig. 5.79). The prementum is light-yellow-brown, slightly longer than wide, the lateral margins are shallowly convex and serrulate, and the ventral surface lacks a longitudinal central sulcus. The ligula is poorly developed, its apical margin is shallowly convex, but straight at the middle third, it has a ventral row of 16–25 short, subquadrate, reddish-brown teeth and dorsal rows of long, yellowish-brown, piliform setae (Fig. 5.79). The labial palp is yellowish-brown, the apical lobe tapers to a blunt apex, and it has a row of 11–16 small, squarely truncate teeth on the mesal margin. The movable hook is reddish-brown, shorter than the palp, sharp, and moderately incurved.

Fig. 5.79 The head, prementum, forelegs, and middle legs in ventral view

Thorax. The prothorax is narrower than the head and setose on the ventral margin of the propleuron. The pronotum is densely covered with scale-like setae, except on the large, dorsolateral, bare, subrectangular areas; the anterior margin is straight, the lateral margins are rounded, and the posterior margin is shallowly convex. The pterothorax is densely setose on the ventrolateral surfaces and light-yellow-brown. The legs are short (Figs. 5.77 and 5.80) (i.e., when fully extended, the hind legs do not reach the posterior margin of S7) and setose, with scale-like setae mainly on the femora and tibiae; the anterior surface of the fore tibia is armed with short, blunt, stout, reddish tubercles (Figs. 5.78 and 5.79); the fore- and middle tibiae have moderately to well-developed burrowing hooks; the pretarsal claws are simple, long, and each has a pulvilliform empodium. The anterior and posterior wing sheaths are strongly divergent and surpass the posterior margin of S3 and reach the basal half of S4, respectively (Fig. 5.77). A transverse fringe of long, white setae is along the posterior margin of the metasternum, or lacking.

Fig. 5.80 Ventral habitus of the nymph of *Lamelligomphus*

Abdomen. The abdomen is more or less fusiform (Figs. 5.77 and 5.80), widest at S5–S6, and light-yellow-brown with a complex pattern of light-brown and reddish-brown spots and dots. The ventral surface is flat, whereas the dorsal surface is convex and has middorsal protuberances on S2–S9, which can be poorly or well-developed (Figs. 5.77 and 5.81), and that on S9 is the largest; short, sharply or bluntly pointed lateral spines are on S6–S9 or S7–S9. Terga 2–10 are granulose including on the posterior margins. The lateral margins of S2–S9 are densely setose. Sterna 1–8 are yellow (Fig. 5.80), 2–8 minutely setose, and 9–10 granulose with long, ventrally directed, stiff setae. Sterna 3–7 and 9 each are divided into three sternites and sternum 8 into five sternites. The caudal appendages (Fig. 5.81) are elongate and sharply pointed, roughly granulose or covered with scale-like setae, and are four times longer than the middorsal length of S10. The epiproct is almost twice longer than its basal width and as long as the paraprocts. The cerci are 0.15× to 0.25× shorter than the epiproct in lateral view, the ventral margin of male's cercus is widely concave, and the tips are slightly reflexed (Fig. 5.81). Cerci of female are straight.

Fig. 5.81 The abdomen in lateral view showing dorsal protuberances and caudal appendages sharply pointed. Note the sinuate ventral margin of male's cercus with the tips slightly reflexed

Measurements (in mm). Total length including caudal appendages 22.0–28.2; abdomen 13.4–17.3; maximum width of head 4.6–5.7; length of hind femur 3.0–4.3.

5.13.3 Remarks

Lamelligomphus was typically included in *Onychogomphus* (*s.l.*) until Chao (1954) split *Melligomphus* and *Nepogomphus* from the Chinese *Onychogomphus*. Since then, there has not been a general consensus in how to treat this complicated group of genera taxonomically (Sasamoto and Kawashima 2009a). Here, we follow Chao's proposal. The nymphs of *Lamelligomphus* inhabit in lowland forest streams in riffles with the substrate composed mainly of gravel.

5.13.4 Distribution

In Thailand, *Lamelligomphus* has been recorded only from the provinces of Nakhon Si Thammarat and Trang (Hämäläinen and Pinratana 1999 [as *Leptogomphus risi* Laidlaw, 1933]). ***New province records:*** Chanthaburi, Kanchanaburi, Nan, Phetchaburi, Prachuap Khiri Khan, and Ratchaburi (Fig. 5.82).

Fig. 5.82 Distribution of *Lamelligomphus*

5.14 *Leptogomphus* Selys, 1878

Fig. 5.83 Dorsal habitus of the nymph of *Leptogomphus*

5.14.1 General Appearance

The nymphs of *Leptogomphus* are characterized by their medium size (20–21 mm), with the body robust and more or less flat, coloration yellowish-brown to reddish-brown, and abdomen with a complex pattern of dark and light spots (Fig. 5.83).

5.14.2 Description

Head. The head is wider than long (Fig. 5.84), but narrower than the pterothorax and abdomen. The labrum is 0.4–0.5 mm long and bare and has the anterior margin setose and flattened ventrally. The clypeus is bare, the frons, vertex, and occiput are densely covered with reddish-brown granules and white, elongate, scale-like setae. The anterior margin of the postfrons has a shallow U-shaped concavity (Fig. 5.84). The antennal scape is subglobose, the pedicel ring-shaped, and both are densely covered with scale-like setae; antennomere 3 is oval (Fig. 5.84), 0.33× longer than its widest part, strongly flattened, slightly reflexed at tip, the dorsal and ventral surfaces are finely granulose, the lateral margins are beset with elongate, claviform, scale-like, reflexed setae, and it has a brush of long, white, stiff setae; antennomere 4 is spherical, strongly reduced, and mostly covered by claviform setae except at the extreme tip.

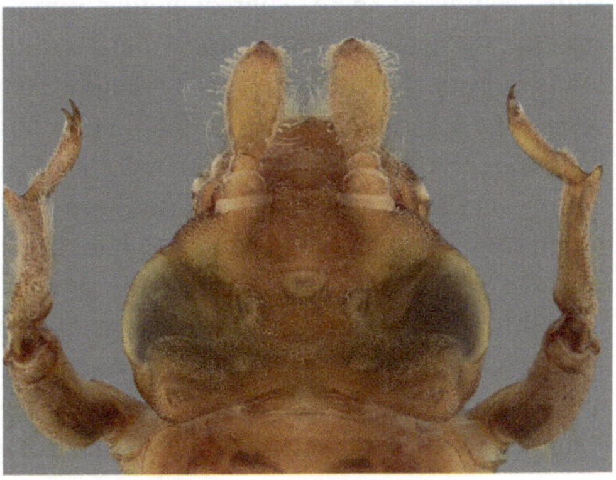

Fig. 5.84 The head and forelegs in dorsal view

The compound eyes are relatively large. The occipital lobes are rounded and slightly bulging (Fig. 5.84). Mandibles have the formula: L 1234 0 a($m^{1,2,3,4}$)b/R 1234 y a($m^{1,2,3,4}$)b, with a > b in both mandibles and noticeably larger in the right mandible. The maxilla's galeolacinia is as described for *Acrogomphus*. The ventral pad of the hypopharynx is subpentagonal and whitish with a large, light-brown, subtriangular sclerite on its ventral surface, the posterior margin is strongly concave and with a low, blunt tubercle on each side of this concavity, and the anterior margin has a row of long, stiff setae. Labium: The prementum-postmentum articulation reaches the posterior margin of the fore coxae. The prementum is slightly longer than wide (Fig. 5.85), the lateral margins are shallowly convex and serrulate, and the basal margin is widely emarginate with a shallow ventromedian sulcus along the basal half. The ligula is poorly developed, straight or very shallowly convex, and it has a ventral row of 13–14 short, square, dark, truncate teeth and a dorsal row of abundant long, piliform setae on the anterior margin (Fig. 5.85). The apical lobe of the labial palp gradually tapers to the apex, ending in a short, conical tooth, and it has a row of 11–13 short, robust, square, reddish-brown teeth on the mesal margin. The movable hook is reddish, slightly longer than the palp, sharp, and moderately incurved.

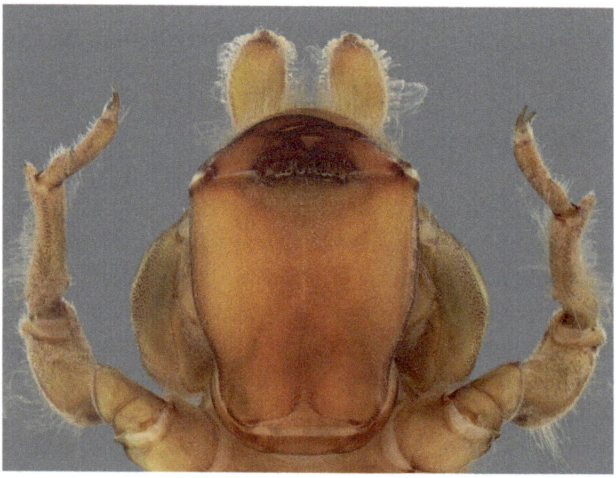

Fig. 5.85 The prementum in ventral view showing the straight ligula

Thorax. The prothorax is narrower than the head (Fig. 5.83) and setose on the ventral margin of the propleuron. The pronotum is densely granulose, except on the large, depressed, bare, subrectangular dorsolateral areas; the anterior margin is straight, the lateral margins are widely rounded, and the posterior margin is shallowly convex. The pterothorax is granulose and densely setose on the ventrolateral areas. The legs are short (Figs. 5.83 and 5.86) (i.e., when fully extended, the hind legs scarcely reach the posterior margin of S8) and setose, with minute granules mainly on the femora and tibiae; the fore- and middle tibiae are armed with short, blunt, stout, reddish tubercles on the lateral surfaces (Figs. 5.84 and 5.85), and have well-developed burrowing hooks; the pretarsal claws are simple and each has a pulvilliform empodium. The anterior and posterior wing sheaths are strongly divergent (Fig. 5.83), their surfaces are covered with minute spines and the margins with abundant delicate setae, and both reach the basal half of S4. A transverse fringe of long, white setae is along the anterior margin of the metasternum.

Fig. 5.86 Ventral habitus of the nymph of *Leptogomphus*

Abdomen. The abdomen is oblong, robust, parallel-sided at the middle (Figs. 5.83 and 5.86), and has a complex color pattern. The ventral surface is flat, whereas the dorsal surface is shallowly convex. Terga 2–10 are finely granulose including on the posterior margins. Middorsal protuberances are on S2–S9, which are vertical on S2–S3, low on S4–S6, vestigial on S7–S9 (Fig. 5.87), and that on S2 is the highest. The lateral margins of S1–S8 are setose, whereas those of S7–S9 have a row of strong, spiniform setae. S7–S9 have short, sharply pointed, robust posterolateral spines, which increase in size and robustness posteriorly; that on S9 is the longest (Fig. 5.87). Sterna 2–7 are yellow-brown (Fig. 5.86), whereas sterna 8–10 are reddish-brown, all of which are finely granulose. Sterna 3, 7, and 8 are each divided into five sternites and sterna 4–6 and 9 into three sternites. The caudal appendages are pyramidal (Fig. 5.87), short, robust, and bluntly pointed, and all are densely and roughly granulose. The epiproct is 0.25× longer than its basal width. The cerci are 0.85× the length of the epiproct, and the paraprocts are the longest of the caudal appendages.

Fig. 5.87 The abdomen and caudal appendages in lateral view

Measurements (in mm). Total length including caudal appendages 20–21; abdomen 12.7–13.0; maximum width of head 5.0–5.1; length of hind femur 4.2–4.3.

5.14.3 Remarks

Inconsistences in the descriptions of the nymphs of several species of *Leptogomphus* were summarized by Xu (2012c), who tentatively suggested to define the *Leptogomphus* nymph based on the body shape flat and narrowly oblong, the third antennomere narrowly oval, the ligula straight with squarish serrations, the wing sheaths strongly divergent, dorsal protuberances present on most of the abdominal segments, and lateral spines on S7–S9. Only three species of *Leptogomphus* have been recorded from Thailand: *L. gestroi* Selys, 1891, *L. inclitus* Selys, 1878, and *L. risi* Laidlaw, 1932 (Hämäläinen and Pinratana 1999). The nymph has not been described for any of these species. Nymphs of other species of *Leptogomphus* have been found in muddy gravel (Wilson 1995), leaf litter (Orr 2003), and in a small stream deeply buried in the sand (Xu 2012c).

5.14.4 Distribution

In Thailand, *Leptogomphus* has been recorded from the provinces of Chiang Mai, Kanchanaburi, Loei, Nakhon Si Thammarat, Tak, Trang (Hämäläinen and Pinratana 1999), Ranong (Hämäläinen 2002), Chiang Rai (Katatani et al. 2004), and more recently from Ratchaburi (Day et al. 2012). ***New province record***: Phayao (Fig. 5.88).

Fig. 5.88 Distribution of *Leptogomphus*

5.15 *Macrogomphus* Selys, 1857

Fig. 5.89 Dorsal habitus of the nymph of *Macrogomphus*

5.15.1 General Appearance

The nymphs of *Macrogomphus* are characterized by their very large size (42–45 mm), with the body moderately hairy, robust, and strongly tapering posteriorly, S9 noticeably elongate and forming a tube together with S10, and coloration light-yellow-brown (Fig. 5.89).

5.15.2 Description

Head. The head is wider than long (Fig. 5.90), but narrower than the thorax and abdomen. The labrum is 0.5 mm long, has very minute, delicate setae, is flattened ventrally, and its anterior margin is densely setose. The clypeus is bare, and the frons, vertex, and occiput are densely covered with minute, delicate setae. The anterior margin of the postfrons is shallowly concave, and a row of small, reddish, peg-like tubercles (apparently some kind of sensilla) is behind each antenna parallel to the mesal margins of the compound eyes. The antennae are densely covered with long, whitish setae on the lateral margins (Fig. 5.90) and have small, scattered, reddish, peg-like tubercles on the dorsal surfaces of antennomeres 1–3, with the setae more abundant on antennomere 3; the scape is globose, the pedicel is subglobose and the shortest, antennomere 3 is cylindrical, five times longer than its widest part, and the largest, and antennomere 4 is reduced, conical, laterally reflexed over the apex of antennomere 3 and has some reddish, peg-like tubercles. The compound eyes are moderately developed. The occipital lobes are rounded and slightly bulging (Fig. 5.90). Mandibles have the formula: L 1234 0 a($m^{1,2,3,4,5,6}$) b/R 1234 y a($m^{1,2,3,4,5}$)b. The maxilla's galeolacinia is as described for *Acrogomphus*. The ventral pad of the hypopharynx is subpentagonal, whitish, and mostly soft, it has abundant long, stiff setae mainly on the lateral margins, and a U-shaped ventrobasal sclerite.

Fig. 5.90 The head, pronotum, and forelegs in dorsal view

Labium: The prementum-postmentum articulation reaches the posterior margin of the fore coxae. The prementum is light-yellow-brown, longer than wide, and parallel-sided along the apical 0.60 (Fig. 5.91). It has a longitudinal, bare, central sulcus along the basal half of the ventral surface (Fig. 5.91), and a large, shallow, setose concavity just below the ligula. The ligula is very poorly developed; the apical margin is sinuate and has a small central concavity and double row of piliform setae (Fig. 5.91). The labial palp is stout and yellowish-red, the apical lobe is sharp and strongly incurved, and the mesal margin has six robust, sharply pointed teeth and a short, square, apical tooth, with the three basal teeth longer (Fig. 5.91). The movable hook is yellowish-red, as long as the palp, sharp, and moderately incurved.

Fig. 5.91 The prementum in ventral view

Thorax. The prothorax is yellow, narrower than the head, and setose on the ventral margin of the propleuron. The pronotum is covered with minute setae, except on the large, bare, excavated areas between the anterior and posterior lobes and the large, dorsal, irregular areas of the posterior lobe. The anterior margin is straight, the lateral margins are straight and divergent, and the posterior margin is sinuate and has a large, central, shallow concavity. The pterothorax is yellow, setose on the ventral margins, and the meso- and metaspiracles are easily visible. The legs are very short (Figs. 5.89 and 5.92) (i.e., when fully extended, the hind legs scarcely reach the posterior margin of S6), strongly setose, and have long, whitish, delicate setae mainly on sides, and short, stiff, yellow setae mainly on the femora and tibiae. The tibiae lack burrowing hooks. The pretarsal claws are simple and chelate-like on the fore- and middle tarsi, but are long and sharp on the hind tarsi, and each has a pulviliform empodium. The wing sheaths are parallel (Fig. 5.89), yellowish, with abundant delicate setae on the margins, minute setae along the principal veins, and the remainder of the surface is mostly bare; the anterior and posterior wing sheaths reach the basal half and posterior margin of S3, respectively.

Fig. 5.92 Ventral habitus of the nymph of *Macrogomphus*

Abdomen. The abdomen is very enlarged and ends in a long tube posteriorly (Figs. 5.89, 5.92, and 5.93). The ventral surface is more or less flat, whereas the dorsal surface is convex. Middorsal protuberances are on S3–S9, which are very low and almost imperceptible on S3–S5, and well-defined on S6–S9; they increase in size and height posteriorly and that on S9 is a strong, posteriorly directed spine (Fig. 5.93). S7–S8 or S7–S9 each ends in a short, lateral spine; the spine is stouter on S7 and the smallest on S9. Terga 2–10 have minute to small, stiff setae, which are more abundant along the midline, and the posterior margins have a row of red, stiff setae. The lateral margins of S2–S8 are densely setose. S9–S10 are cylindrical and reddish-brown; S9 is very long, more than three times longer than its basal margin, and as long as S6+S7+S8, and has a longitudinal carina mostly along the basal half. Sterna 1–8 are light-yellow to yellowish-brown (Fig. 5.92), and sterna 2–10 are covered with small, delicate setae along the midline. Sterna 3 and 7–8 are each divided into five sternites and sterna 4–6 into three sternites. The caudal appendages are very short and beset with small, stiff setae (Fig. 5.93). The cerci and paraprocts are triangular and the same length. The epiproct is hood-like, broadly rounded at the apex, has a small, conical, dorsal, subapical protuberance, its ventral surface is widely and deeply concave, and is slightly shorter than the cerci and paraprocts.

Fig. 5.93 The abdomen and caudal appendages in lateral view. Notice the large dorsal spine on S9

Measurements (in mm). Total length including caudal appendages 42–45; abdomen (ventral) 28–31; segment 9 (lateral) 8.0–8.5; maximum width of head 5–6; length of hind femur (lateral) 4.8–5.0.

5.15.3 Remarks

The nymph of *Macrogomphus* resembles that of the monotypic Chinese genus *Labrogomphus*, both belonging to Gomphinae. Both genera appear similar to the Neotropical *Aphylla* and *Phyllocycla* (Gomphoidinae) by the elongate posterior end of the abdomen, although in *Macrogomphus* and *Labrogomphus*, the elongation is of S9, whereas in *Aphylla* and *Phyllocycla* it is of S10. By the elongation of S9, the nymph of *Macrogomphus* is easily distinguished from all other nymphs inhabiting Thailand. Five species of *Macrogomphus* have been recorded from Thailand: *Macrogomphus albardae* Selys, 1878, *M. kerri* Fraser, 1932, *M. matsukii* Asahina, 1986, *M. parallelogramma* (Burmeister, 1839), and *M. thoracicus* McLachlan, 1884 (Hämäläinen and Pinratana 1999). Nymphs of *Macrogomphus* have been reported to inhabit sluggish lowland forest streams in deep silt (Orr 2005).

5.15.4 Distribution

In Thailand, *Macrogomphus* has been recorded from the provinces of Chaiyaphum, Chanthaburi, Chiang Mai, Kanchanaburi, Lampang, Nakhon Nayok, Nakhon Ratchasima, Nakhon Si Thammarat, Phrae, Prachuap Khiri Khan, Ranong, Saraburi, Songkhla, Tak, Trang (Hämäläinen and Pinratana 1999), Chiang Rai (Katatani et al. 2004), Phangnga (Ferro et al. 2009), and more recently from Ratchaburi (Day et al. 2012). ***New province records***: Mae Hong Son, Phitsanulok, and Satun (Fig. 5.94).

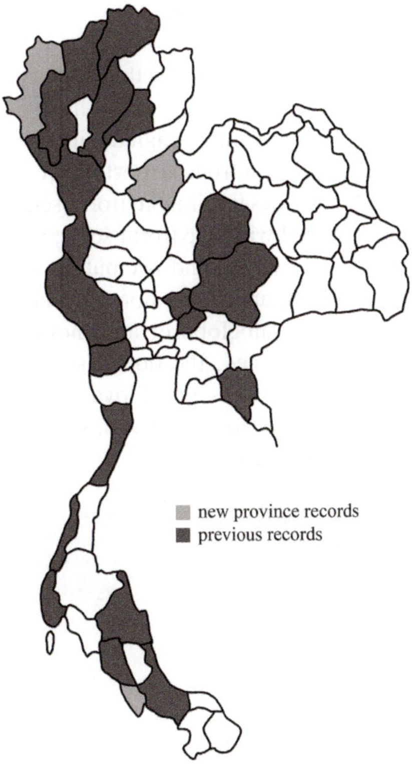

Fig. 5.94 Distribution of *Macrogomphus*

5.16 *Megalogomphus* Campion, 1923

Fig. 5.95 Dorsal habitus of the nymph of *Megalogomphus*

5.16.1 General Appearance

The nymphs of *Megalogomphus* are characterized by their large size (36 mm length), with the body very robust, enlarged, and setose, eyes small, burrowing hooks strong, and coloration light-yellow-brown (Fig. 5.95).

5.16.2 Description

Head. The head is wider than long (Fig. 5.96). The labrum is 0.9 mm long, mostly bare, flattened ventrally, and it has a dense brush of long setae along the frontal and lateral margins intermingled with rows of claviform setae. The clypeus is bare and rugose. The antefrons is mostly bare and with the apical margin slightly raised at the middle and beset with claviform setae, the postfrons is widely concave at the middle, covered with claviform setae, and has a V-shaped bare area in the middle concavity. The vertex and occiput are densely covered with claviform setae, and the lateral ocelli are raised along the mesal margin. The antennal scape is globose and the pedicel subglobose; antennomere 3 is the largest (Fig. 5.96), oval-shaped, strongly flattened, the dorsal surface is covered with abundant claviform setae, except on a bare, central, oval-shaped area, and the ventrolateral margins have a dense brush of long, reflexed, stiff setae; antennomere 4 is small, conical, and has claviform setae (Fig. 5.96).

Fig. 5.96 The head in dorsal view

The compound eyes are small and cap the anterolateral angles of the head (Figs 5.95 and 5.96). The occipital lobes are large, rounded, and bulging (Fig. 5.96). Mandibles have the formula: L 1234 0 a(m ^{1,2,3,4,5})b/R 1234 y a(m^{1,2,3,4,5})b. The maxillae are densely setose; the maxilla's galeolacinia is as described for *Acrogomphus*. The ventral pad of the hypopharynx is subpentagonal, whitish, entirely soft, almost as long as wide, and has a row of subapical setae forming a semicircle. Labium: The prementum-postmentum articulation reaches the anterior margin of the middle coxae (Fig. 5.98). The prementum is light-yellow-brown (Fig. 5.97), slightly longer than its widest part, subrectangular, and parallel-sided along the apical 0.60; the ventral surface lacks a median sulcus, although a large shallow concavity is evident, and transverse rows of long, white, delicate setae are just below the ligula. The ligula is poorly developed, its apical margin is straight (Figs. 5.4 and 5.97), and it has a ventral row of approximately 22 rectangular, dark, truncate teeth and dorsal rows of long, reddish-brown, piliform setae on the apical margin. The labial palp is reddish-yellow, stout, and the apical lobe is strongly incurved, sharp, and has a row of seven large, stout, sharp, incurved, reddish-brown teeth (Figs. 5.4 and 5.97). The movable hook is reddish-brown, longer than the palp, sharp, and strongly incurved.

Fig. 5.97 The prementum in ventral view

Thorax. The prothorax is narrower than the head (Fig. 5.95) and setose on the ventral margin of the mostly smooth propleuron. The pronotum is covered with claviform setae mostly on the margins, whereas the remainder surface is smooth; the anterior lobe is narrow and raised, the lateral margins of the posterior lobe are poorly developed, slanting, and rounded, and the posterior margin is shallowly convex. The pterothorax is light-yellow-brown, densely setose on the ventrolateral areas; most of the episterna have claviform setae and most of the epimera are smooth. The legs are short (Figs. 5.95 and 5.98) (i.e., when fully extended, the hind legs scarcely surpass the posterior margin of S7) and setose with claviform setae mainly on the femora and tibiae; burrowing hooks are strongly developed on the fore- and middle tibiae (Fig. 5.98); the hind tarsi are as long as the hind tibiae; the pretarsal claws are simple, long, and each has a pulvilliform empodium. The wing sheaths are divergent (Fig. 5.95), yellow-brown, and have claviform setae and hair-like setae on the margins, whereas the remainder of the surface is mostly bare; the anterior and posterior wing sheaths reach the basal half and posterior margin of S4, respectively.

Fig. 5.98 Ventral habitus of the nymph of *Megalogomphus*

Abdomen. The abdomen is enlarged and tapers posteriorly (Figs. 5.95 and 5.98). The ventral surface is flat, whereas the dorsal surface is convex. The coloration is light-yellow-brown on S1–S3, with a complex pattern of light-brown and reddish-brown spots and dots on S4–S8, and an irregular, transverse, light-brown spot on the middle third of S6 and apical half of S9. Middorsal protuberances are low, blunt, and moderately developed on S2–S6, and large, blunt, carinate, and spine-like on S7–S9 (Fig. 5.99); they increase in size and height posteriorly and the spine on S9 is as long as S10. S7–S9 end in strong lateral spines which increase in size and robustness posteriorly (Figs. 5.95 and 5.99). Terga 1–10 have long, white, delicate setae on the lateral margins; the remainder of the surface is beset with a mixture of short, stiff, brown setae, minute spiniform and claviform setae, and the posterior margins have small, claviform setae. Sterna 1–7 are yellow and 8–10 yellowish-brown (Fig. 5.98); all have abundant short, delicate setae. Sterna 3–9 are each divided into three sternites. The caudal appendages are yellow, straight, sharply pointed (Fig. 5.99), and mostly covered with spiniform setae; they are longer than the dorsal length of S9 and twice longer than S10. The epiproct and paraprocts are the same length and the cerci are slightly shorter.

Fig. 5.99 The abdomen and caudal appendages in lateral view

Measurements (in mm). Total length including caudal appendages 36–40; abdomen (ventral) 23–27; maximum width of head 7.5–8.5; length of hind femur (lateral) 5.5–6.0.

5.16.3 Remarks

The nymph of *Megalogomphus* is a shallow burrower (Orr 2003) that inhabits fast water upstream of gravel beds (Orr 2005) and fine and coarse sand and gravel, sometimes several inches beneath the substrate (Wilson 1995). In Thailand, only *Megalogomphus sumatranus* (Krüger, 1899) and *Megalogomphus* sp. have been recorded (Hämäläinen and Pinratana 1999; Katatani et al. 2004). By the nymphal morphology, there is no a particular close resemblance of *Megalogomphus* to other genera that occur in Thailand.

5.16.4 Distribution

In Thailand, *Megalogomphus* has been recorded from the provinces of Kanchanaburi, Songkhla, Trang, Yala (Hämäläinen and Pinratana 1999), Chiang Rai (Katatani et al. 2004), Chumphon (Ferro et al. 2009), and more recently Ranong, Ratchaburi, and Surat Thani (Day et al. 2012). ***New province records:*** Chiang Mai, Nakhon Ratchasima, Nakhon Si Thammarat, Phangnga, Phatthalung, and Phetchaburi (Fig. 5.100).

Fig. 5.100 Distribution of *Megalogomphus*

5.17 *Microgomphus* Selys, 1857

Fig. 5.101 Dorsal habitus of the nymph of *Microgomphus*

5.17.1 General Appearance

The nymphs of *Microgomphus* are characterized by their small size (18–20 mm), with the body robust, strongly flattened, minutely scaly, and coloration variegated-reddish-brown to brown (Fig. 5.101).

5.17.2 Description

Head. The head is wider than long and densely granulose (Fig. 5.102). The labrum, frons, and occiput are densely covered with claviform setae, whereas the clypeus is bare. The labrum is 0.5 mm long, and the anterior margin is flattened ventrally, and densely covered with short, stiff setae. The anteromesal part of the postfrons is shallowly concave to U-shaped. The antennae are light-yellowish-brown; the scape is cylindrical and the pedicel globose; antennomere 3 is claviform (Fig. 5.102), strongly flattened, the largest, its lateral margins have rows of scale-like, claviform setae, and the dorsal and ventral surfaces are mostly bare with scattered claviform setae; antennomere 4 is subconical, granulose, and strongly reduced (Fig. 5.102).

Fig. 5.102 Dorsal view of the head

The compound eyes are moderately developed and have a longitudinal, strong, serrate carina below (Fig. 5.103). The occipital lobes are rounded and slightly bulging. Mandibles have the formula: L 1234 0 a($m^{1,2,3,4,5,6 \text{ or } 1–7}$)b/R 1234 y a($m^{1,2,3 \text{ or } 1–5 \text{ or } 1–6}$)b, with a > b in both mandibles and noticeably larger in the right mandible. The maxilla's galeolacinia is as described for *Acrogomphus*. The ventral pad of the hypopharynx is subpentagonal, whitish, soft, the posterior margin is widely concave, and the anterior margin has a row of long, stiff setae. Labium: The prementum-postmentum articulation reaches the middle of the fore coxae (Fig. 5.104). The prementum is yellowish dorsally, reddish-yellow ventrally, subquadrate, slightly wider than long, and the sides are straight, serrate, slightly divergent apically, and convergent basally (Fig. 5.103); the ventral surface is granulose, it has a longitudinal median sulcus along the basal half, which widens basally, and the basal margin is sinuate. The ligula is poorly developed, the apical margin is convex, and it has a ventral row of approximately 18 short, truncate, reddish teeth (Fig. 5.103) and dorsal rows of whitish, piliform setae on the apical margin. The labial palp is yellow, the apical lobe is widely rounded, and the mesal margin bears a row of 12–15 small, squarely truncate teeth. The movable hook is reddish-yellow, shorter than the palp, sharp, and moderately incurved.

Fig. 5.103 The head in ventral view showing the prementum and subocular carinae

Thorax. The prothorax is narrower than the head (Fig. 5.101). The ventral margin of the propleuron has a tuft of long, white, delicate setae. The pronotum is mostly granulose, except on the large, depressed, bare, subrectangular dorsolateral areas; the anterior margin is almost straight and raised, the posterior margin is widely convex, and the lateral margins appear as large, blunt, cylindrical, scaly, slightly reflexed tubercles. The pterothorax is granulose and sparsely setose. The legs are long (Figs. 5.101 and 5.104) (i.e., when fully extended, the hind legs reach the posterior margin of S10), slender, granulose, and with abundant claviform setae; the coxae, trochanters, and femora are reddish-brown with a yellow subapical ring on the femora; the tibiae are yellow with a wide, central, dark band and lack burrowing hooks; the hind tibia is shorter than the femur; the two basal tarsomeres are yellow and the apical tarsomeres yellowish-brown; the pretarsal claws are light-brown basally, dark-brown distally, and have a row of minute denticles and comb of stiff, white setae on the apical half of the ventral margin; the middle and hind pretarsal claws are the largest and are strongly hooked apically; all have a pulvilliform empodium.

Fig. 5.104 Ventral habitus of the nymph of *Microgomphus*

The anterior and posterior wing sheaths are parallel, surpassing the posterior margin of S3 and reaching the basal half of S4, respectively (Fig. 5.101). The sternum is finely granulose, and the prosternum has a short, blunt, conical tubercle close to the articulation of the fore coxae.

Abdomen. The abdomen is large, parallel-sided, and widest at S7 (Figs. 5.101 and 5.104). The ventral surface is flat, whereas the dorsal surface is slightly convex and may have or not middorsal protuberances (Fig. 5.105), but when present they are on S4–S9, S5–S9, or only S9. The lateral margins of S2–S9 have a keel with a row of regularly spaced, strong spines intermingled with short, stiff, flat, white setae. The keels and spines increase in size and robustness posteriorly; S3–S6 end in small spines, S7–S9 end in large, strong spines, which are divergent on S7–S8 and directed posteriorly on S9; S10 is laterally concealed by the spines of S9. Terga 2–10 are finely granulose including on the posterior margins and have a complex, variegated color pattern.

Fig. 5.105 The abdomen and caudal appendages in lateral view

The sternum is finely granulose and reddish-brown to brown (Fig. 5.104). Sterna 3–6 and 9 are each divided into three sternites and sterna 7–8 into five sternites; the median sutures are convergent apically on 2, parallel on 3–8, and divergent apically on 9; the posterior margins are straight on sterna 1–7, sinuate on 8, and convex on 9–10; sterna 8 and 9 are fused. The female gonapophyses are vestigial and conical. The caudal appendages are pyramidal, densely granulose, and shorter than the dorsal length of S9+S10 (Fig. 5.105). The epiproct is dark-brown with the tip yellow and blunt. The cerci are yellowish-brown, sharply pointed, and with the tips divergent. The paraprocts are obtusely pointed with the tips straight, light-brown dorsally, and reddish-brown ventrally. The paraprocts are the longest of the caudal appendages, and the epiproct and cerci are the same length.

Measurements (in mm). Total length including caudal appendages 18.1–20.3; abdomen 12.0–12.8; maximum width of head 4.2–4.8; length of hind femur 5.0–5.2.

5.17.3 Remarks

The nymph of *Microgomphus chelifer* Selys, 1859, was described by Ngiam (2010) as *Heliogomphus* cf. *retroflexus* Ris (Ngiam et al. 2011). The main difference between the nymph described here and that of *M. chelifer* is the lack of dorsal protuberances on S4–S8. Boonsoong and Chainthong (2014b) described the nymph of *Microgomphus svihleri* (Asahina, 1970) (as *M. thailandica* Asahina, 1981). Recently, the adult of a new species, *M. farrelli* Makbun and Fleck, 2018, was described together with its nymph. Makbun and Fleck (2018) considered *M. thailandica* as a junior synonym of *M. svihleri*. The nymph of *Microgomphus* inhabits leaf litter in forest pools (Ngiam 2010) in first order streams (Boonsoong and Chainthong 2014b). Four species of *Microgomphus* have been recorded in Thailand: *M. chelifer*, *M. farrelli*, *M. jurzitzai* Karube, 2000, and *M. svihleri* (Hämäläinen and Pinratana 1999; Kitagawa and Katatani 2005; Makbun and Fleck 2018).

5.17.4 Distribution

In Thailand, *Microgomphus* has been recorded from the provinces of Chanthaburi, Chiang Mai, Nakhon Nayok, Nakhon Ratchasima, Phrae, Ratchaburi, Yala (Hämäläinen and Pinratana 1999), Kanchanaburi (Hämäläinen 2002), Chumphon, Loei, Songkhla (Ferro et al. 2009), Nan (Katatani et al. 2010), Surat Thani (Day et al. 2012), and more recently Mae Hong Son (Makbun and Fleck 2018). ***New province records:*** Kamphaeng Phet, Phangnga, Phatthalung, Phetchaburi, Phitsanulok, Prachuap Khiri Khan, Sa Kaew, Suphan Buri, Trang, Trat, and Uttaradit (Fig. 5.106).

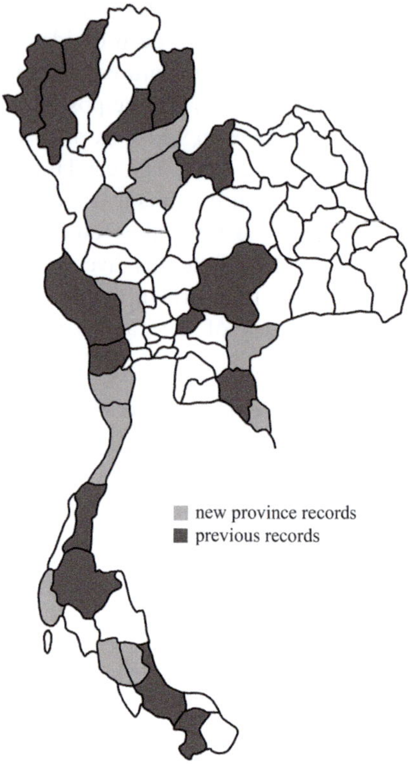

Fig. 5.106 Distribution of *Microgomphus*

5.18 *Nepogomphus* Fraser, 1934

Fig. 5.107 Dorsal habitus of the nymph of *Nepogomphus*

5.18.1 General Appearance

The nymphs of *Nepogomphus* are characterized by their small size (14–17 mm), with the body robust and beset with short, stiff setae, abdomen with blunt middorsal protuberances, and coloration yellowish-brown to reddish-brown (Fig. 5.107).

Fig. 5.108 The head in dorsal view

5.18.2 Description

Head. The head is wider than long and densely setose with erect, scale-like setae (Fig. 5.108). The labrum is 0.5 mm long, the anterior margin is flattened ventrally, and it has scale-like and bristle-like setae. The clypeus is bare. The frons has the anteromesal margin shallowly concave, and the anterior and lateral margins have long, stiff setae. The antennal scape is globose, the pedicel is short and subglobose, and both are covered with scale-like setae; antennomere 3 is the largest, spindle-shaped, moderately depressed dorsoventrally, the dorsal surface has scale-like setae, and the lateral margins are covered with scale-like setae and long stiff setae; antennomere 4 is subconical, strongly reduced, and mostly bare. The compound eyes are rather small. The occipital lobes are rounded and bulging (Fig. 5.108). Mandibles have the formula: L 1234 0 $a(m^{1,2,3,4,5,6,7,8})b$/R 1234 y $a(m^{1,2,3,4,5})b$, with $a > b$ in both mandibles and noticeably larger in the right mandible. The maxilla's galeolacinia is as described for *Acrogomphus*. The ventral pad of the hypopharynx is subpentagonal, whitish, mostly soft, its posterior margin is widely concave, and the anterior margin has a row of long, stiff setae.

Fig. 5.109 The prementum, forelegs, and middle legs in ventral view

Labium: The prementum-postmentum articulation reaches the anterior margin of the middle coxae. The prementum is quadrate and as long as wide, and the lateral margins are shallowly convex and serrulate (Fig. 5.109). The ligula is moderately developed (Fig. 5.5), the apical margin is strongly convex, and has a ventral row of approximately 17 short, rectangular, dark, truncate teeth and dorsal rows of long, grayish, piliform setae. The apical lobe of the labial palp tapers slightly, and the mesal margin has a row of 11–12 small, square, brown teeth. The movable hook is almost as long as the palp, sharp, and moderately incurved (Fig. 5.5).

Fig. 5.110 Ventral habitus of the nymph of *Nepogomphus*

Thorax. The prothorax is narrower than the head (Fig. 5.107). The ventral margin of the propleuron has long, white setae. The pronotum is mostly setose with the setae short, stiff, reflexed, and resembling small hooks, except on the large, bare, subcircular dorsolateral areas; the anterior and posterior margins are shallowly sinuate and the lateral margins rounded. The pterothorax is setose with setae as described for the pronotum, and a tuft of long setae on the ventrolateral surface. The legs are short (Figs. 5.107 and 5.110) (i.e., when fully extended, the hind legs scarcely reach the posterior margin of S8) and densely setose; the fore- and middle tibiae have moderately developed burrowing hooks, the anterior surfaces have long, stiff setae and sparse, short, robust, conical, blunt spines; the pretarsal claws are simple and each has a pulvilliform empodium. The wing sheaths are divergent (Fig. 5.107), most of the surfaces are covered with stiff, hook-shaped setae, and the ventral margin of the posterior wing sheath has short, erect, stiff setae; the anterior and posterior wing sheaths reach the basal half and posterior margin of S4, respectively.

Fig. 5.111 The abdomen and caudal appendages in lateral view. Notice the large, blunt, dorsal protuberances

Abdomen. The abdomen is elongate, gradually widening on S2–S5, and then tapers posteriorly; its ventral surface is flat, whereas the dorsal surface is convex. Terga 2–10 are setose and with a bare, circular, sublateral area on each side. Middorsal protuberances are well-developed on S2–S9 and increase in size and robustness posteriorly (Fig. 5.111); they are almost vertical on S2–S4 and slant gradually on S5–S9. S7–S9 are keeled and have large, lateral, acutely pointed, and moderately incurved spines that increase in size and robustness posteriorly (Fig. 5.111). The sternum is setose. Sterna 3–9 are each divided into three sternites. The caudal appendages are elongate, setose, acutely pointed, and almost as long as the dorsal length of terga 9+10 (Fig. 5.111). The dorsal margin of the epiproct and lateral margins of the cerci and paraprocts are shallowly concave, and coarsely granulose on the cerci. The paraprocts are the longest of the caudal appendages, the epiproct is longer than its basal width, and the cerci are the shortest.

Measurements (in mm). Total length including caudal appendages 13.8–17.0; abdomen 7.7–9.5; maximum width of head 3.6–4.0; length of hind femur 2.8–3.0.

5.18.3 Remarks

There has been some controversy regarding whether this is a distinct genus or simply a member of *Onychogomphus* (Sasamoto and Kawashima 2009). Based upon nymphal morphology, we support the opinion of Zhao (1990, cited by Sasamoto and Kawashima 2009) that it is a valid genus, and the nymphal morphology suggests it might be more similar to *Nihonogomphus*. Nymphs are found in small, clear forest streams in sand and silt banks (Orr 2005). Only *Nepogomphus walli* (Fraser, 1924) has been recorded in Thailand (Hämäläinen and Pinratana 1999); the nymph was described by Yang and Davies (1993) and Sasamoto and Kawashima (2009).

5.18.4 Distribution

In Thailand, *Nepogomphus* has been recorded from the provinces of Chanthaburi, Chiang Mai, Chiang Rai, Kanchanaburi, Krabi, Nakhon Nayok, Tak, and Yala (Hämäläinen and Pinratana 1999; Hämäläinen 2002). ***New province records***: Chumphon, Loei, Mae Hong Son, Nakhon Si Thammarat, Nan, Narathiwat, Phangnga, Phayao, Phetchabun, Phetchaburi, Phrae, Prachuap Khiri Khan, Songkhla, Trang, and Uttaradit (Fig. 5.112).

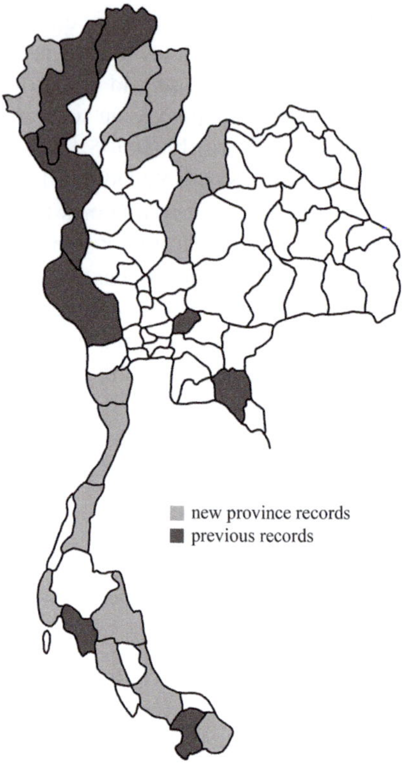

Fig. 5.112 Distribution of *Nepogomphus*

5.19 *Nihonogomphus* Oguma, 1926

Fig. 5.113 Dorsal habitus of a young nymph of *Nihonogomphus*

5.19.1 General Appearance

The nymphs of *Nihonogomphus* are characterized by their body robust and oblong, coarsely granulose, hairy on the sides, and coloration light-yellow-brown (Fig. 5.113).

Fig. 5.114 The head in dorsal view

5.19.2 Description

Head. The head is wider than long (Fig. 5.114), but narrower than the thorax and abdomen. The labrum is 0.5 mm long, its anterior margin is flattened ventrally, and densely setose. The clypeus is bare, and the frons, vertex, and occiput are granulose. The antennae are densely covered with small, scale-like setae on the dorsal surface, and the lateral margins have long, white, stiff setae (Figs. 5.114 and 5.115); the scape is subcylindrical, the pedicel annular and the shortest, antennomere 3 is the largest, subcylindrical, flattened dorsoventrally, and has a circular tuft of setae on the apical margin, and antennomere 4 is a hemispherical rudiment.

Fig. 5.115 The head and prementum in ventral view

The compound eyes are moderately developed. The occipital lobes are rounded and bulging (Fig. 5.114). Mandibles have the formula: L 1234 0 a($m^{1,2,3,4,5,6,7,8}$)b/R 1234 y a($m^{1,2,3,4}$)b. The maxilla's galeolacinia is as described for *Acrogomphus*. The ventral pad of the hypopharynx is subpentagonal, whitish, mostly soft and bare, and has a transverse basal sclerite that extends ventrolaterally to each side of the midline as a triangular branch along the basal half of the pad. Labium: The prementum-postmentum articulation reaches the posterior margin of the fore coxae. The prementum is yellow-brown, slightly longer than wide, has the sides very finely serrulated, is strongly convex along the basal half, and nearly parallel-sided along the apical half (Fig. 5.115); the basal margin is strongly emarginated and there is no ventral sulcus. The ligula is poorly developed and convex (Fig. 5.115), and its apical margin has a dorsal double row of piliform setae and ventral row of 22 small, truncate, reddish teeth. The labial palp is yellow basally and yellow-brown apically, the apex is rounded, and the mesal margin is shallowly concave and beset with 14–15 small, truncate teeth (Fig. 5.115). The movable hook is yellowish-red, shorter than the palp, sharp, and moderately incurved.

Fig. 5.116 Ventral habitus of a young nymph of *Nihonogomphus*

Thorax. The prothorax is narrower than the head and setose on the ventral margin. The pronotum is covered with minute, scale-like setae except on the large, dorsolateral, bare, irregular areas; the anterior margin is straight, the lateral margins are rounded, and the posterior margin is widely convex. The pterothorax is granulose and densely setose on the ventrolateral surface. The legs are short (Figs. 5.113 and 5.116) (i.e., when fully extended, the hind legs hardly reach the anterior margin of S7), strongly setose, granulose, and with golden setae on the fore- and middle legs; the burrowing hooks are short on the fore- and middle tibiae; the pretarsal claws are simple and each has a pulvilliform empodium. The anterior and posterior wing sheaths are strongly divergent and have minute, stiff setae on the margins, whereas the remainder surface is mostly bare; both probably reach the anterior margin of S4.

Abdomen. The abdomen is oblong and concolorous yellowish-brown without a color pattern (Figs. 5.113 and 5.116). The ventral surface is more or less flat and the dorsal surface convex. Short, dorsal protuberances on S2–S9 increase in size but diminish in height posteriorly (Fig. 5.117). S2–S9 end in lateral spines which increases in size and robustness posteriorly (Figs. 5.113 and 5.116). Terga 2–10 and the sternum are densely granulose. The lateral margins of S2–S10 are setose and serrulate. Sterna 3–9 are each divided into three sternites and sterna 2–7 each have a central, low, mound-like protuberance. The female gonapophyses are vestigial. The caudal appendages are long, pyramidal, sharply pointed, densely granulose, and nearly as long as the length of S9+S10 (Fig. 5.117). The epiproct and cerci have the tips decurved, whereas the paraprocts have the tips reflexed. All of them are the same length.

Fig. 5.117 The abdomen and caudal appendages in lateral view

Measurements (in mm). Based on a young nymph, probably F-3: Total length including caudal appendages 17.5; abdomen (ventral) 12.5; maximum width of head 3.4; length of hind femur (lateral) 2.8.

5.19.3 Remarks

Only *Nihonogomphus pulcherrimus* (Fraser, 1927) and *Nihonogomphus* sp. have been recorded in Thailand (Hämäläinen and Pinratana 1999; Katatani et al. 2010). Chao (1984) described the first nymph of the genus, *N. lieftincki* Chao, 1954, from China and reported them inhabiting river banks. Apparently, the nymphs of this genus are characterized by a full set of dorsal protuberances and lateral spines on abdominal segments 2–9. By the nymphal morphology, *Nihonogomphus* appears more similar to *Nepogomphus* and to a lesser extent to *Onychogomphus*.

5.19.4 Distribution

In Thailand, *N. pulcherrimus* has been recorded only from Kanchanaburi Province (Hämäläinen and Pinratana 1999) and *Nihonogomphus* sp. from Nan Province (Katatani et al. 2010). ***New province records***: Chiang Mai and Phetchaburi (Fig. 5.118).

Fig. 5.118 Distribution of *Nihonogomphus*

5.20 *Nychogomphus* Carle, 1986

Fig. 5.119 Dorsal habitus of the nymph of *Nychogomphus duaricus*

5.20.1 General Appearance

The nymphs of *Nychogomphus* are characterized by their medium size (24 mm), with the body stout, bulky, and setose on the sides, integument finely granulose, antennomere 3 widening distally, and coloration uniformly light-brown to dark-brown (Fig. 5.119).

5.20.2 Description

Head. The head is wider than long (Fig. 5.120), wider than the prothorax, but narrower than the pterothorax and abdomen, and densely granulose (Figs. 5.119 and 5.120). The labrum is 0.7 mm long, its anterior margin is flattened ventrally, and it has a dense brush of piliform setae. The clypeus is bare. The frons and vertex are mostly flat dorsally. The antefrons is very short and granulose, and the postfrons is widely concave on the anterior margin and laterally produced as a wide, rounded lobe. The ocelli are hidden by minute scales making them hardly visible. The antennal scape is globose, the pedicel subglobose, antennomere 3 is enlarged and gradually widens to a truncate apex (Fig. 5.120), and antennomere 4 is a conical rudiment that caps the apex of antennomere 3. The compound eyes are large. The occiput is small, and the occipital lobes are rounded, finely granulose, and bulging. The posterior margin of the head is widely concave. Mandibles have the formula: L 1234 0 a($m^{1\text{-}8 \text{ or } 1\text{-}10}$)b/R 1234 y a($m^{1\text{-}6 \text{ or } 1\text{-}8}$)b, with a > b in both mandibles. The maxilla's galeolacinia is as described for *Acrogomphus*. The ventral pad of the hypopharynx is subpentagonal, whitish, mostly soft, has abundant long, stiff setae mainly on the lateral margins, and a U-shaped ventrobasal sclerite.

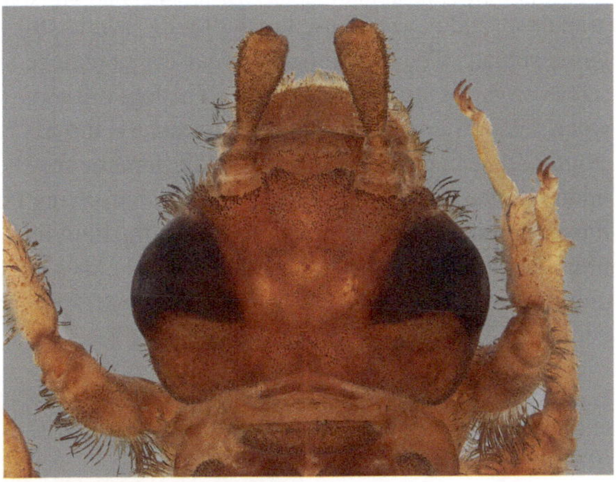

Fig. 5.120 Dorsal view of the head

Labium: The prementum-postmentum articulation reaches the posterior margin of the fore coxae. The prementum is subrectangular, longer than wide, the sides are shallowly convex at the middle, and the ventral surface lacks a median sulcus (Fig. 5.121). The ligula is poorly developed, representing 45% the maximum width of the prementum, its apical margin is convex, and it has a ventral row of 24–31 short, subquadrate, reddish-brown teeth, and dorsal rows of whitish, piliform setae. The labial palp is mostly yellowish-brown; the apical lobe is reddish-brown, widely rounded, and its mesal margin has a row of 15–16 small, truncate teeth. The movable hook is reddish-brown, 0.70× the length of the palp, sharp, and moderately incurved.

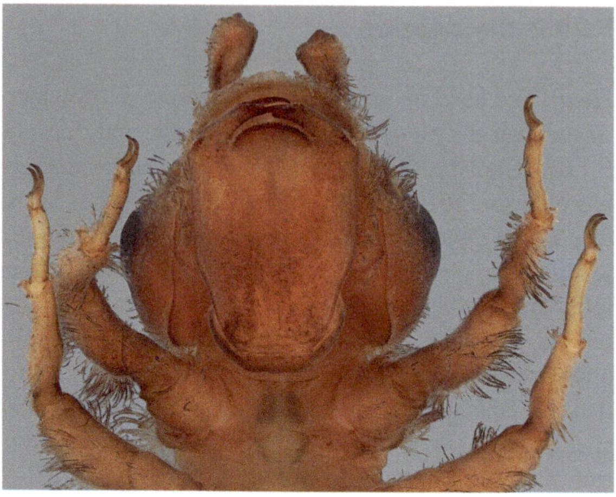

Fig. 5.121 The head, prementum, and fore- and middle legs in ventral view

Thorax. The prothorax is narrower than the head (Fig. 5.119). The propleuron is mostly bare, but its ventral margin has a dense brush of long, white, stiff setae. The pronotum anterior lobe is short and thick, the posterior lobe is wide and has a small mound-like tubercle at the middle, the lateral margins are widely rounded, densely granulose, and have a dense brush of long, white, stiff setae, and the posterior margin is sinuate. The pterothorax is coarsely granulose and light-yellow, but densely setose on the ventrolateral surfaces. The legs are short (Figs. 5.119 and 5.122) (i.e., when fully extended, the hind legs scarcely surpass the posterior margin of S5), stout, and setose; the fore- and middle femora are short and bulky, the fore- and middle tibiae have moderately developed burrowing hooks and numerous reddish-brown tubercles on the anterior surfaces; the hind femur is slightly longer than the tibia; the tarsi are creamy pale; the pretarsal claws are long, simple, and each has a pulvilliform empodium. The anterior and posterior wing sheaths are strongly divergent (Fig. 5.119) and reach the basal half of S4.

Fig. 5.122 Ventral habitus of the nymph of *Nychogomphus duaricus*

Abdomen. The abdomen is more or less fusiform, widest at S5, and strongly setose on the sides (Fig. 5.122). The ventral surface is flat, whereas the dorsal surface is convex and has mound-like middorsal protuberances on S2–S8, and a low, short, blunt spine on S9 (Fig. 5.123).

Fig. 5.123 The abdomen and caudal appendages in lateral view

Lateral spines are lacking on S2–S6, but the lateral margins of S7–S9 are serrated and end in short spines (Fig. 5.123). Sterna 2–6 are densely covered with small, spiniform setae, sterna 7–9 are beset with long, stiff setae, and sternum 10 is granulose. Sterna 3–7 and 9 are each divided into three sternites and sternum 8 into five sternites. The female gonapophyses apparently are lacking. The caudal appendages are densely granulose, sharply pointed, and longer than the middorsal length of S9+S10 (Fig. 5.123). The epiproct and paraprocts are pyramidal, and the cerci are digitiform with the mesal margins sinuate. The epiproct is 0.40× longer than its basal width and slightly longer than the paraprocts and cerci, which are the same length.

Measurements (in mm). Total length including caudal appendages 22–24; abdomen 13.4–15.0; maximum width of head 4.6–4.7; length of hind femur 3.3–3.5.

5.20.3 Remarks

The nymphs of *Nychogomphus* inhabit "headwater streams, where they were found in the sandy gravel substrates"(Chainthong and Boonsoong 2016). The nymph of this genus exhibits strong resemblance to that of *Onychogomphus* and to a lesser extent to *Lamelligomphus*.

5.20.4 Distribution

In Thailand, *Nychogomphus* has been recorded from the provinces of Chiang Mai, Kanchanaburi, Nakhon Nayok, Tak (Hämäläinen and Pinratana 1999 [as *Onychogomphus duaricus*), Phetchabun (Day et al. 2012), Ratchaburi (Chainthong and Boonsoong 2016), and more recently Bueng Kan, Kalasin, Sakon Nakhon (Buppachat et al. 2020) . ***New province records***: Chumphon, Mae Hong Son, Nan, Prachuap Khiri Khan, Surat Thani (Fig. 5.124).

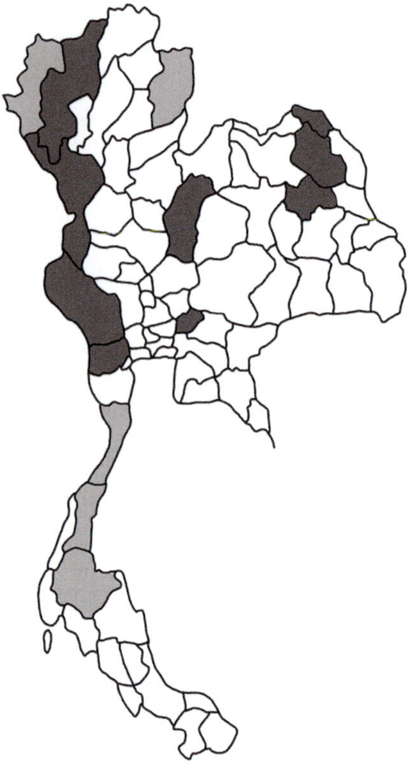

Fig. 5.124 Distribution of *Nychogomphus*

5.21 *Onychogomphus* Selys, 1854

Fig. 5.125 Dorsal habitus of the nymph of *Onychogomphus louissiriusi*

5.21.1 General Appearance

The nymphs of *Onychogomphus* (*s.l.*) are largely heterogeneous, but in Thailand they are characterized by their small size (17.4–17.8 mm), with the body stout and bulky, integument mostly granulose and with very few long hair-like setae, antennomere 3 strongly hypertrophied, and coloration light-yellow to light-brown (Fig. 5.125).

5.21.2 Description

Head. The head is wider than long (Fig. 5.126), wider than the prothorax, but narrower than the pterothorax and abdomen, and densely granulose. The labrum has a dense brush of piliform setae on the anterior margin. The clypeus is short and very finely granulose. The frons and vertex are mostly flat dorsally. The antefrons is produced anteriorly as a triangle with its tip nipple-shaped (Fig. 5.126), the postfrons is widely concave on the anterior margin and laterally produced as a wide, rounded lobe. The antennal scape is twice wider than long and granulose, and the pedicel is short and irregular in shape, antennomere 3 is highly hypertrophied, subrectangular, scoop-shaped, strongly granulose with its dorsal (external) surface convex and highly sculptured, its apex is directed mesad in such a manner that both antennae form a frontal "shield" (Figs. 5.126 and 5.127); antennomere 4 is a mere hemispheric rudiment positioned on the anterodorsal surface of antennomere 3. The compound eyes are relatively small. The occiput is large and the occipital lobes are rounded, coarsely granulose, and bulging. The posterior margin of the head is widely concave.

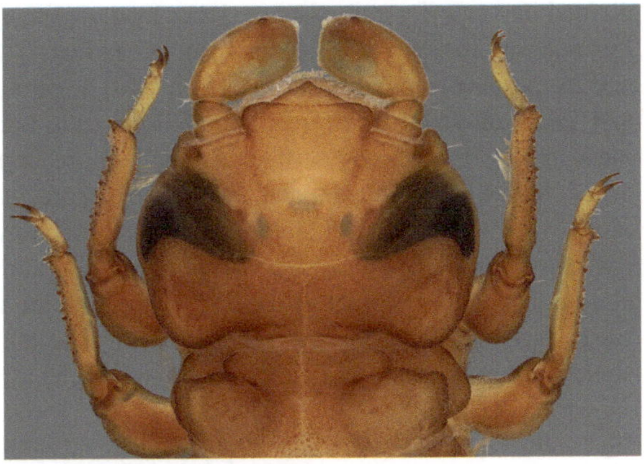

Fig. 5.126 The head, pronotum, and fore- and middle legs in dorsal view

Mandibles have the formula: L1234 0 a(m^{1-9})b/R1234 y a(m^{1-7})b. The maxilla's galeolacinia is as described for *Acrogomphus*. The ventral pad of the hypopharynx is subpentagonal, the basal third has a transverse, rectangular, light-amber plate, and the distal third has long, stiff setae arranged in a semicircle. Labium: The prementum-postmentum articulation reaches the posterior margin of the fore coxae. The prementum is subquadrate with the sides shallowly convex (Fig. 5.127), and it has a weak, longitudinal, median sulcus along the basal 0.70 of the ventral surface. The ligula is moderately developed, representing 40% of the maximum width of the prementum, its apical margin is convex, and it has a ventral row of 18–20 short, subquadrate, reddish-brown teeth (Fig. 5.127), and dorsal rows of whitish, piliform setae. The labial palp is mostly yellowish-brown, its apical lobe is reddish-brown and broadly rounded, and its mesal margin has a row of 11–12 small, squarely truncate teeth. The movable hook is reddish-brown, half the length of the palp, sharp, and moderately incurved.

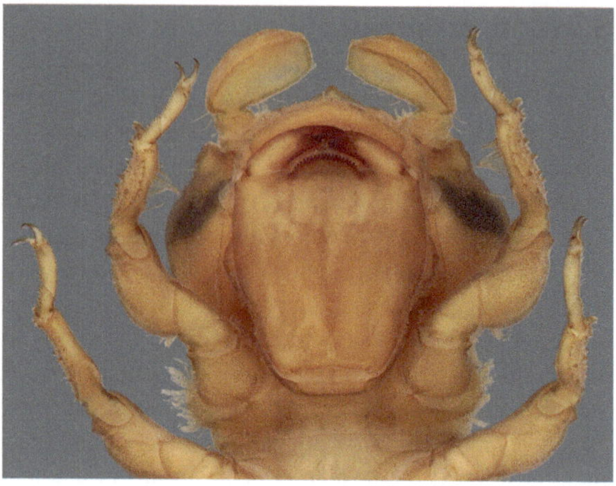

Fig. 5.127 The head, prementum, and fore- and middle legs in ventral view

Thorax. The prothorax is narrower than the head. The propleuron is densely granulose, and its ventral border has long, stiff setae. The pronotal disc is partially divided into anterior and posterior lobes by anterolateral narrow grooves; the anterior margin of the posterior lobe is sinuate and has a small tubercle at the middle, and the posterior margin is strongly convex. The pterothorax is coarsely granulose and light-yellow; the mesinfraepimeron lacks setae, but the ventral part of the metapleuron has a tuft of long, stiff setae. The legs are stout and short (Figs. 5.126 and 5.127) (i.e., when fully extended, the hind legs scarcely surpass the posterior margin of S5); the fore- and middle femora are short and bulky, the fore- and middle tibiae have moderately developed burrowing hooks and numerous reddish-brown tubercles on the anterior surfaces (Fig. 5.126); the tarsi are mostly granulose on the dorsal surface. The anterior and posterior wing sheaths are strongly divergent and slightly surpass the anterior margin of S4 and reach the basal half of S4, respectively.

Fig. 5.128 Ventral habitus of the nymph of *Onychogomphus louissiriusi*

Abdomen. The abdomen is more or less fusiform, widest at S5–S6 (Fig. 5.128), and lacks setae; the ventral surface is flat, whereas the dorsal surface is convex with middorsal protuberances as follows: on S2 as a blunt, anteriorly directed cone, on S3–S9 as very low and mound-like protuberances, and are almost imperceptible on S3–S6 (Fig. 5.129).

Fig. 5.129 The abdomen and caudal appendages in lateral view

Lateral spines are lacking. The sternum is finely granulose. Sterna 3–7 and 9 are each divided into three sternites and sternum 8 into five sternites; the median sutures of the sterna are convergent posteriorly on 2, parallel on 3, and slightly gradually divergent posteriorly on 4–9, the posterior margins are straight on sterna 1–9 and sinuate on 10. The female gonapophyses are very small, conical, and bluntly pointed, with a wide, U-shaped gap between them. The caudal appendages are densely and roughly granulose and twice as long as the middorsal length of S10 (Fig. 5.129). The epiproct is roundly pointed, slightly longer than its basal width, and as long as the paraprocts. The paraprocts and cerci are acutely pointed. The cerci are 0.12× shorter than the epiproct and paraprocts.

Measurements (in mm). Total length including caudal appendages 17.4–17.8; abdomen 11.1–11.4; maximum width of head 4.8–4.9; length of hind femur 2.8–3.0.

5.21.3 Remarks

Onychogomphus (*s.l.*) is a large, heterogeneous and complicated group of gomphids. It is distributed widely in Asia and Africa, with some species occurring in Europe and Russia (Davies and Tobin 1985). Chao (1954) split the Chinese *Onychogomphus* into several genera including *Lamelligomphus*, *Melligomphus*, and *Nepogomphus*. Based upon the nymphal and adult morphology, we think *Onychogomphus thienemanni* Schmidt, 1934, from Malaysia, and the recently described *O. (Siriusonychogomphus) louissiriusi* Fleck, 2020, together could well represent a new genus. Fleck (2020) considered these last two species to be closely related to the Bornean species *O. marijanmatoki* Dow, 2014. The nymphs of *Onychogomphus* inhabit small, shallow, rocky, pristine streams of first to fourth orders, with rapidly flowing cool water and high oxygen content. The microhabitat is rough gravel, cobbles, and coarse or fine sand riffles with sparsely scattered boulders and much or a moderate amount of detritus (50–70%) (Novelo-Gutiérrez and Che Salmah 2013).

5.21.4 Distribution

In Thailand, *Onychogomphus* has been recorded from the provinces of Chiang Mai, Khon Kaen, Sakon Nakhon, Songkhla, Tak (Hämäläinen and Pinratana 1999, 2000),

Chiang Rai (Katatani et al. 2004), Phrae (Katatani et al. 2010), Phetchabun, (Day et al. 2012), and more recently Surat Thani (Fleck 2020). ***New province records***: Chumphon, Kamphaeng Phet, Krabi, Loei, Mae Hong Son, Nakhon Ratchasima, Nakhon Si Thammarat, Nan, Narathiwat, Pattani, Phangnga, Phatthalung, Phayao, Satun, Sa Kaew, Trang, Uttaradit, and Yala (Fig. 5.130).

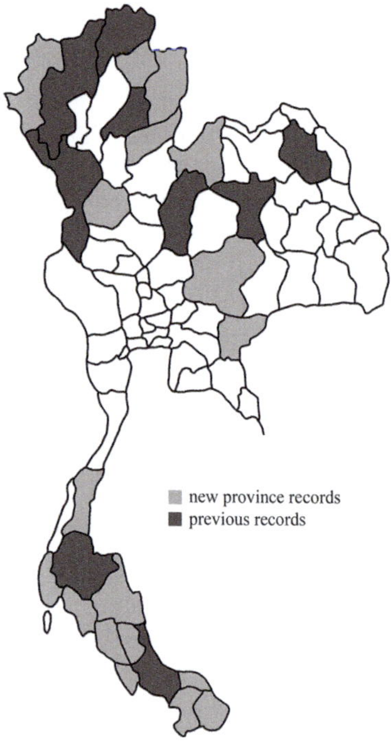

Fig. 5.130 Distribution of *Onychogomphus*

5.22 *Orientogomphus* Chao & Xu, 1987

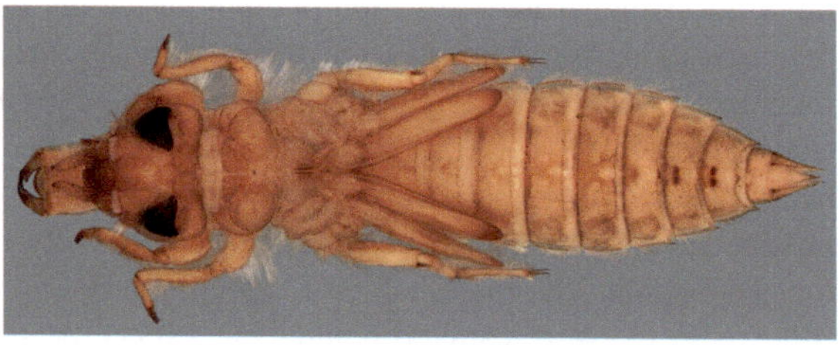

Fig. 5.131 Dorsal habitus of the nymph of *Orientogomphus minor*

5.22.1 General Appearance

The nymphs of *Orientogomphus* are characterized by their medium size (23–25 mm), with the body hairy, enlarged, tapering posteriorly (Fig. 5.131), and coloration light-yellow-brown.

5.22.2 Description

Head. The head is wider than long, but narrower than the thorax and abdomen. The labrum is 0.5 mm long and bare, and its anterior margin is flattened ventrally and densely setose. The clypeus is glabrous, the frons is densely covered with robust, stiff, erect setae, and the anterior margin of the postfrons is concave. The vertex and occiput have small, spiniform setae. The antennae are densely covered with robust, yellow, stiff setae on the dorsal surface, and long golden setae on the lateral margins; the scape is globose and the pedicel subglobose; antennomere 3 is the largest, wider at the base and tapers to the apex, the ventral surface is flat and has a tuft of long setae on the apical margin; antennomere 4 is an elongate cone reflexed over the apex of antennomere 3 and has long setae. The compound eyes are moderately developed. The occipital lobes are rounded and bulging. The posterior margin of the head is widely concave (Fig. 5.132). Mandibles have the formula: L 1234 0 a($m^{1,2,3,4,5,6}$)b/R 1234 y a($m^{1,2,3}$)b. The maxilla's galeolacinia is as described for *Acrogomphus*. The ventral pad of the hypopharynx is subpentagonal, whitish, mostly soft and bare, almost as long as wide, and has a transverse row of long setae on the distal third. Labium: The prementum-postmentum articulation reaches the posterior margin of the fore coxae. The prementum is light-yellow-brown, longer than wide, the sides are moderately convex at midlength, and it has a vestigial longitudinal central sulcus on the basal half of the ventral surface with a row of long, stiff setae on each side. The ligula is well-developed, its distal margin is strongly convex and very slightly pointed at the middle, and it has a double row of piliform setae. The labial palp is grayish-yellow, and the mesal margin is slightly serrate and rounded at the tip. The movable hook is brown, shorter than the palp, sharp, and moderately incurved.

Fig. 5.132 The head in dorsal view

Thorax. The prothorax is as wide as the head and setose on the ventral margin. The pronotum is narrower than the head, bulky, rounded dorsolaterally, and is covered with small, spiniform setae, except on the large, middorsal, bare, and subcircular areas; the anterior lobe comprises two large, triangular sclerites separate by a middorsal suture; the posterior lobe has the lateral margins rounded, and the posterior margin is widely convex but shallowly concave at the middle. In lateral view, the dorsal surface of the pronotum appears flat and continuous. The pterothorax is densely setose on the ventrolateral surface. The legs are short (i.e., when fully extended, the hind legs hardly reach the posterior margin of S6) and strongly setose, mainly on the fore- and middle legs; burrowing hooks are lacking; the middle coxae are closer to each other than are the fore- and hind coxae; the pretarsal claws are chelate-like on the fore- and middle tarsi, but long and sharp on the hind tarsi, and each has a pulvilliform empodium. The wing sheaths are strongly divergent and mostly bare, but with marginal setae; the anterior and posterior wing sheaths reach the posterior margin of S4 and basal half of S5, respectively.

Fig. 5.133 Ventral habitus of the nymph of *Orientogomphus minor*

Abdomen. The abdomen is elongate, widest at S6, and gradually tapers posteriorly (Fig. 5.133). The ventral surface is more or less flat, whereas the dorsal surface is convex and creamy-pale.

Fig. 5.134 The abdomen and caudal appendages in lateral view

The dorsal midline has a pair of small, brown spots on S8–S9, protuberances as mere domes on S2–S5, short triangular projections of the posterior margins which increase in size posteriorly on S6–S9 (Fig. 5.134). The lateral margins of S2–S7 are densely setose, and those of S7–S9 have strong posterolateral spines of the same size. Terga 2–10 are covered with minute, spiniform setae including on the posterior margins. Sterna 1–5 are creamy-pale and 6–10 light-yellow. Sterna 2–10 are covered with abundant minute setae, laterosternites 3–6 have rows of long, delicate setae, and 7–9 have these setae extending along the entire width. Sterna 3–9 are each divided into three sternites. The female gonapophyses are vestigial. The caudal appendages are sharply pointed, straight, and as long as the middorsal length of S9+S10 (Fig. 5.134). The paraprocts are slightly longer than the epiproct, and the cerci are slightly shorter than the epiproct.

Measurements (in mm). Total length including caudal appendages 23.3–25.0; abdomen (ventral) 15.3–16.1; maximum width of head 4.2; length of hind femur (lateral) 2.7.

5.22.3 Remarks

By its general morphology, the nymph of *Orientogomphus* appears closely related to *Acrogomphus* and *Paragomphus*. *Orientogomphus minor* (Laidlaw, 1931) was recorded in Thailand by Hämäläinen and Pinratana (1999) as *Acrogomphus minor*. However, Wilson (2008) transferred this species to *Orientogomphus* based on an analysis of adult morphology. The nymphs can be found in pristine, small, forest streams with abundant litter and sand banks. The nymph of *O. minor* was recently described by Chainthong and Boonsoong (2022).

5.22.4 Distribution

In Thailand, *Orientogomphus* has been recorded from the provinces of Phatthalung (Hämäläinen and Pinratana 1999 [as *Acrogomphus minor]*), Chumphon, Prachuap Khiri Khan, Ratchaburi, and Surat Thani (Ferro et al. 2009). ***New province records***: Chiang Mai, Kanchanaburi, Mae Hong Son, Phayao, Phetchaburi, Phitsanulok, Satun, and Yala (Fig. 5.135).

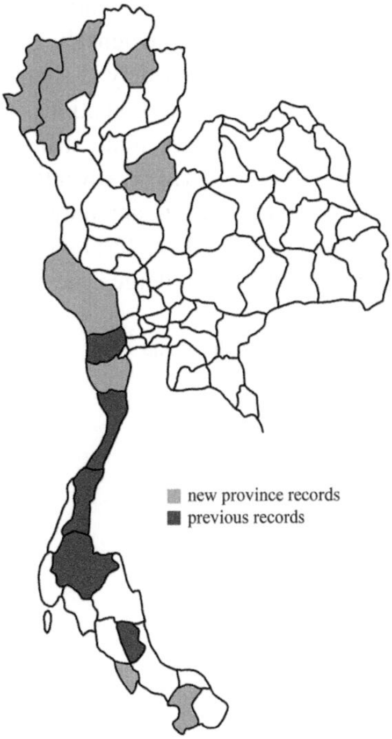

Fig. 5.135 Distribution of *Orientogomphus*

5.23 *Paragomphus* Cowley, 1934

Fig. 5.136 Dorsal habitus of the nymph of *Paragomphus capricornis*

5.23.1 General Appearance

The nymphs of *Paragomphus* are characterized by their medium size (23–28 mm), with the body robust and oblong, mostly granulose, and setose on the sides, the low number of *m* denticles in both mandibles, ligula strongly convex, middle coxae closer to each other, and coloration yellow-brown (Fig. 5.136).

5.23.2 Description

Head. The head is wider than long (Fig. 5.137), but narrower than the thorax and abdomen. The labrum is 0.4 mm long, has very minute spiniform setae, and its anterior margin is flattened ventrally and densely setose. The clypeus is bare; the frons vertex and occiput are densely granulose with small, stiff setae. The anterior margin of the postfrons is widely U-shaped at the middle, but is laterally produced as short, blunt, setose conical projections. The antennal scape is globose, the pedicel annular, and both have short, stout, stiff setae; antennomere 3 is the largest, subcylindrical, moderately flattened dorsoventrally, setose on the sides, and has a tuft of long setae on the apical margin; antennomere 4 is finger-like, reflexed over the apex of antennomere 3 (Fig. 5.137), has long setae, and is 0.40x the length of antennomere 3. The compound eyes are large. The occipital lobes are rounded and bulging (Fig. 5.137).

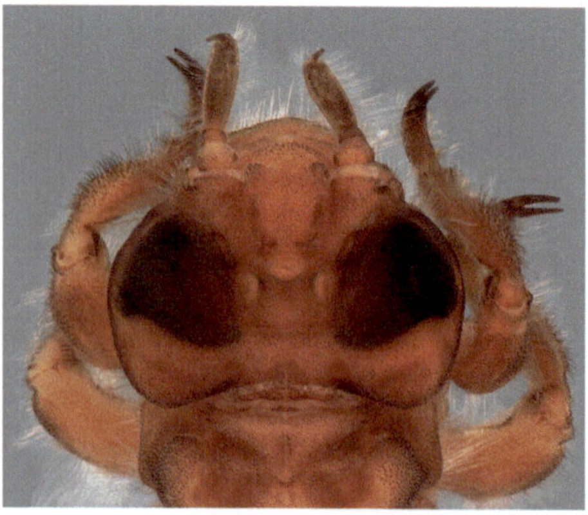

Fig. 5.137 The head and pronotum in dorsal view

Mandibles have the formula: L 1234 0 a($m^{1,2}$)b/R 1234 y a(m^{1})b, with b > a in the left mandible and a > b in the right mandible. The maxilla's galeolacinia is as described for *Acrogomphus*. The ventral pad of the hypopharynx is subpentagonal, whitish, soft, bare, and almost as long as wide. Labium: The prementum-postmentum articulation reaches the posterior margin of the fore coxae. The prementum is light-yellow, subrectangular, longer than wide (Fig. 5.138); its sides are shallowly convex and divergent, its basal margin is widely emarginated, and there is no ventral sulcus. The ligula is well-developed, its apical margin is convex, slightly angulate at the middle (Fig. 5.138), and it has a dorsal double row of piliform setae and ventral row of approximately 60 small, quadrate teeth. The apical lobe of the labial palp is finger-like, bluntly pointed, and the mesal margin is shallowly concave and very finely serrulated. The movable hook is shorter than the palp, obtusely pointed, and moderately incurved (Fig. 5.138).

Fig. 5.138 The head, prementum, and forelegs in ventral view

Thorax. The prothorax is narrower than the head (Figs. 5.136 and 5.137), coarsely granulose, and setose on the ventral margin. The pronotum is densely granulose, except on the large, dorsolateral, bare, semicircular areas; the anterior and posterior margins are sinuate, and the lateral margins are rounded (Fig. 5.137). The pterothorax is densely setose on the ventrolateral surface, and the mesepimeron is strongly convex. The legs are very short (Figs. 5.136 and 5.139) (i.e., when fully extended, the hind legs hardly reach the anterior margin of S6), strongly setose, and have white setae mainly on the fore- and middle legs; burrowing hooks are lacking; the middle coxae are closer to each other than are the fore- and hind coxae (Fig. 5.139); the pretarsal claws are simple and chelate-like on the fore- and middle legs, but long and sharp on the hind legs (Figs. 5.138, 5.139, and 5.140), and each has a pulvilliform empodium. The wing sheaths are strongly divergent (Fig. 5.136) and have the margins serrulated and setose, whereas the remainder of the surfaces are mostly bare; the anterior and posterior wing sheaths surpass the anterior margin and reach the basal half of S4, respectively.

Fig. 5.139 Ventral habitus of the nymph of *Paragomphus capricornis*

Abdomen. The abdomen is lanceolate and tapers posteriorly (Figs. 5.136 and 5.139). Coloration is yellow-brown with a complex pattern of dark spots, although S6 is mostly reddish-brown. The ventral surface is more or less flat and the dorsal surface is convex.

Fig. 5.140 The abdomen and caudal appendages in lateral view

Middorsal protuberances are small on S2–S3 and S7–S9, and vestigial on S4–S6 (Fig. 5.140). S4–S9 end in lateral spines that increase in size and robustness posteriorly, with those of S8–S9 the largest. Terga 2–10 are densely granulose, and the posterior margins of S2–S6 have a sparse row of red, long, stiff setae; the lateral margins of S2–S10 are densely setose and finely serrulated. The sternum is mostly yellow and granulose, but sterna 8–10 are reddish-brown (Fig. 5.139). Sterna 3–9 are each divided into three sternites. The caudal appendages are pyramidal, sharply pointed, covered with minute, spiniform setae, and are as long as S9+S10 (Fig. 5.140). The epiproct is the longest of the caudal appendages, and the paraprocts and cerci are the same length.

Measurements (in mm). Total length including caudal appendages 23.1–25.2; abdomen (ventral) 16.5–17.1; maximum width of head 4.5–4.6; length of hind femur (lateral) 3.0.

5.23.3 Remarks

Only *Paragomphus capricornis* (Förster, 1914) has been recorded in Thailand (Hämäläinen and Pinratana 1999). This species inhabits clear streams with cobble and gravel substrates in open country or light forest (Orr 2005). Other species apparently inhabit the shallow, sandy bottom near the banks (Bedjanič 2013) or in sandy patches in rapids (Suhling and Martens 2007). Wilson (1995) illustrated and briefly described the nymph of *P. capricornis* from Hong Kong. Some species have a very fast nymphal development (as few as 60 days), such as *P. genei* (Selys, 1841) in Namibia (Suhling and Martens 2007). The nymph here diagnosed is characterized by the low number of *m* denticles in both mandibles and the strongly convex ligula, as compared to those of other genera. By the nymphal morphology, *Paragomphus* appears more similar to *Orientogomphus* and to a lesser extent to *Acrogomphus*.

5.23.4 Distribution

In Thailand, *Paragomphus* is widespread with records from the provinces of Chanthaburi, Chiang Mai, Kanchanaburi, Krabi, Loei, Mae Hong Son, Nakhon Nayok, Phetchaburi, Prachuap Khiri Khan, Ranong, Ratchaburi, Songkhla, Tak, Trang, Yala (Hämäläinen and Pinratana 1999, 2000), Chiang Rai (Katatani et al. 2004), Nakhon Si Thammarat (Ferro et al. 2009), Nan, Phrae (Katatani et al. 2010), and more recently Surat Thani (Day et al. 2012). ***New province records:*** Chumphon, Nakhon Ratchasima, Phatthalung, Phetchabun, and Phitsanulok (Fig. 5.141).

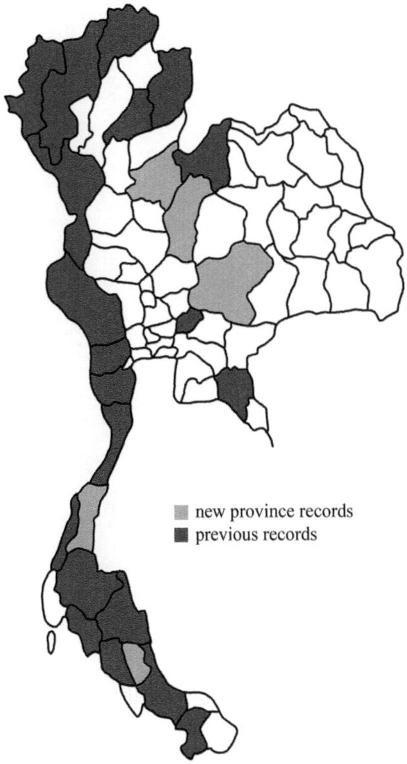

Fig. 5.141 Distribution of *Paragomphus*

5.24 *Phaenandrogomphus* Lieftinck, 1964

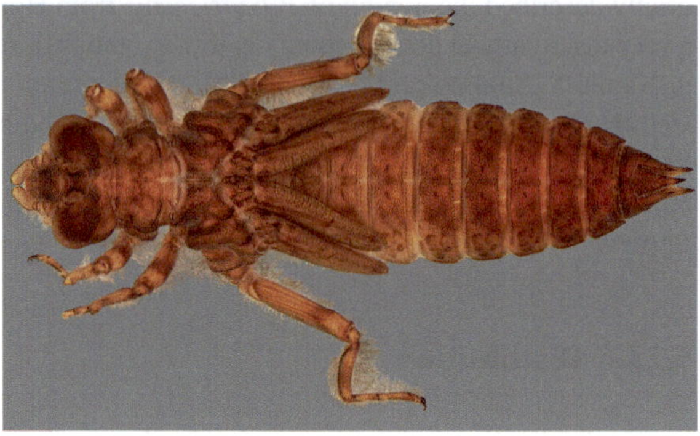

Fig. 5.142 Dorsal habitus of the nymph of *Phaenandrogomphus tonkinicus* (After Novelo and Sites 2019, with permission of Magnolia Press)

5.24.1 General Appearance

The nymphs of *Phaenandrogomphus* are characterized by their small to medium size (17–25 mm), with the body robust and granulose, compound eyes bulging anterolaterally, abdomen lanceolate, and coloration yellow-brown to reddish-brown (Fig. 5.142).

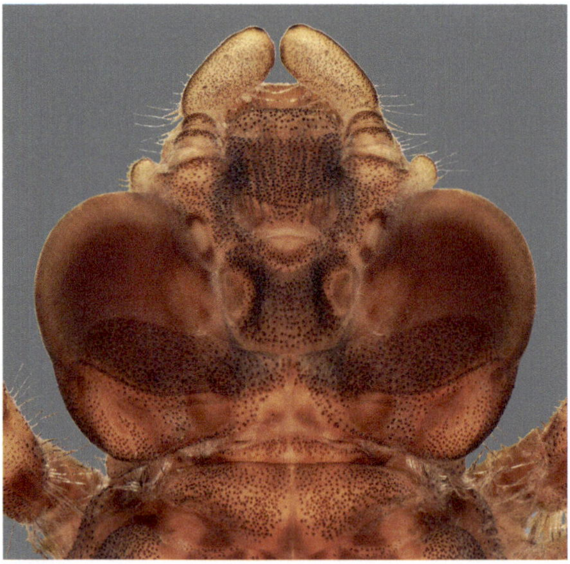

Fig. 5.143 The head in dorsal view (After Novelo and Sites 2019, with permission of Magnolia Press)

5.24.2 Description

Head. The head is wider than long (Fig. 5.143), but narrower than the pterothorax and abdomen. The labrum is 0.5–0.6 mm long and has the anterior margin flattened ventrally and setose. The clypeus is bare, and the labrum, frons, vertex, and occiput are mostly coarsely granulose. The antefrons sometimes has a thickened anterior margin and can be slightly produced as a short shelf; the anterior margin of the postfrons is wide and shallowly concave at the middle, and the anterolateral portions are rounded, slightly to strongly produced anteriorly, and beset with short, stiff setae. The antennae are coarsely granulose or scaly, and the scape is subglobose and pedicel annular; antennomere 3 is the largest, flattened, fusiform or with the mesal margin strongly convex at the middle and the lateral margin convex, its dorsal and ventral surfaces are densely granulose or covered with claviform setae, the lateral margins have a few or abundant long, white, stiff, reflexed setae; antennomere 4 is strongly reduced to a conical rudiment or is vestigial (Figs. 5.143 and 5.144), mostly scaly, dome-like, and not visible in dorsal view. The compound eyes are large and have the posterior margin concave (Fig. 5.143). A moderate to strongly developed longitudinal carina is on the ventral part of the head below the compound eye (Fig. 5.144). The occipital lobes are rounded (Fig. 5.143) and sometimes with a posterior mound-like tubercle.

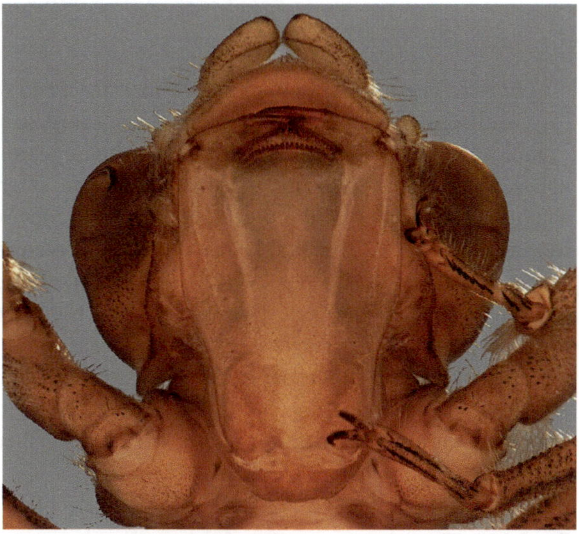

Fig. 5.144 The head and prementum in ventral view (After Novelo and Sites 2019, with permission of Magnolia Press)

Mandibles have the formula: L 1234 y a($m^{1,2,3,4,5,6}$)b/R 1234 y a($m^{1,2,3}$)b, with a > b in both mandibles; the *m* denticles are sharp in the right mandible, but truncate in the left mandible. The maxilla's galeolacinia is as described for *Acrogomphus*. The ventral pad of the hypopharynx is subpentagonal, whitish, mostly soft, and has a basal, finger-like sclerite on each side of the midline connected to each other by a basal, transverse, denticulate, sclerotized bridge; its posterior margin is sinuate, and the apical half has abundant long, stiff setae. Labium: The prementum-postmentum articulation reaches the posterior margin of the fore coxae (Fig. 5.145). The prementum is subrectangular and longer than wide (Fig. 5.144); the lateral margins are finely serrate, convex along the apical half, and sinuate and divergent along the basal half; the basal margin is widely emarginate and there is no ventromedian sulcus. The ligula is poorly developed and shallowly convex (Fig. 5.144), its anterior margin has a dorsal row of abundant long, piliform setae and ventral row of 17–30 short, red-brown, truncate teeth. The apical lobe of the labial palp is widely rounded at the tip, and the mesal margin is nearly straight to shallowly concave and has a row of 10–12 very short, square teeth. The movable hook is reddish, shorter than the palp, sharp, and slightly incurved.

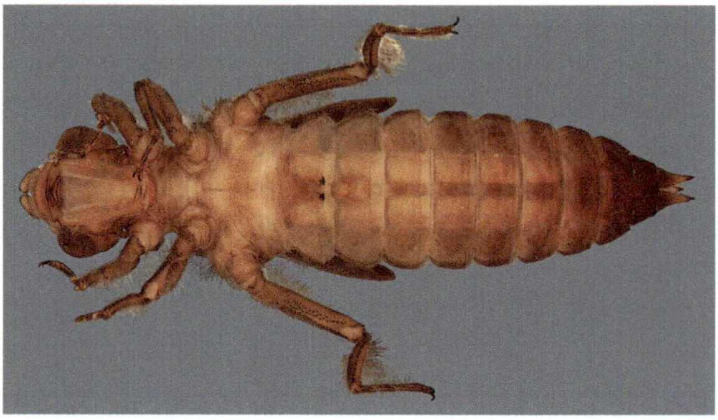

Fig. 5.145 Ventral habitus of the nymph of *Phaenandrogomphus tonkinicus* (After Novelo and Sites 2019, with permission of Magnolia Press)

Thorax. The prothorax is narrower than the head (Fig. 5.142), coarsely granulose, and setose on the ventral margin of the propleuron. The pronotum is strongly granulose and has minute setae, except on the large, depressed, bare, irregular, dorsolateral areas; the anterior margin is sinuate, the lateral margins are flattened and shallowly concave, and the posterior margin is convex. The pterothorax is granulose, bare on the sutures, and setose on the ventrolateral margin. The legs are short (Figs. 5.142 and 5.145) (i.e., when fully extended, the hind legs hardly reach the posterior margin of S6) and setose; burrowing hooks are poorly to well-developed; the external (anterior) surfaces of the fore- and middle tibiae are tuberculate; the pretarsal claws are simple and each has a pulvilliform empodium. The wing sheaths are strongly divergent (Fig. 5.142), their surfaces are mostly smooth, but the margins are granulose; the anterior and posterior wing sheaths reach the basal half and posterior margin of S4, respectively.

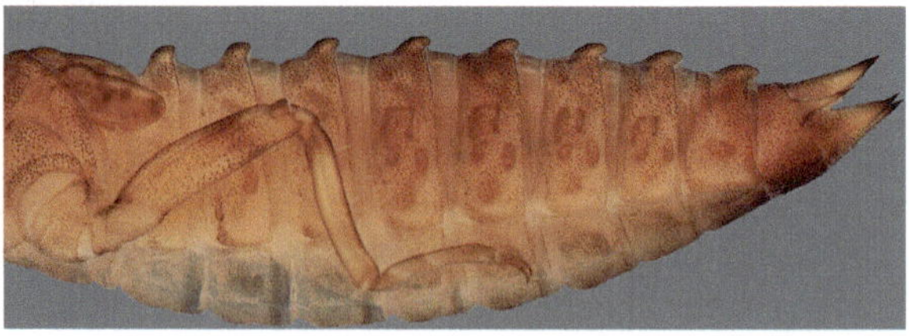

Fig. 5.146 The abdomen and caudal appendages in lateral view (After Novelo and Sites 2019, with permission of Magnolia Press)

Abdomen. The abdomen is lanceolate, robust, and has a complex color pattern (Figs. 5.142 and 5.145); its ventral surface is flat, whereas the dorsal surface is convex. Terga 2–10 are coarsely granulose, including on the posterior margins, and have sparse, long, white setae; low, blunt, middorsal protuberances are on S2–S9 (Fig. 5.146). S1–S9 lateral margins are shallowly convex, setose, and finely serrulated; S8–S9 or only S9 end in a small, bluntly pointed, lateral spine. The sternum is yellowish-brown to reddish-brown (Fig. 5.145), densely granulose, and with or without small setae. Sterna 3–7 and 9 are each divided into three sternites and sternum 8 into five sternites. The female gonapophyses are vestigial. The caudal appendages are pyramidal and sharply pointed (Fig. 5.146), granulose, and the same length, or the cerci are the shortest. The epiproct and paraprocts are as long as the dorsal length of S9+S10 (Fig. 5.146). The male's epiproct has a pair of small, rounded, dorsal tubercles at the basal 0.60–0.70.

Measurements (in mm). Total length including caudal appendages 17–25; abdomen 10.7–17.0; maximum width of head 4.0–4.8; length of hind femur 3.0–4.3.

5.24.3 Remarks

Three species of *Phaenandrogomphus* have been recorded from Thailand: *P. asthenes* Lieftinck, 1964, *P. dingavani* (Fraser, 1924) [as *Onychogomphus dingavani*], and *P. tonkinicus* (Fraser, 1926) (Hämäläinen and Pinratana 1999). *P. dingavani* was transferred to the genus *Scalmogomphus* Chao, 1990 by Karube (2015), but here we decided to follow the World Odonata List (Paulson et al. 2023). The nymphs inhabit clear forest streams where they burrow into sand and gravel (Orr 2005). Lieftinck (1964) considered this genus allied to *Onychogomphus* and *Acrogomphus* based on adult characteristics, although he also recognized strong differences in both adults and nymphs among the three genera. Based upon nymphal features (Novelo-Gutiérrez and Sites 2019b), we support Lieftinck's concept of a close relationship with *Onychogomphus*, and to a lesser extent with *Amphigomphus*, but not with *Acrogomphus*. The nymph of *Acrogomphus* is quite different in many respects.

5.24.4 Distribution

In Thailand, *Phaenandrogomphus* has been recorded from the provinces of Chanthaburi, Chiang Mai, Kanchanaburi, Ranong, Tak, Yala (Hämäläinen and Pinratana 1999), Chiang Rai (Katatani et al. 2004), Nan, Phrae (Katatani et al. 2010), and recently Chumphon, Kamphaeng Phet, Loei, Mae Hong Son, Nakhon Si Thammarat, Narathiwat, Phangnga, Phatthalung, Phayao, Phetchaburi, Prachuap Khiri Khan, Ratchaburi, Songkhla, Surat Thani, Trang, Udon Thani, and Uttaradit (Novelo-Gutiérrez & Sites 2019). ***New province record***: Nakhon Ratchasima (Fig. 5.147).

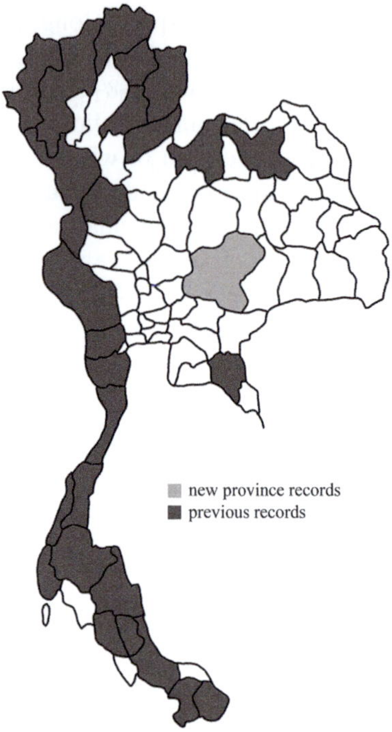

Fig. 5.147 Distribution of *Phaenandrogomphus*

5.25 *Sieboldius* Selys, 1854

Fig. 5.148 Dorsal habitus of a young nymph of *Sieboldius* showing the expanded abdomen

5.25.1 General Appearance

The nymphs of *Sieboldius* are characterized by their very large size (41 mm), with the body leaf-like, fore- and middle legs darker, abdomen strongly flattened and wide, and coloration yellow-brown to reddish-brown (Fig. 5.148).

5.25.2 Description

Head. The head is wider than long and wider than the prothorax, but narrower than the pterothorax and abdomen (Fig. 5.148). The labrum is 1.0 mm long and flattened ventrally. The clypeus is trapezoidal. The antefrons is produced medially as a frontal shelf with its distal margin strongly convex (Fig. 5.149); the anterior margin of the postfrons has a deep, V-shaped concavity. The antennal scape is globose, the pedicel is annular, antennomere 3 is the largest, greatly expanded and strongly flattened (Fig. 5.149), and all margins bear rows of closely set, scale-like setae, and antennomere 4 is conical, strongly reduced, and mostly covered by scale-like setae except at the tip. The compound eyes are relatively small. The occiput has a dorsal, submedian, mound-like tubercle on each side, a large, posterolateral, subconical tubercle on the occipital lobes (Fig. 5.149), and a lateral tubercle on the ventral margin of the occiput (Fig. 5.150); the occipital lobes are rounded.

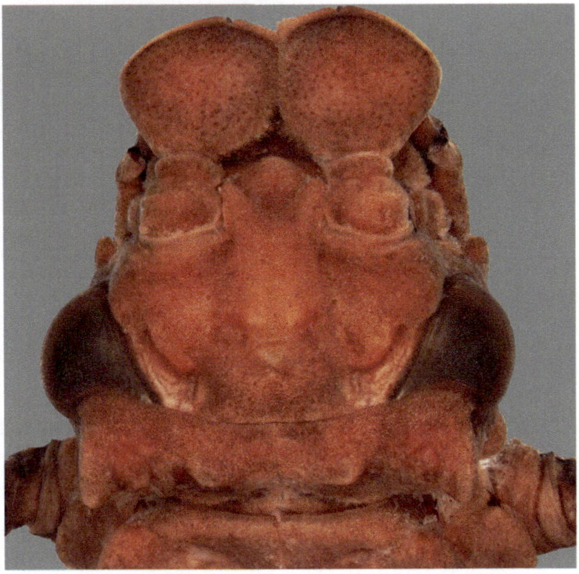

Fig. 5.149 The head in dorsal view

Mandibles have the formula: L 1234 0 a($m^{1,2,3,4,5,6}$)b/R 1234 y a($m^{1,2,3,4,5,6}$)b, with a > b in both mandibles, but noticeably larger in the right mandible. The maxillae are mostly covered with long, golden setae; the maxilla's galeolacinia is as described for *Acrogomphus*. The ventral pad of the hypopharynx is subpentagonal, whitish, mostly soft, its basal third has a transverse, chitinous plate, and the soft area has long, abundant setae. Labium: The prementum-postmentum articulation reaches the midlength of the prosternum (Fig. 5.151). The prementum is yellow-brown, slightly wider than long, and its lateral margins are straight (Fig. 5.150); the ventral surface has a vestigial median sulcus and a large, central, shallow concavity just below the ligula. The ligula is poorly developed, very shallowly convex, and has a ventral row of 15–16 short, square, dark, truncate teeth (Fig. 5.150); dorsally it has a row of abundant long, yellowish, piliform setae on the anterior margin, the surface has four longitudinal tufts of long, golden setae intermingled with glabrous, shining areas. The labial palp is dark-reddish-brown, its apical lobe is widely rounded, the apical and mesal margins are serrulate, and the lateral margin is smooth. The movable hook is black, almost as long as the palp, sharp, and moderately incurved.

Fig. 5.150 The head and prementum in ventral view

Thorax. The prothorax is narrower than the head. The pronotum anterior margin is straight, the lateral margins are rounded and strongly reflexed, and the posterior margin is shallowly convex. The pronotal disc is partially divided into anterior and posterior lobes by narrow anterolateral grooves. The pterothorax is finely granulose, and the meso- and metaspiracles are evident. The legs are long (Figs. 5.148 and 5.151) (i.e., when fully extended, the hind legs surpass the caudal appendages); the fore- and middle legs are darker than the hind legs (Fig. 5.148); burrowing hooks are absent; the pretarsal claws are simple and each has a pulvilliform empodium. The anterior and posterior wing sheaths are parallel and almost reach and surpass the posterior margin of S4, respectively (Fig. 5.148). The sternum is finely granulose, and the pro- and mesosterna each have a large, ventrally directed, blunt, subconical tubercle close to the articulation of their respective coxae.

Abdomen. The abdomen is uniformly colored, leaf-like, wide, strongly flattened, and the sides are gradually divergent posteriorly then strongly converge at S9–S10 (Figs. 5.148 and 5.151).

Fig. 5.151 Ventral habitus of a young nymph of *Sieboldius*

Terga 2–10 are finely granulose, but the posterior margins are smooth. Middorsal protuberances are on S2–S5 with those on S2–S4 spine-like, on S5 a rounded, longitudinal ridge is on the posterior half, and on S6–S10 is a longitudinal, very low, rounded ridge. The lateral margins of S2–S10 have abundant strongly close-packed, scale-like setae, and those of S2–S9 end in a large spine. S7 is the widest and S9 the longest. The sternum is finely granulose, yellow-brown anteriorly and reddish-brown posteriorly. Sterna 3–9 are each divided into three sternites, and spiracles are evident on S2–S8. The caudal appendages are enlarged, robust, densely covered with scale-like setae, and are slightly shorter than the dorsal length of S9+S10 (Fig. 5.152). The epiproct and paraprocts are bluntly pointed; the cerci are sharply pointed and very short, scarcely more than one third as long as the epiproct. The epiproct is longer than its basal width, and the paraprocts are the longest of the caudal appendages (Fig. 5.152).

Fig. 5.152 The abdomen in lateral view showing the caudal appendages with tips rounded

Measurements (in mm). Total length including caudal appendages 41; abdomen (ventral) 28.5; maximum width of head 8.3; length of hind femur (lateral) 14.7.

5.25.3 Remarks

This huge nymph is easily recognized by its strongly flattened, leaf-like body. *Sieboldius* and the North American monotypic genus *Hagenius* compose the subfamily Hageniinae. The nymphs of *Sieboldius* inhabit leaf packs in gentle riffles

(Orr 2005). Two species of *Sieboldius* have been recorded from Thailand: *S. japponicus* Selys, 1854 and *S. nigricolor* (Fraser, 1924) (Hämäläinen and Pinratana 1999). The nymph of *S. japponicus* was described by Lieftinck (1932).

5.25.4 Distribution

In Thailand, *S. nigricolor* is a northern species that has been recorded from only Chiang Mai Province (Hämäläinen and Pinratana 1999). In contrast, *S. japponicus* is a southern species recorded from the peninsular provinces of Songkhla, Trang, and Yala (Hämäläinen and Pinratana 1999), and more recently from Surat Thani (Day et al. 2012). ***New province record***: Narathiwat (Fig. 5.153).

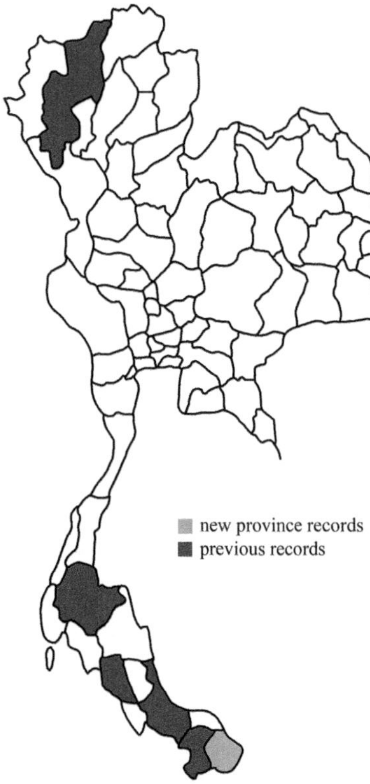

Fig. 5.153 Distribution of *Sieboldius*

5.26 *Sinictinogomphus* Fraser, 1939

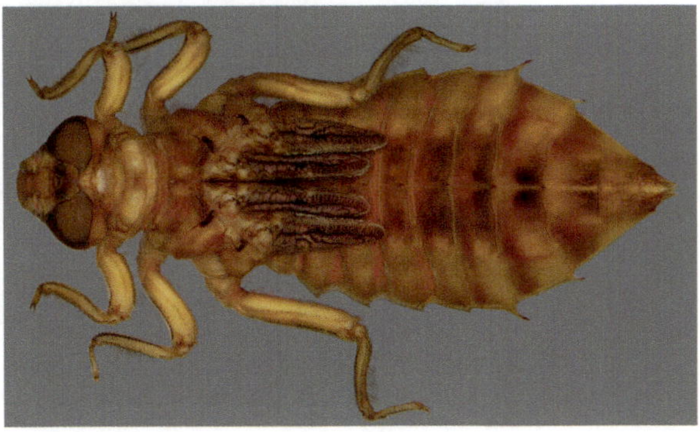

Fig. 5.154 Dorsal habitus of the nymph of *Sinictinogomphus clavatus*

5.26.1 General Appearance

The nymphs of *Sinictinogomphus* are characterized by their very large size (45 mm), with the body mostly bare and robust, labial palp strongly incurved and sharply pointed, abdomen strongly convex dorsally and flattened ventrally, and coloration light-yellowish-brown (Fig. 5.154).

5.26.2 Description

Head. The head is wider than long, but narrower than the pterothorax and abdomen. The labrum, postclypeus, frons, vertex, and occiput are slightly rugose, and the anteclypeus is bare. The labrum is 1.0 mm long, and its anterior margin is slightly flattened ventrally and covered with delicate setae. The antefrons is moderately swollen in the middle, the postfrons is flat, and the vertex is mound-like with three large ocelli. The genae protrude anteroventrally as small bluntly pointed tusks. The antennal scape is globose, the pedicel barrel-shaped, antennomere 3 is the largest and distinctly longer than wide, cylindrical, and straight, and antennomere 4 is strongly reduced. The compound eyes are large (Fig. 5.154). The occiput is very short and the occipital lobes are small, pale, slightly bulging, and have vertical rows of small, spiniform setae separated by bare areas. Mandibles have the formula: L 1234 0 a(m^{1-4})b/R 1234 y a(m^{1-4})b, with a > b in both mandibles; in the left mandible, the tooth "b" has four, smaller, thin, extra teeth on its mesal margin, and three teeth in the right mandible; the teeth of the molar crest are arranged in a semicircle in both mandibles. The maxilla's galeolacinia has seven teeth; the three dorsal teeth are moderately incurved and subequal in length and robustness, the four ventral teeth are strongly incurved and subequal in size and robustness, and the apical tooth is the largest. The ventral pad of the hypopharynx is subpentagonal, has a large, triangular, yellowish, sclerotized plate covering most of its surface, the remainder is soft, the anterior margin has a row of long, stiff setae, and the posterior margin is sinuate (Fig. 5.155).

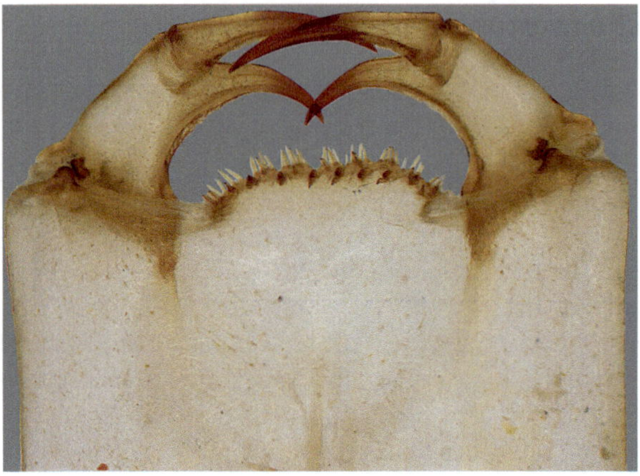

Fig. 5.155 Detail of the prementum showing the ligula and palps in dorsal view

Labium: The prementum-postmentum articulation reaches the anterior margin of the mesosternum. The prementum is yellow, rectangular, and longer than wide; the lateral margins are setose and sinuate; there is no median sulcus on the ventral surface; a transverse, depressed, narrow, bare area is along the basal margin. The ligula is convex and moderately developed, and the apical margin is armed with a dorsal row of eight short, stout, triangular, reflexed, reddish teeth and a ventral row of yellowish, elongate, spiniform setae (Fig. 5.155). The labial palp is yellow, its reddish apical lobe is strongly incurved and sharply pointed, and the mesal margin is sharp, smooth and strongly concave (Fig. 5.155). The movable hook is reddish-brown, shorter than the palp, sharply pointed, and moderately incurved.

Fig. 5.156 Ventral habitus of the nymph of *Sinictinogomphus clavatus* (labium detached)

Thorax. The prothorax is narrower than the head (Fig. 5.154). The ventral margin of the propleuron is granulose with abundant long, delicate setae; the proepisternum is greatly reduced, and the proepimeron is rounded and bulging. The pronotum anterior lobe is thick, raised, and finely granulose, the posterior lobe is subrectangular (Fig. 5.154), largely smooth and with the lateral margins rounded and granulose, and the posterior margin is sinuate. The ventral margins of the meso- and metapleura are setose. The legs are entirely light-yellow and relatively short (Figs. 5.154 and 5.156) (i.e., when fully extended, the hind legs scarcely surpass the basal half of S8), and have abundant delicate, long, golden setae. The tibiae lack burrowing hooks, but have large brushes of long, reddish-brown setae on the lateral surfaces. The tarsal formula is 2–2–2; the pretarsal claws are simple, short, reddish, and lack pulvilliform empodia. The anterior and posterior wing sheaths are parallel and mostly bare, but with the margins serrulated; they reach the posterior margin of S4 and basal half of S5, respectively. The sternum is mostly smooth; the meso- and metasterna have a short, longitudinal, low, rounded, and finely granulose ridge close to the articulation of the middle and hind coxae.

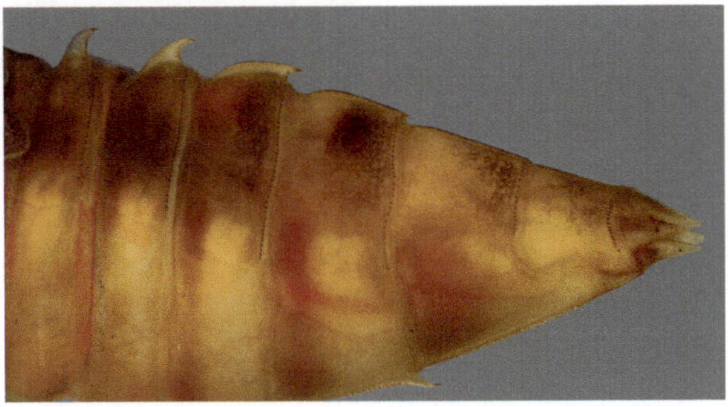

Fig. 5.157 Abdominal segments 5–10 and caudal appendages in lateral view

Abdomen. The abdomen is oval in shape and widest at the posterior margin of S7 (Figs. 5.154 and 5.156). Coloration is reddish-yellow on the lateral third and reddish-brown on the middle third. The ventral surface is flat (Fig. 5.156), whereas the dorsal surface is convex and reaches its highest point at S5 (Fig. 5.157). Dorsal protuberances are present on S2–S9 and are large vertical spines on S2–S3, hook-like on S4–S7 (Fig. 5.157), and short spines on S8–S9. The lateral margins of S3–S9 have strongly serrated keels which end in large and strong spines on S7–S9 (Fig. 5.158); the spines are sharp and divergent on S7–S8, parallel on S9, and those on S7 are the largest; S10 is laterally concealed by the spines of S9 (Fig. 5.158). S8 and S9 are fused. Terga 2–10 are finely granulose, and the posterior margins have a row of minute, truncate, spiniform setae. The sternum is mostly yellow, although reddish-yellow on the sides (Fig. 5.156), and mostly smooth, although sterna 8–9 are very finely granulose. Sterna 3–6 and 9 are each divided into three sternites and sterna 7–8 into five sternites; the median sutures are convergent apically on 2, parallel on 3–5 and 8–9, and slightly divergent on 6–7; the posterior margins are straight on sterna 2–5 and 9 and sinuate on S1, S6–S8, and S10. The female gonapophyses are rudimentary (Fig. 5.158), conical, and widely divergent. The caudal appendages are shorter than the dorsal length of S10 (Fig. 5.157), pyramidal, setose, and pale. The epiproct and paraprocts are obtusely pointed. The epiproct is the longest of the caudal appendages. The cerci are sharply pointed and 0.90× the length of the epiproct. The paraprocts are 0.80× the length of the epiproct, and are the shortest.

Fig. 5.158 Sterna 5–10 showing the serrate lateral margins and posterolateral spines

Measurements (in mm). Total length including caudal appendages 45; abdomen 30; maximum width of head 9; length of hind femur 9

5.26.3 Remarks

The nymph of *Sinictinogomphus* appears similar to that of *Ictinogomphus*, although the former is larger, mostly smooth, has a smaller occiput, and the mesal margin of the apical lobe of the labial palp is smooth and sharp. Only one species has been recorded from Thailand: *S. clavatus* (Fabricius, 1775) (Hämäläinen and Pinratana 1999 [as *Ictinogomphus clavatus phaleratus* (Selys, 1854)]). This species inhabits lowland ponds.

5.26.4 Distribution

In Thailand, *Sinictinogomphus* has been recorded from only the provinces of Khon Kaen (Hämäläinen and Pinratana 1999) and Chiang Rai (Katatani et al. 2004) (Fig. 5.159).

Fig. 5.159 Distribution of *Sinictinogomphus*

5.27 *Stylogomphus* Fraser, 1922

Fig. 5.160 Dorsal habitus of the nymph of *Stylogomphus*

5.27.1 General Appearance

The nymphs of *Stylogomphus* are characterized by their small size (17.0–17.5 mm), with the body elongate and slightly flattened, integument granulose, antennomere 3 large and triangular, and coloration yellowish-brown to brown (Fig. 5.160).

5.27.2 Description

Head. The head is wider than long and wider than the prothorax, but narrower than the pterothorax and abdomen (Fig. 5.160). The labrum is 0.5 mm long, its surface is bare, and the anterior margin is flattened ventrally and densely covered with long, stiff setae. The clypeus is bare. The antefrons is granulose and produced over the clypeus forming a short shelf; the anterior margin of the postfrons is deeply U-shaped, forming a sharply pointed horn on each anterolateral corner (Fig. 5.161). The antennal scape is subcylindrical and the pedicel is ring-shaped; antennomere 3 is the largest, triangular, paddle-like, and concave dorsally (Fig. 5.161), its margins are thick, the lateral margin is strongly convex with short, close-packed, scale-like setae, 2–3 long, reflexed, stiff setae are on the basal third, the mesal margin is straight and has a row of enlarged, claviform setae, and the apical margin is rounded and thick; antennomere 4 is hemispherical and vestigial. The compound eyes are moderately developed and strongly concave on the posterior margin (Fig. 5.161); a longitudinal carina covered with short, spiniform setae is below each compound eye (Fig. 5.162). The occipital lobes are rounded, bulging, and granulose (Fig. 5.161).

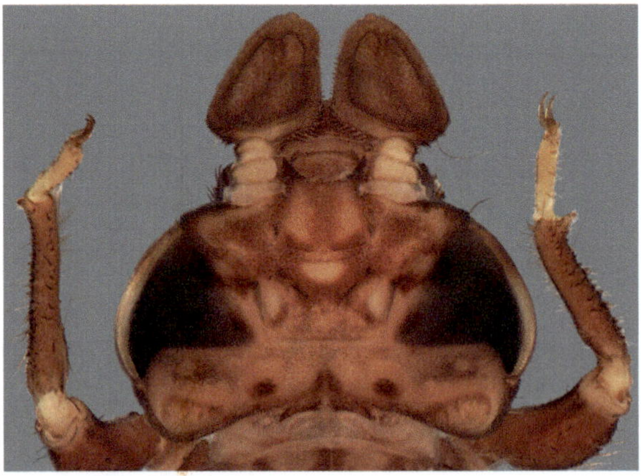

Fig. 5.161 The head and forelegs in dorsal view

Mandibles have the formula: L 1234 0 a($m^{1,2,3}$)b/R 1234 y a($m^{1,2,3,4}$)b, with a > b in both mandibles. The maxilla's galeolacinia is as described for *Acrogomphus*. The ventral pad of the hypopharynx is subpentagonal and yellowish, the basal half is lightly sclerotized, the apical half is soft and whitish, the posterior margin is straight, and a row of short, transverse, delicate setae is on the apical third. Labium: The prementum-postmentum articulation almost reaches the posterior margin of the fore coxae. The prementum is reddish-yellow, longer than wide, the sides are nearly straight along the apical 0.70, but narrowed and convergent along the basal 0.30 (Fig. 5.162); the ventral surface has a poorly developed, longitudinal, median sulcus; the basal margin is deeply notched. The ligula is poorly developed (Fig. 5.162), its apical margin is nearly straight, and it has a ventral row of 3–4 short, subconical, reddish teeth and dorsal rows of whitish, piliform setae. The labial palp has the apical lobe broadly rounded and its mesal margin has a row of 8–10, triangular, sharply pointed teeth. The movable hook is concolorous, shorter than the palp, sharp, and moderately incurved.

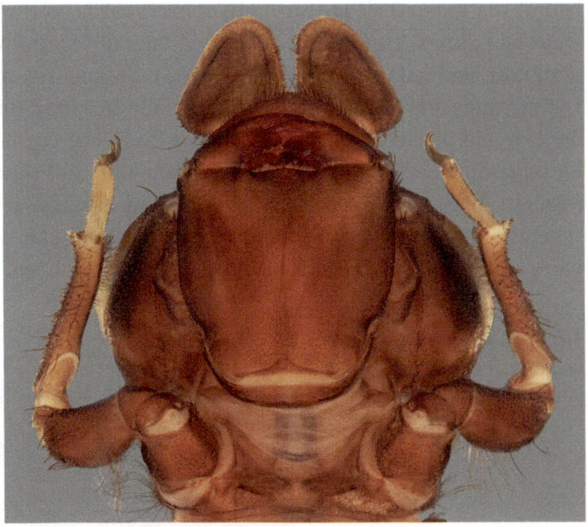

Fig. 5.162 The head, prementum, and prothorax in ventral view

Thorax. The prothorax is narrower than the head, and the ventral margin of the propleuron is setose and finely granulose. The pronotum is mostly granulose, except on the large, depressed, bare, dorsolateral areas; its anterior margin is almost straight, the posterior margin is widely convex, and the lateral margins are rounded and bulging. The pterothorax is granulose. The legs are short (Figs. 5.160 and 5.163) (i.e., when fully extended, the hind legs reach the basal half of S7), granulose, with abundant setae, fully reddish-brown, but the tarsi are pale. The fore- and middle tibiae have burrowing hooks and multiple tubercles on the lateral surfaces (Fig. 5.161); the femur and tibia of the hind legs are the same length; the pretarsal claws are pale and each has a pulvilliform empodium. The anterior wing sheaths are parallel, the posterior wing sheaths are divergent basally and convergent apically, and they almost reach the basal half and posterior margin of S4, respectively (Fig. 5.160). The sternum is very finely granulose. The prosternum has a transverse, thick ridge set with a row of short, robust, stiff setae.

Fig. 5.163 Ventral habitus of the nymph of *Stylogomphus*

Abdomen. The abdomen is more or less parallel-sided and widest at S6; the ventral surface is flat, whereas the dorsal surface is convex and lacks protuberances (Fig. 5.164). Terga 1–10 are granulose, including on the lateral and posterior margins, and have a complex, mottled pattern. S8–S9 end in a short, bluntly pointed spine; the lateral margins of S9 are strongly convergent posteriorly (Fig. 5.160). S10 is very short and almost completely concealed laterally by the spines of S9. The sternum is granulose and reddish-brown (Fig. 5.163). Sterna 3–4 and 7–8 are each divided into five sternites and sterna 5–6 and 9 into three sternites; the median sutures are parallel on 2–4 and 9, and divergent apically on 5–8. The caudal appendages are pyramidal, bluntly pointed (Fig. 5.164), densely granulose, and shorter than the dorsal length of S9+S10. The male's epiproct is nearly as long as its basal width and has a pair of small tubercles at 0.66 of its length. The cerci are triangular with the tips convergent. The paraprocts are the longest of the caudal appendages, and the cerci are the shortest.

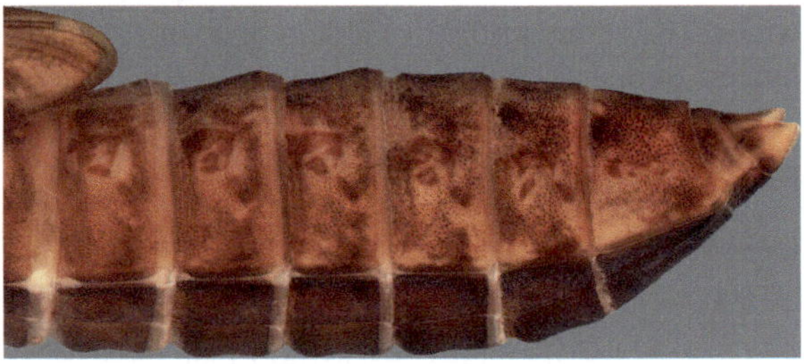

Fig. 5.164 The abdomen and caudal appendages in lateral view

Measurements (in mm). Total length including caudal appendages 17.0–17.5; abdomen 10.7–11.0; maximum width of head 4.0–4.2; length of hind femur 3.1–3.2.

5.27.3 Remarks

Two species of *Stylogomphus* have been recorded in Thailand: *S. malayanus* Sasamoto, 2001, and *S. thongphaphumensis* Chainthong, Sartori and Boonsoong, 2020 (Chainthong et al. 2020). The nymphs are readily recognizable by the shape of the third antennal segment and sterna 3–4 and 7–8 each divided into five sternites. Nymphs of *Stylogomphus* inhabit clear forest streams (Orr 2005) in beds of sand and gravel (Needham et al. 2014), "… often amongst tree roots in upland streams with steep gradient dominated by boulder and cobble substrate" (Wilson 1995).

5.27.4 Distribution

The first record of *Stylogomphus* from Thailand was provided by Matsuki (1988) based on two exuviae collected from Doi Suthep National Park, Chiang Mai Province. Recently, Novelo-Gutiérrez and Sites (2019b) published the first records of *Stylogomphus* for the provinces of Chiang Mai, Chiang Rai, Kamphaeng Phet, Kanchanaburi, Mae Hong Son, Nan, Phetchaburi, Prachuap Khiri Khan, Tak, and Uttaradit. ***New province record:*** Phangnga (Fig. 5.165).

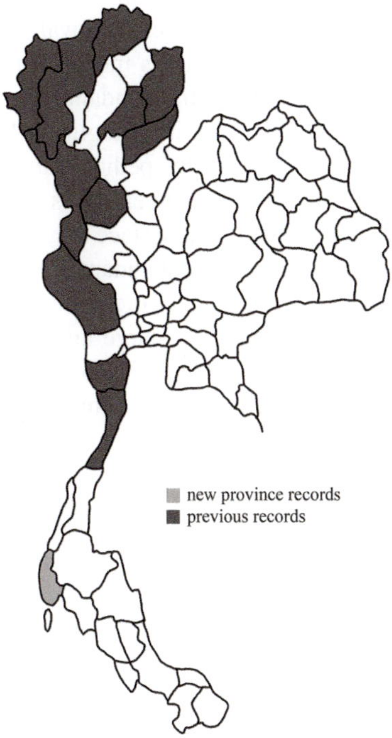

Fig. 5.165 Distribution of *Stylogomphus*

5.28 Genus Undetermined

Fig. 5.166 Dorsal view of the exuvia of an unknown gomphid genus

5.28.1 General Appearance

The nymphs of this undetermined genus are characterized by their medium to large size (27–31 mm), with the body robust, labial palp strongly incurved with a stout end hook, ligula with two small teeth, abdomen flattened, and coloration yellow-brown (Fig. 5.166).

5.28.2 Description

Head. The head is wider than long, but narrower than the pterothorax and abdomen (Fig. 5.166). The labrum is 0.7 mm long, covered abundantly with minute spines, except on a large, central, oval area; its anterior margin is flattened ventrally and setose. The clypeus is mostly bare, the frons, vertex, and occiput appear coarsely granulose and with abundant elongate, scale-like setae, with those on the frons the longest. The anterior margin of the postfrons is shallowly concave. The antennal scape is globose, the pedicel subglobose, and both are covered with elongate scale-like setae; antennomere 3 is the largest, fusiform, strongly flattened, slightly reflexed, its dorsal surface is granulose with elongate, scale-like setae, and the lateral margins have abundant long, white, delicate setae; antennomere 4 is strongly reduced to a conical rudiment and mostly covered by short, scale-like setae. The compound eyes are large and have a tuft of long, reflexed, delicate setae on the anteroventral margin (Fig. 5.167). The occipital lobes are rounded, slightly bulging, and have long, delicate, white setae (Fig. 5.167).

Fig. 5.167 The head and forelegs in dorsal view. Notice the large burrowing hooks

Mandibles have the formula: L 1234 0 a($m^{1,2,3,4,5,6}$)b/R 1234 y a($m^{1,2,3,4}$)b, with a = b in both mandibles. The maxilla's galeolacinia are as described for *Acrogomphus*. The ventral pad of the hypopharynx is subpentagonal, whitish, and mostly soft; it has a basal triangular sclerite on each side of the midline, the posterior margin is sinuate, and the anterior margin has a row of long, stiff setae. Labium: The prementum-postmentum articulation reaches the posterior margin of the fore coxae (Fig. 5.169). The prementum is quadrate, finely serrate, the lateral margins are straight and parallel in the distal half, and the basal margin is widely emarginate and ridged (Fig. 5.168); the ventral surface has a large medial sulcus along the basal 0.60. The ligula is poorly developed and straight (Fig. 5.168), its anterior margin is finely serrulated, and it has a dorsal row of abundant long, piliform setae and two ventral small teeth, one on each side of the midline. The apical lobe of the labial palp is reddish-brown, strongly incurved, ends in a stout, sharp hook (Fig. 5.168), and the mesal margin has a row of 10–11 short, robust, rectangular teeth. The movable hook is reddish-brown, longer than the palp, sharp, and strongly incurved.

Fig. 5.168 The head, prementum, and forelegs in ventral view

Thorax. The prothorax is narrower than the head and setose on the ventral margin of the propleuron. The pronotum is granulose, except on the large, depressed, bare, irregular dorsolateral areas; the anterior margin is straight, the lateral margins are rounded and bulging, and the posterior margin is sinuate. The pterothorax is granulose, but setose on the ventral margins. The hind legs are short (Figs. 5.166 and 5.169) (i.e., when are fully extended, the tips of the claws scarcely reach the posterior margin of S7) and setose; the burrowing hooks are large and well-developed on the fore- and middle tibiae; the pretarsal claws are simple and each has a pulvilliform empodium. The wing sheaths are parallel (Fig. 5.166), their surfaces are sparsely granulose, and the margins have delicate and long setae; the anterior and posterior wing sheaths slightly surpass the posterior margin of S4 and almost reach the basal half of S5, respectively.

Fig. 5.169 The exuvia of an unknown gomphid genus in ventral view

Abdomen. The abdomen is wide, robust, parallel-sided on S4–S7 and then tapers posteriorly (Figs. 5.166 and 5.169); its ventral surface is flat, whereas the dorsal surface is shallowly convex and lacks dorsal protuberances (Fig. 5.170). The tergum is finely granulose including on the posterior margins, but is setose on the sides. Terga 4–7 have a longitudinal suture along the midline. S7–S9 each end in a short, lateral, sharply pointed spine, which increases in size and robustness posteriorly. Sterna 1–5 are light-yellowish-brown, sterna 6–10 are light-reddish-brown (Fig. 5.169), and all are minutely granulose. Sterna 3–8 are each divided into five sternites and sternum 9 into three sternites. The female gonapophyses are conical and rudimentary. The caudal appendages are short, robust, granulose, and as long as the dorsal length of S9 (Fig. 5.170). The epiproct and paraprocts are pyramidal; the paraprocts are slightly longer than the epiproct. The epiproct is 0.2× longer than its basal width. The male's epiproct has a pair of large, rounded tubercles at the basal 0.70. The cerci are triangular and the shortest of the caudal appendages, 0.60× the length of the epiproct.

Fig. 5.170 The abdomen and caudal appendages in lateral view

Measurements (in mm). Total length including caudal appendages 27.0–30.5; abdomen 18.0–20.5; maximum width of head 6.5; length of hind femur 5.0–5.1.

5.28.3 Remarks

The nymph of this undetermined genus closely resembles the general appearance of *Asiagomphus*, with the ligula poorly developed, the apical lobe of the labial palp strongly incurved and with a stout end hook; however, it is easily distinguished from *Asiagomphus* by the presence of two small ventral teeth on the anterior margin of the ligula, the lack of middorsal protuberances, and sterna 3–8 divided into five sternites (*Asiagomphus* has one minute tooth on the anterior margin of the ligula, small middorsal protuberances on S8–S9, and only sterna 7–8 divided into five sternites). Also, it resembles *Burmagomphus* in having a middorsal longitudinal suture on S4–S7 and lateral spines on S7–S9 (*Burmagomphus* has a longitudinal suture on S5–S8). The nymphs here described were collected in a shallow highland creek in Chiang Mai Province.

5.28.4 Distribution

This is the first record of this apparently unknown genus in Thailand. Collection data are Chiang Mai Province, Doi Inthanon National Park, creek at Twin Pagodas, 1 May 2003, 2119 m (18°33'N 98°28'E), coll: UMC and CMU teams (Fig. 5.171).

Fig. 5.171 Distribution of the unknown genus

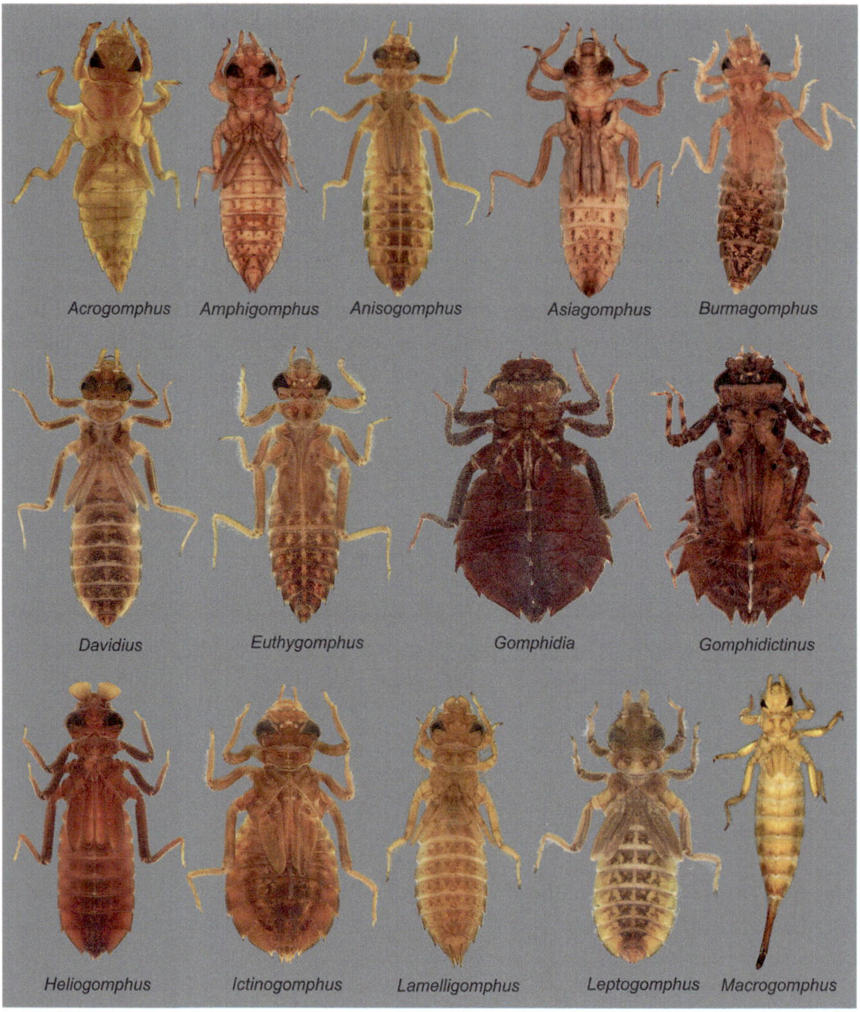

Plate 5.1. Dorsal habitus of Gomphidae nymphs of Thailand. The nymph of *Asahinagomphus* is unknown

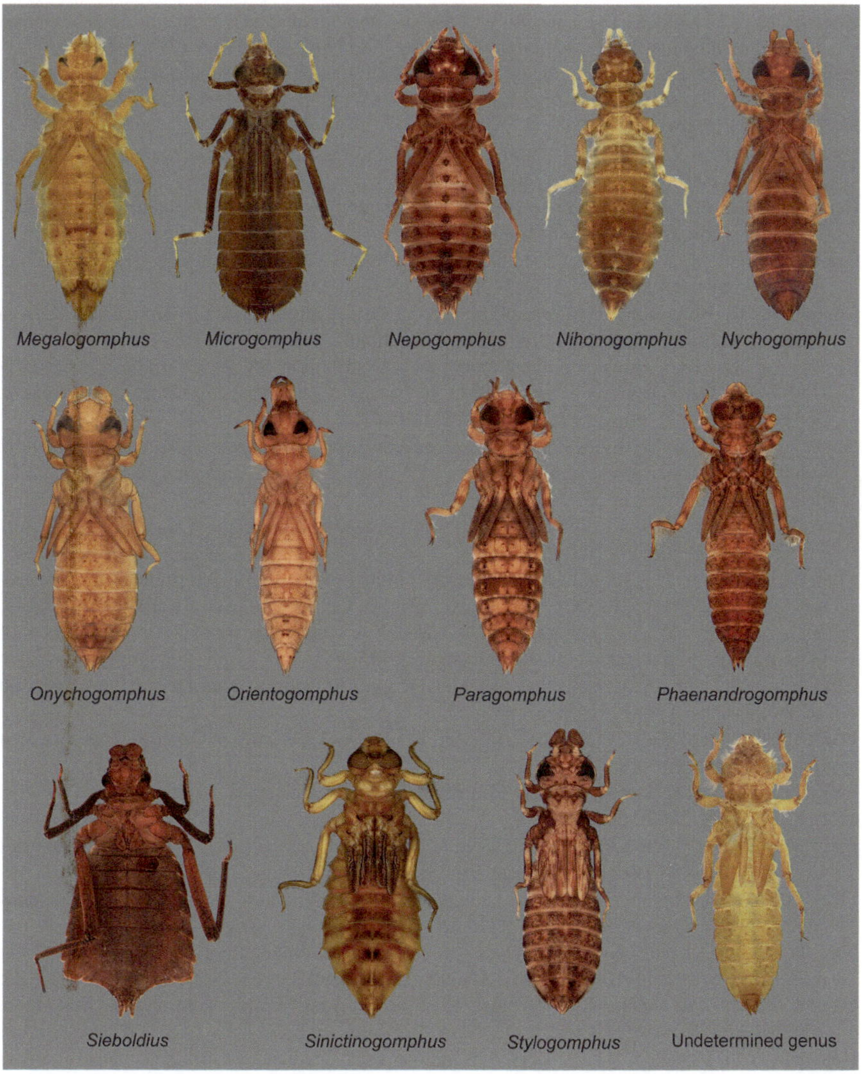

Plate 5.2. Dorsal habitus of Gomphidae nymphs of Thailand. The nymph of *Mattigomphus* is unknown

References

Aoki T. Larval development, emergence and seasonal regulation in *Asiagomphus pryeri* (Selys) (Odonata: Gomphidae). Hydrobiologia. 1999;394:179–92.

Bedjanič B. *Paragomphus campestris* spec. nov. a new endemic dragonfly from Sri Lanka (Anisoptera: Gomphidae). Odonatologica. 2013;42(1):45–53.

Belle J. Higher classification of the South-American Gomphidae (Odonata). Zool. Med. Leiden. 1996;70:297–324.

Boonsoong B, Chainthong D. Description of the final-instar larva of *Heliogomphus selysi* Fraser (Odonata: Gomphidae). Zootaxa. 2014a;3764(4):482–8.

Boonsoong B, Chainthong D. Description of the last stadium larva and female of *Microgomphus thailandica* Asahina, 1981 (Odonata: Gomphidae). Zootaxa. 2014b;3811(2):271–9.

Buppachat N, Inkaew P, Thitiarchagul T, et al. Additions to the Odonata fauna of Thailand (Aeshnidae, Devadattidae, Euphaeidae, Gomphidae, Libellulidae, Platycnemididae). Agrion. 2020;24(3):184–91.

Bybee S, Kalkman V, Erickson RJ, et al. Phylogeny and classification of Odonata using targeted genomics. Mol Phylogenet Evol. 2021;160:1–15.

Carle FL. The classification, phylogeny and biogeography of the Gomphidae (Anisoptera). I. Classification. Odonatologica. 1986;15:275–326.

Carle FL, Kjer KM, May ML. A molecular phylogeny and classification of Anisoptera (Odonata). Arthropod Syst Phylog. 2015;73:281–301.

Chainthong D, Boonsoong B. Description of two final stadium *Onychogomphus* larvae from Thailand (Odonata: Gomphidae). Zootaxa. 2016;4066(5):561–70.

Chainthong D, Boonsoong B. Taxonomy and Distribution of the Gomphid Dragonfly *Orientogomphus minor* (Laidlaw, 1931) (Odonata: Gomphidae) in Thailand. Diversity. 2022;14(291):1–12. https://doi.org/10.3390/d14040291.

Chainthong D, Sartori M, Boonsoong B. *Stylogomphus thongphaphumensis* (Odonata: Anisoptera: Gomphidae), a new gomphid dragonfly and the first record of *S. malayanus* Sasamoto, 2001 from Thailand. Zootaxa. 2020;4763(2):231–45.

Chao H-F. Classification of Chinese dragonflies of the family Gomphidae (Odonata). Part 3. Acta Entomol. Sinica. 1954;4:213–75.

Chao H-F. Notes on the gomphid dragonflies from Fujian Province, with descriptions of a new species and the nymphs of two known species (Odonata, Gomphidae). Wuyi Sci J. 1984;4:151–7.

Davies DAL, Tobin P. The Dragonflies of the World: A systematic list of the extant species of Odonata, vol. 2. SIO Rapid Comm: Anisoptera; 1985.

Day L, Farrell D, Günther A, et al. New provincial records of Odonata from Thailand mostly based on photographs. Agrion. 2012;16(1):16–25.

Dumont HJ, Vierstraete A, Vanfleteren JR. A molecular phylogeny of the Odonata (Insecta). Syst Entomol. 2010;35:6–18.

Ferro ML, Sites RW, Vitheepradit A. Contributions to the faunistics of Odonata in Thailand. Insecta Mundi. 2009;104:1–24.

Ferro ML, Sites RW. Description of the larva of *Gomphidictinus perakensis* (Laidlaw) (Odonata: Gomphidae), with distributional notes. Proc Ent Soc Washington. 2006;108(1):76–81.

Fleck G. *Onychogomphus (Siriusonychogomphus) louissiriusi*, a new species and new subgenus from Thailand (Odonata: Anisoptera: Gomphidae). Faunitaxys. 2020;8(7):1–9.

Fleck G, Ullrich B, Brenk M, et al. A phylogeny of anisopterous dragonflies (Insecta, Odonata) using mtRNA genes and mixed nucleotide/doublet models. J Zool Syst Evol Res. 2008;46(4):310–22.

Hämäläinen M. The species list of Thai dragonflies increase steadily – an update. Malangpo. 2002;19:176–9.

Hämäläinen M, Pinratana BA. Atlas of the dragonflies of Thailand. Distribution maps by provinces. Bangkok: Brothers of Saint Gabriel in Thailand; 1999.

Hämäläinen M, Pinratana BA. Additions and corrections to dragonfly lists of five protected areas in Thailand. Malangpo. 2000;17:156–7.

Katatani N, Kitagawa K, Muraki A, Matuzawa T. Notes on Thai Odonata, Part 3. The Odonata from Chiang Rai. Aeschna. 2004;41:59–70.

Katatani N, Kitagawa K, Muraki A, et al. Notes on Thai Odonata, Part 5. The Odonata from Nan and Phrae. Aeschna. 2010;46:48–58.

Kitagawa K, Katatani N. Notes on Thai Odonata, Part 1. The Odonata from Chanthaburi in March 1998. Aeschna. 2002;39:33–42.

Kiyoshi T, Sota T. Differentiation of the dragonfly genus *Davidius* (Odonata: Gomphidae) in Japan inferred from mitochondrial and nuclear gene genealogies. Zool Sci. 2006;23:1–8.

Kitagawa K, Katatani N. Notes on Thai Odonata, Part 4. The Odonata from Kanchanaburi. Aeschna. 2005;42:23–30.

Kosterin OE. Reconsideration of the genera *Merogomphus* Martin, 1904, and *Anisogomphus* Selys, 1857, including erection of a new genus, with a new species and discussion of additional specimens from Cambodia. Zootaxa. 2016;4171(1):51–76.

Kosterin OE, Makbun N, Dawwrueng P. *Burmagomphus asahinai* sp. nov., a new species from Cambodia and Thailand, with a description of the male of *B. gratiosus* Chao, 1954. Int J Odonatol. 2012;15(4):275–92.

Kosterin OE, Vikhrev NE. Some new provincial records of Odonata made in Thailand in 2007–2009 and a new record for Vietnam. Agrion. 2009;13(2):75–9.

Kosterin OE, Vikhrev N, Zinchenko V, Korshunov A. More dragonfly records from Thailand. Agrion. 2011;15(1):12–6.

Lieftinck MA. Notes on the larvae of two interesting Gomphidae (Odon.) from the Malay peninsula. Bulletin of the Raffles Museum, Straits Settlements. 1932;7:102–15.

Lieftinck MA. Studies on Oriental Gomphidae (Odon.), with descriptions of new and interesting larvae. Treubia. 1941;18:233–53.

Lieftinck MA. Some Gomphidae and their larvae, chiefly from the Malay Peninsula (Odonata). Zool Ver. 1964;69:1–38.

Lieftinck MA, Lien JC, Maa TC. Catalogue of Taiwanese dragonflies (Insecta: Odonata). Taichung: Asian Ecol. Soc; 1984.

Makbun N. *Anisogomphus yingsaki* (Odonata: Gomphidae) sp. nov., a new gomphid species from Thailand. Zootaxa. 2017;4306(3):437–43.

Makbun N, Fleck G. Description of *Microgomphus farrelli* sp. nov. (Odonata: Anisoptera: Gomphidae) based on adults of both sexes and larvae from Northern Thailand. Zootaxa. 2018;4422(3):442–50.

Matsuki K. Larvae of the genus *Stylogomphus* from Thailand. Gekkan-Mushi. 1988;212:38–9.

Michalski J. A manual for the identification of the dragonflies and damselflies of New Guinea, Maluku, and the Solomon Islands. Morristown: Kanduanum Books; 2012.

Needham JG, Westfall MJ Jr, May ML. Dragonflies of North America. The Odonata (Anisoptera) fauna of Canada, the Continental United States, Northern Mexico and the Greater Antilles. Gainesville: Scientific Publishers; 2014.

Ngiam RWJ. *Heliogomphus* cf. *retroflexus* Ris, 1912, (Odonata: Anisoptera: Gomphidae), a possible new record for Singapore. Nature in Singapore. 2010;3:221–5.

Ngiam RWJ, Sun SW, Sek JY. An update of *Heliogomphus* cf. *retroflexus* Ris, 1912 with notes on *Microgomphus chelifer* Selys, 1858 in Singapore (Odonata: Anisoptera: Gomphidae). Nature in Singapore. 2011;4:95–9.

Novelo-Gutiérrez R, Che Salmah MR. Two interesting larvae of *Onychogomphus* from Malaysia (Anisoptera: Gomphidae). Odonatologica. 2013;42(1):31–8.

Novelo-Gutiérrez R, Sites RW. The larva of *Amphigomphus somnuki* Hämäläinen, 1996 and the first records of the genus *Stylogomphus* Fraser, 1922 for Thailand (Odonata: Gomphidae). Zootaxa. 2019;4555(1):121–6.

Novelo-Gutiérrez R, Sites RW. The larvae of *Phaenandrogomphus* Lieftinck, 1964 in Thailand, including the description of *P. tonkinicus* (Fraser, 1926) with larval diagnosis and new province records of *P. asthenes* Lieftinck, 1964 (Odonata: Gomphidae). Zootaxa. 2019b;4700(3):377–84.

Orr AG. A guide to the dragonflies of Borneo. Their identification and biology. Kota Kinabalu, Sabah: Natural History Publications (Borneo); 2003.

Orr AG. Dragonflies of peninsular Malaysia and Singapore. A pocket guide. Kota Kinabalu, Sabah: Natural History Publications (Borneo); 2005.

Paulson D. Dragonflies and damselflies of the West. Princeton: Princeton University Press; 2009.

Paulson DR, Schorr M, Abbott J, et al. World Odonata List. OdonataCentral: University of Alabama; 2023. https://www.odonatacentral.org/app/#/wol/. Accessed 18 Aug 2023.

Sasamoto A. *Anisogomphus yanagisawai* sp. nov., a new gomphid dragonfly from northern Thailand (Odonata: Anisoptera: Gomphidae). Zootaxa. 2015;3904(3):421–6.

Sasamoto A, Kawashima I. Description of the last instar larva of *Amphigomphus nakamurai* Karube, 2001 (Anisoptera, Gomphidae, Onychogomphinae) from Northern Vietnam, Indo-China. Tombo, Matsumoto. 2007;49(1/4):5–9.

Sasamoto A, Kawashima I. Description of the last instar larva of *Nepogomphus walli* (Fraser, 1924) (Anisoptera, Gomphidae, Onychogomphinae) from West Malaysia, southeastern Asia. Tombo, Osaka. 2009;52:7–12.

Suhling F, Martens A. Dragonflies and damselflies of Namibia. Windhoek: Gamsberg Macmillan Publishers; 2007.

Tennessen KJ. Dragonfly nymphs of North America. Springer, Cham: An identification guide; 2019.

Wilson KDP. The gomphid dragonflies of Hong Kong, with descriptions of two new species (Anisoptera: Gomphidae). Odonatologica. 1995;24(3):319–40.

Wilson KDP. Family Gomphidae (Clubtails). In: Silsby J. Collinwood: Dragonflies of the world. CSIRO Publishing; 2001. p. 135–41.

Wilson KDP. Crepuscular activity of *Orientogomphus minor* (Laidlaw) comb. nov. from Thailand and clarification of the taxonomic status of closely related species. Echo. 2008;5:15–21.

Xu Q-H. Description of the last instar larva of *Amphigomphus hansoni* Chao, with notes on the systematic status of the genus *Amphigomphus* Chao (Anisoptera: Gomphidae). Odonatologica. 2012a;41(1):55–9.

Xu Q-H. A description of the final stadium larva of *Leptogomphus elegans* Lieftinck, with a discussion of taxonomic characters of the larvae of the genus *Leptogomphus* Selys (Odonata: Gomphidae). Int J Odonatol. 2012b;15(1):25–9.

Yang B, Davies DAL. Gomphid dragonflies of Yunnan, China, with descriptions of new species and some views on the origin of the group (Anisoptera: Gomphidae). Odonatologica. 1993;22(1):45–62.

Yang B, Davies DAL. Two new species and one new subspecies of Gomphidae from southwestern China, with descriptions of larvae and distribution records (Anisoptera). Odonatologica. 1996;25(3):283–96.

Zhao X-F. The Gomphid Dragonflies of China (Odonata: Gomphidae). Fuzhou: The Science and Technology Publishing House; 1990.

Chlorogomphidae

6

Contents

6.1 Key to Genera of Chlorogomphidae of Thailand 240
6.2 *Chlorogomphus* Selys, 1854 241
  6.2.1 General Appearance 241
  6.2.2 Description 241
  6.2.3 Remarks 245
  6.2.4 Distribution 245
References 247

The Chlorogomphidae is a small family of very large and spectacular dragonflies confined largely to the mountainous regions of tropical Asia (Orr 2003). It has long been considered a subfamily of Cordulegastridae (Karube 2002), but recent molecular studies (Dumont et al. 2010, Fleck 2011, Dijkstra et al. 2013, Carle et al. 2015, Bybee et al. 2021) provided sufficient evidence to support the family rank (Zhang and Tong 2013). The family comprises 3 widely recognized genera with 55 species: *Chlorogomphus* Selys, 1854 (47 species); *Chloropetalia* Carle, 1995 (4 species); and *Watanabeopetalia* Karube, 2002 (4 species) (Paulson et al. 2023). However, Carle (1995) considered the family to be represented by eight genera, six of which he erected: *Aurorachlorus* Carle, *Chlorogomphus* Selys, *Chloropetalia* Carle, *Eorogomphus* Carle, *Indorogomphus* Carle, *Neorogomphus* Carle, *Orogomphus* Selys, and *Sinorogomphus* Carle.

Adult chlorogomphids inhabit submontane and montane well-vegetated forests, with most species preferring open and shallow streams (Fig. 6.1), although some occupy shaded, narrow streams or seepages (Zhang and Tong 2013). Nymphs can be found at stream edges, at pools, or at the bases of waterfalls (Orr 2005) where they burrow into the sand, mud, or gravel (Orr 2003). Nymphs usually are crepuscular, becoming active at dawn and dusk, and are fierce predators, preying on small aquatic insects, fish, and shrimp (Zhang and Tong 2013).

R. Novelo-Gutiérrez, R. W. Sites, *The Dragonfly Nymphs of Thailand (Odonata: Anisoptera)*, https://doi.org/10.1007/978-3-031-33712-3_6

Fig. 6.1 Stream at Namtok Ton Prai at Khao Lampi-Hat Thai Mueang National Park in Phangnga Province, a typical habitat of *Chlorogomphus*

Two genera of Chlorogomphidae inhabit Thailand:

Chlorogomphus Selys with six species recorded: *C. arooni* Asahina, 1981; *Chlorogomphus miyashitai* Karube, 1995; *Chlorogomphus piaoacensis* Karube, 2013; *C. yokoii* Karube, 1995; and two undetermined species

Watanabeopetalia Karube with only one species: *W. atkinsoni* (Selys, 1878) (Hämäläinen and Pinratana 1999, Hämäläinen 2002, as *Chlorogomphus atkinsoni*, Yanagisawa and Sasamoto 2017)

6.1 Key to Genera of Chlorogomphidae of Thailand

1 Ligula median process with tips sharply pointed (Fig. 6.2a); lateral margin of prementum serrulate (Fig. 6.2a); mesal margin of protibia distinctly sinuate; caudal appendages as long as S9 + S10 or shorter . . . . ***Chlorogomphus***

1’ Ligula median process with tips rounded (Fig. 6.2b); lateral margin of prementum with short spiniform setae (Fig. 6.2b); mesal margin of protibia more or less straight; caudal appendages about 0.1× longer than S9 + S10 . . . . ***Watanabeopetalia***

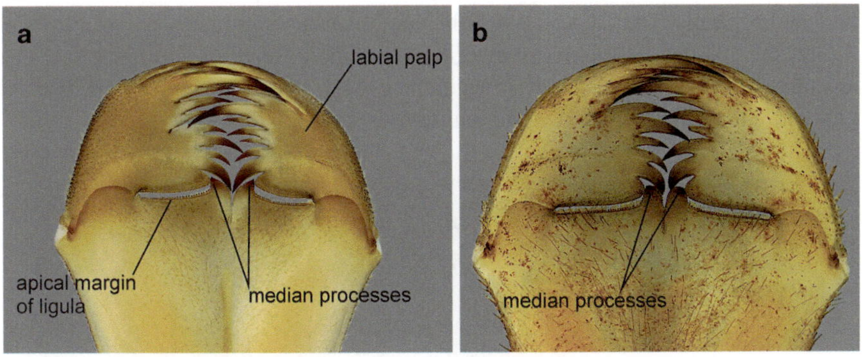

Fig. 6.2 Ligula and palps of *Chlorogomphus* (**a**) and *Watanabeopetalia* (**b**)

6.2 *Chlorogomphus* Selys, 1854

Fig. 6.3 Dorsal habitus of the nymph of *Chlorogomphus*

6.2.1 General Appearance

The nymphs of *Chlorogomphus* and *Watanabeopetalia* are very similar; thus, they are treated together here. They are characterized by their large size (32.0–44.5 mm), body setose and tapering caudally, labial palps strongly developed, legs relatively short, caudal appendages very sharply pointed, and coloration light-yellow to yellow-brown (Fig. 6.3). They differ mainly in the shape of the ligula lobes: sharply pointed in *Chlorogomphus* and rounded in *Watanabeopetalia*.

6.2.2 Description

Head. The head is almost twice as wide as long (Fig. 6.3) and covered with small, cup-like setae, scale-like setae, and spiniform setae (Fig. 6.4), with several bare, smooth areas in a complex arrangement. The labrum has long setae on the distal half including along the anterior margin, which is widely concave at the middle and not flattened ventrally. The clypeus is mostly bare; the frons is produced as a trapezoidal frontal shelf, wider at the anterior margin and mostly covered by cup-like setae. The antennae are very short and seven-segmented with the scape ring-shaped, pedicel globose, and third antennomere the longest.

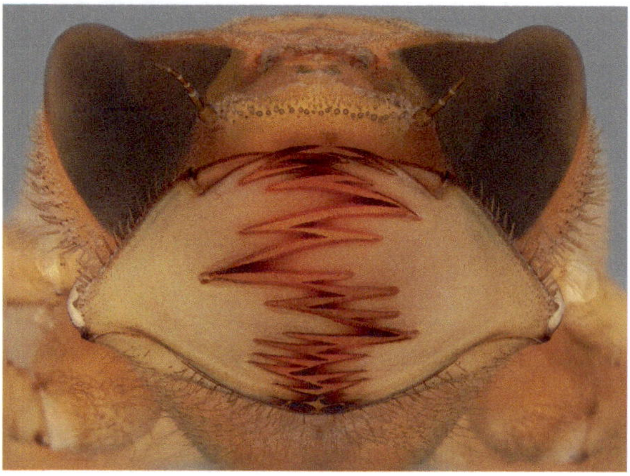

Fig. 6.4 The head of *Chlorogomphus* in frontal view, showing the irregularly jagged apical lobes of the labium

The compound eyes are moderately developed and slightly protruding dorsally. The occiput is very large and strongly bulging laterally with the lateral margins covered with elongate, scale-like setae intermingled with spiniform setae and fine setae on the ventral surface; the dorsal surface is sparsely covered with cup-like setae; an occipital carina continues along the ventral surface of the head. Mandibles have the following formula: L 12'234 0 a($m^{1,2,3,4,5}$)b/R 1234 y ab; the right mandible lacks a molar crest but has tooth "a" very large and tooth "b" smaller; the molar crest of the left mandible is movable, with tooth "b" not in line with *m* denticles, but lower. The maxilla's galeolacinia has seven teeth; the three dorsal teeth are slightly incurved and of similar lengths and robustness, and the four ventral teeth are of different sizes. The ventral pad of the hypopharynx is wide and subrectangular and has abundant stiff setae throughout.

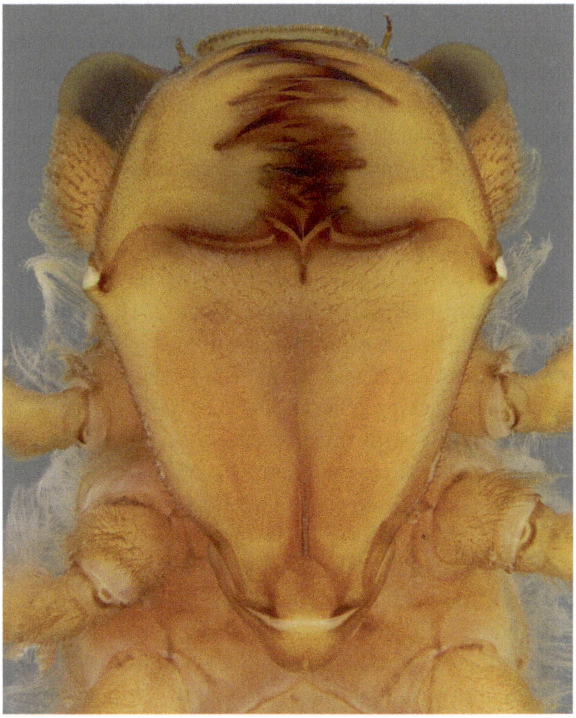

Fig. 6.5 The prementum of *Chlorogomphus* in ventral view

Labium. The prementum-postmentum articulation reaches the middle of the metacoxae. The prementum is subtriangular with the lateral margins widely divergent (Fig. 6.5) and spiniform setae on the apical 0.70; the ventral surface has a longitudinal, bare, central sulcus in the basal half and is covered with abundant long setae; 6 + 5 or 6 + 4 premental setae are on each side of the midline; the ligula is poorly developed and has a V-shaped median process on the apical margin and a row of small, dark teeth intermingled with piliform setae on the apical margin except on the median process. The apical margin of the ligula is shallowly concave and the tips of the median process are sharply pointed in *Chlorogomphus* (Figs. 6.2a and 6.5), whereas the apical margin is straight and the tips of the median process are rounded in *Watanabeopetalia* (Fig. 6.2b). The labial palps are well developed, covering the anterior part of the head as a mask (Fig. 6.4); the apical lobes are divided into 11–12 long, sharply pointed hooks of different sizes and robustness and 1 internal rounded lobe, all of which are finely serrulated on the margins; the dorsal margin of each palp has a tight row of approximately 30 short, stiff setae; the ventral margin is shallowly convex in *Chlorogomphus* and straight in *Watanabeopetalia*; the movable hook is long, slender, and sharply pointed, with short setae, and a row of 4–5 stout spiniform setae is on the dorsomesal margin. There are four palpal setae.

Thorax. The thorax has a large pronotal disc with the lateral margins strongly convex and is fringed with cup-like setae, elongate scale-like setae, and short, fine setae (Fig. 6.3). The dorsum of the pterothorax has spiniform or scale-like setae, and the lateral surface and ventral margins are strongly setose. The legs are setose and relatively short (i.e., when fully extended, the hind legs reach the posterior margin of S8); the mesal margin of the fore tibia is noticeably sinuate in *Chlorogomphus* (Fig. 6.3) and almost straight in *Watanabeopetalia*; the legs are yellow with two dark, incomplete rings on the femora and tibiae; the pretarsal claws are simple, and each has a pulvilliform empodium. The anterior and posterior wing sheaths are yellow and strongly divergent, reaching the posterior margin of S4 and basal half of S5, respectively (Fig. 6.3).

Abdomen. The abdomen is cylindrical, strongly tapering caudally (Fig. 6.3), yellow on S1–S6, and yellow with dark spots on S7–S10; the dark maculation is more abundant in *Watanabeopetalia*. Terga 2–10 are covered with spiniform and scale-like setae. The posterior margins of S2–S10 each has a row of long, scale-like setae or spiniform setae and long, white, fine setae. Dorsal protuberances and lateral spines are lacking. Sterna 1–7 are yellow and 8–10 yellow-brown, and all are densely setose with setae relatively short except on the posterior margins of 6–10 where they are long and stout (Fig. 6.6). Sterna 2–7 are divided into five sternites each (lateralmost sternite is very thin on 7), sternum 8 is divided into three sternites, and the sterna of S9–S10 form a continuous ring with the respective terga. The female gonapophyses are vestigial and those of the male are lacking. A pair of blunt, conical, midventral rudiments are visible on sternum 2 of the male. The caudal appendages (Fig. 6.6) are darker and 0.10× longer than S9 + S10 in *Watanabeopetalia* and as long as or shorter than S9 + S10 in *Chlorogomphus*. The epiproct and paraprocts are pyramidal, the same length, covered abundantly by stout, spiniform, or scale-like setae (Fig. 6.6), with the tips long, reddish-brown, acutely pointed, and slightly recurved. The male's epiproct has one mound-like tubercle on each side of the midline at the basal 0.30. The cerci are conical, very short, 0.10× the length of the epiproct, with the tips moderately recurved (male) or straight (female), and sparsely covered with scale-like setae.

Fig. 6.6 The end of the abdomen in lateral view, showing the cercus, epiproct, and paraproct

Measurements (in mm). The total length including caudal appendages is 32.0–44.5; abdomen (ventral) 21.0–32.0; maximum width of the head 7.8–10.5; and length of hind femur (lateral) 5.2–8.0.

6.2.3 Remarks

Whether Chlorogomphidae deserves the rank of family (Carle 1995) or should be treated as a subfamily of Cordulegastridae (Karube 2002) has been controversial. However, recent molecular studies based on mtRNA (Fleck et al. 2008, Fleck 2011) support the family rank of Chlorogomphidae. In addition, several larval features emphasized here but not mentioned in other studies (i.e., Zhang and Tong 2010, 2013) support the family rank of Chlorogomphidae (those of Cordulegastridae in parentheses): antennae short (long); right mandible without molar crest (with molar crest and *m* denticles 1–4); movable hook of labial palp with a dorsomesal row of 4–5 stout spiniform setae (movable hook smooth); abdominal segment 9 that is a continuous ring (tergum and sternum separated); sterna 2–7 divided into five sternites, with lateralmost sternite very thin on 7, and sternum 8 into three sternites (2–8 in five sternites); male genitalia rudiments visible only on sternum 2 (visible on sterna 2–3); without posterolateral spines on the abdomen (with posterolateral spines on S8–S9 in *Anotogaster* and S9 in *Neallogaster* and with or without in *Cordulegaster*); and female gonapophyses vestigial (well-developed).

6.2.4 Distribution

In Thailand, the species of *Chlorogomphus* and *Watanabeopetalia* have been recorded from the provinces of Chiang Mai (*C. arooni*, *C. miyashitai*, *C. piaoacensis*, *Chlorogomphus* sp., and *W. atkinsoni*), Phrae (*C. arooni*), Chiang Rai and Nan (*C. yokoii*), and Chiang Rai and Tak (*Chlorogomphus* sp.) in the northern region, as well as Chanthaburi (*Chlorogomphus* sp.) in the southeastern region, and Ranong and Surat Thani (*C. arooni*) in the peninsular region (Hämäläinen and Pinratana 1999, 2000, Katatani et al. 2004). ***New province records***: Loei, Mae Hong Son, Phangnga, Phatthalung, Phayao, Phetchabun, Phetchaburi, Phitsanulok, Prachuap Khiri Khan, Satun, and Uttaradit (Figs. 6.7 and 6.8).

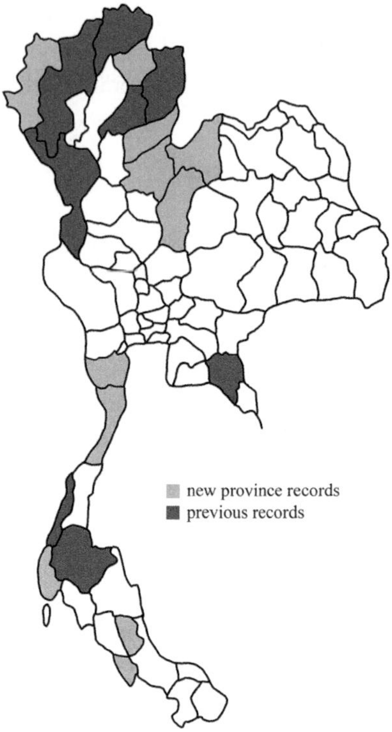

Fig. 6.7 Distribution of *Chlorogomphus*

Fig. 6.8 Distribution of *Watanabeopetalia*

References

Bybee S, Kalkman V, Erickson RJ, et al. Phylogeny and classification of Odonata using targeted genomics. Mol Phylogenet Evol. 2021;160:1–15.

Carle FL. Evolution, taxonomy and biogeography of ancient Gondwanian libelluloids, with comments on anisopteroid evolution and phylogenetics. Odonatologica. 1995;24(4):383–424.

Carle FL, Kjer KM, May ML. A molecular phylogeny and classification of Anisoptera (Odonata). Arthropod Syst Phylog. 2015;73:281–301.

Dijkstra K-DB, Bechly G, Bybee SM, et al. The classification and diversity of dragonflies and damselflies (Odonata). In: Zhang Z-Q (ed) Animal biodiversity: an outline of higher-level classification and survey of taxonomic richness (Addenda 2013). Zootaxa. 2013;3703:36–45.

Dumont HJ, Vierstraete A, Vanfleteren JR. A molecular phylogeny of the Odonata (Insecta). Syst Entomol. 2010;35:6–18.

Fleck G. Phylogenetic affinities of Petaluridae and basal Anisoptera families (Insecta: Odonata). Stutt Beit Nat A, NS. 2011;4:83–104.

Fleck G, Ullrich B, Brenk M, et al. A phylogeny of anisopterous dragonflies (Insecta, Odonata) using mtRNA genes and mixed nucleotide/doublet models. J Zool Syst Evol Res. 2008;46(4):310–22.

Hämäläinen M. The species list of Thai dragonflies increase steadily—an update. Malangpo. 2002;19:176–9.

Hämäläinen M, Pinratana BA. Atlas of the dragonflies of Thailand. Distribution maps by provinces. Bangkok: Brothers of Saint Gabriel in Thailand; 1999.

Karube H. *Watanabeopetalia* gen. nov., a new genus of the dragonflies (Odonata, Cordulegastridae, Chlorogomphinae). Spec Bull Japanese Soc Coleopterol Tokyo. 2002;5:67–85.

Katatani N, Kitagawa K, Muraki A, Matuzawa T. Notes on Thai Odonata, part 3. The Odonata from Chiang Rai Aeschna. 2004;41:59–70.

Orr AG. A guide to the dragonflies of Borneo. Their identification and biology. Sabah: Natural History Publications (Borneo). Kota Kinabalu; 2003.

Orr AG. Dragonflies of peninsular Malaysia and Singapore. A pocket guide. Kota Kinabalu, Sabah: Natural History Publications (Borneo); 2005.

Paulson DR, Schorr M, Abbott J et al. (Coordinators). World Odonata list. Odonata Central, University of Alabama; 2023. https://www.odonatacentral.org/app/#/wol/. Accessed 18 July 2023.

Yanagisawa T, Sasamoto A. New records of *Chlorogomphus miyashitai* Karube, 1995 and *Ch. piaoacensis* Karube, 2013 from northern Thailand, with notes on the characteristics of *Ch. piaoacensis*. Tombo. 2017;59:71–3.

Zhang H-M, Tong X-L. Chlorogomphinae dragonflies of Guizhou Province (China), with first descriptions of *Chlorogomphus tunti* Needham and *Watanabeopetalia usignata* (Chao) larvae (Anisoptera: Cordulegastridae). Odonatologica. 2010;39(4):341–52.

Zhang H, Tong X. Descriptions of the final instar larvae of seven Chinese Chlorogomphidae species, with taxonomic notes on adults (Odonata: Anisoptera). Zootaxa. 2013;3620(2):223–44.

Cordulegastridae 7

Contents

7.1 *Anotogaster* Selys, 1854 250
7.1.1 General Appearance 250
7.1.2 Description 251
7.1.3 Remarks 254
7.1.4 Distribution 254
References 255

The Cordulegastridae is a family of very large dragonflies widely distributed in the temperate and cold regions of the world, mainly in Eurasia (Paulson 2009), but with a few representatives in some tropical mountainous areas. Some authors consider this family almost cosmopolitan, but with no representatives having been reported from smaller islands nor from most of the African continent (Needham et al. 2014). Indeed, it is also completely absent from Oceania, including Australia (Theischinger and Hawking 2006), New Zealand (Rowe 1987, Marinov and Ashbee 2019), the Muller Range of central-western Papua New Guinea (Kalkman et al. 2013), Java, Sumatra (Lieftinck 1954), and Borneo (Orr 2003). The higher classification within this family has been revised several times. Fraser (1929) recognized four genera: *Chlorogomphus* Selys, 1854 (now placed in its own family, Chlorogomphidae); *Anotogaster* Selys, 1854; *Allogaster* Selys, 1878; and *Cordulegaster* Leach, 1815. Carle (1983) recognized *Zoraena* Kirby, 1890 and *Taeniogaster* Selys, 1854, and Lohmann (1992) conducted a phylogenetic study introducing several new genera and resurrecting others, a proposal that has not been widely accepted because the analysis was based on continuous characters or characters with intermediate character states (Tennessen 2004). To date, only 3 genera containing 52 species are widely accepted in the family: *Anotogaster* (14 spp.), *Cordulegaster* (29 spp.), and *Neallogaster* Cowley, 1934 (9 spp.) (Paulson et al. 2023). Cordulegastridae together with Chlorogomphidae and Neopetaliidae have been proposed to constitute the

R. Novelo-Gutiérrez, R. W. Sites, *The Dragonfly Nymphs of Thailand (Odonata: Anisoptera)*, https://doi.org/10.1007/978-3-031-33712-3_7

superfamily Cordulegastroidea (Dijkstra et al. 2013), with Cordulegastridae and Neopetaliidae being sister groups (Fleck 2011).

Fig. 7.1 Stream at Phu Hin Rong Kla National Park, Mhun Daeng Noi, in Phitsanulok Province, a typical habitat of *Anotogaster gregoryi*

Adult cordulegastrids inhabit small, montane, forest streams (Fig. 7.1). They have a ground color of black or dark-brown marked with bright citron-yellow stripes on the thorax and yellow lines, spots, and dots on the abdomen (Fraser 1936). Nymphs live in small, shallow creeks and streams. They are fierce predators, preying on aquatic insects, tadpoles, and small fishes.

7.1 *Anotogaster* Selys, 1854

Fig. 7.2 Dorsal habitus of the nymph of *Anotogaster gregoryi* (after Novelo and Sites 2019, with permission of Magnolia Press)

7.1.1 General Appearance

The nymphs of *Anotogaster* are characterized by their very large size (41–51 mm), body setose and tapering caudally (Figs. 7.2 and 7.6), labial palps large, legs relatively short, caudal appendages sharply pointed, and coloration yellow to reddish-brown or dark-brown.

Fig. 7.3 The head in dorsal view (after Novelo and Sites 2019, with permission of Magnolia Press)

7.1.2 Description

Head. The head is wider than long and covered mainly with small scale-like setae and long, stout, stiff setae on the posterior margin of the occipital lobes. The labrum has long and short, stiff setae and a dense brush of long setae along the widely concave distal margin. The clypeus is mostly bare and the postclypeus has a dense row of long, white setae on the anterior margin.

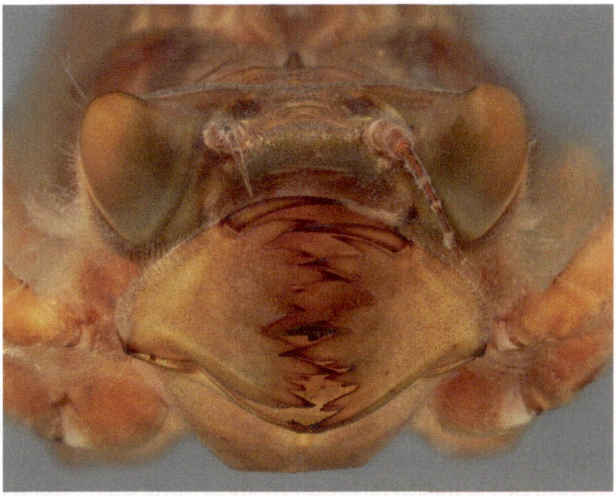

Fig. 7.4 The head in frontal view, showing irregularly jagged apical lobes of the labium and the compound eyes dorsally projected (after Novelo and Sites 2019, with permission of Magnolia Press)

The anterior surface of the frons is large, vertical, and with the anterior margin shallowly convex and beset with a row of long, stout, stiff setae. The antennae are short and seven-segmented (Fig. 7.3), with the scape and pedicel barrel-shaped, the flagellomeres slender and cylindrical, and the antennomere 3 the longest. The vertex is dorsally biconvex and covered with small scale-like setae (Fig. 7.3). The compound eyes are poorly developed and protruding dorsally (Figs. 7.3 and 7.4). The occiput is very large and bulging laterally (Fig. 7.3), with lateral and posterolateral margins covered with short and long, stout, stiff setae; the dorsal surface is densely covered with scale-like setae. Mandibles have the formula L 1234 0 a($m^{1,2,3}$)b/R 1234 y a($m^{1,2,3,4}$)b, with a > b in both mandibles. The maxilla's galeolacinia has seven teeth; the three dorsal teeth are slightly incurved and of similar lengths and robustness; the four ventral teeth are of different sizes. The ventral pad of the hypopharynx is wide, subrectangular, with abundant long setae mainly on the lateral margins, and with a row of stout, stiff setae along the posterior margin directed apically.

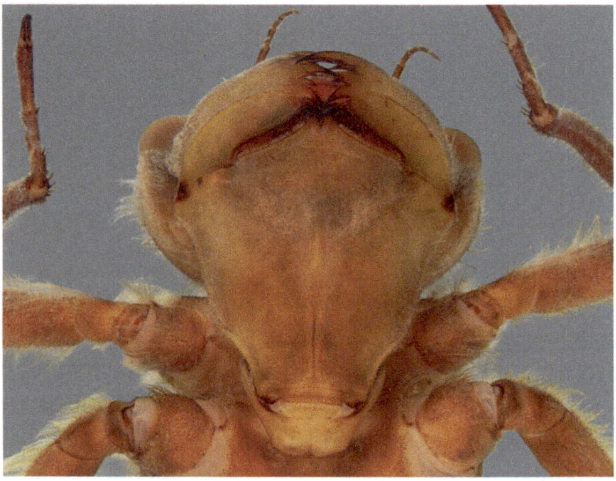

Fig. 7.5 The prementum in ventral view (after Novelo and Sites 2019, with permission of Magnolia Press)

Labium. The prementum-postmentum articulation almost reaches the posterior margin of the mesocoxae. The prementum is subpentagonal with the lateral margins sinuate and divergent distally (Fig. 7.5); it is concave and convergent basally, with minute spiniform setae, which increase in size apically to the base of the palpal articulation; a longitudinal, bare, central sulcus is in the basal third of the ventral surface; the ventral surface is covered moderately with delicate, long, white setae; 5 + 4, 5 + 5, or 6 + 5 premental setae are on each side of the midline; the ligula is well-developed and has a V-shaped median process on the apical margin and a dense row of small, piliform setae except at the median process; the median process is serrulated on the mesal margin and ends in a short spine. The labial palps are strongly developed, covering the anterior part of the head as a mask; the apical lobes are divided into 10–11 long, sharply pointed hooks of different sizes and robustness and 1 rounded lobe, all of which are finely serrulated on the margins; the dorsal margin of each palp has rows of short, stiff setae; the movable hook is long, slender, sharply pointed, and smooth. There are five palpal setae.

Thorax. The thorax is densely covered with minute scale-like setae, and the ventral margin of the pleuron is strongly setose. The prothorax has a large, transverse pronotal disc beset with long, stiff, white setae on its rounded lateral margins; the anterior margin is more or less straight; the posterolateral margins are sinuate; the middle part of the posterior margin is produced caudally as a "pull flap."

Fig. 7.6 Ventral habitus of *Anotogaster gregoryi* (after Novelo and Sites 2019, with permission of Magnolia Press)

The legs are setose and relatively short (e.g., when fully extended, the hind legs barely reach the posterior margin of S8); the posterior margin of the fore tibia is moderately sinuate; the apices of the tibiae and the ventral surfaces of the tarsi have strong, robust spiniform setae; the pretarsal claws are simple and each has a pulvilliform empodium. The anterior and posterior wing sheaths (Fig. 7.2) are yellow and parallel or slightly divergent, reaching the basal half and posterior margin of S4, respectively.

Abdomen. The abdomen is cylindrical, strongly tapering posteriorly (Figs. 7.2 and 7.6), yellow on S1–S3, and reddish-yellow with dark spots on S4–S10. All terga are covered with small, yellow-brown setae and long, fine, golden setae on the sides. The posterior margins of S2–S9 each has a row of small, yellow-brown setae intermingled with long, white, fine setae and scattered, bristle-like setae; the posterior margin of S10 has short, spiniform setae. Dorsal protuberances are lacking (Fig. 7.7). The lateral margin of S9 ends in a short, convergent, acute spine. Sterna 1–8 are orange-yellow and 9–10 reddish-brown (Fig. 7.6); all sterna are covered abundantly with small, spiniform setae; the posterior margins of sterna 1–6 have a row of short, hair-like setae and on 7–10 short, spiniform setae; sterna 2–3 are each divided into three sternites and sterna 4–8 into five sternites, sternum 9 is undivided, and sternum 10 forms a continuous ring with the respective tergum. The female gonapophyses are well-developed; the ventral gonapophyses are the largest and surpass well beyond the posterior margin of sternum 9. The caudal appendages are reddish-brown with the tips darker (Fig. 7.7) and with abundant minute, spiniform setae. The epiproct and paraprocts are pyramidal, the same length, with abundant long setae on the margins, and with the tips long, acutely pointed, and slightly decurved. The male epiproct has a dorsal mound-like tubercle on the basal 0.20. The paraprocts have a row of six large, stout, spiniform setae along the midlateral surface. The cerci are conical and 0.30× as long as the epiproct.

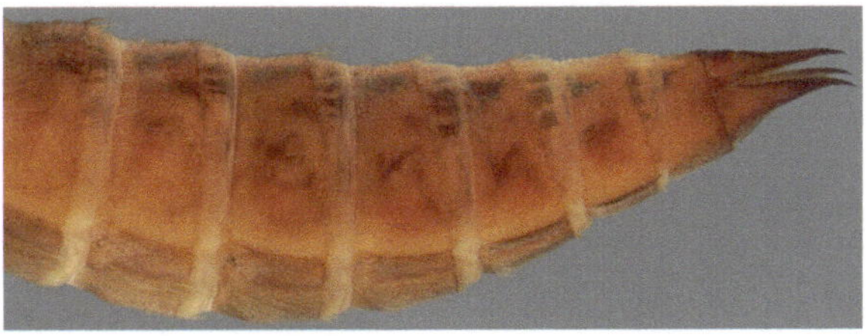

Fig. 7.7 The abdomen of *Anotogaster gregoryi* in lateral view, showing the cercus, epiproct, and paraproct (after Novelo and Sites 2019, with permission of Magnolia Press)

Measurements (in mm). The total length including caudal appendages is 41.0–51.0; abdomen (ventral, including appendages) 29.0–34.2; maximum width of the head 9.2–9.7; and length of hind femur (lateral) 6.5–7.3.

7.1.3 Remarks

Anotogaster gregoryi Fraser, 1924, is the only cordulegastrid recorded from Thailand. The nymphs are found buried in soft beds of mud, silt, or sand, with the tops of the eyes and antennae above the substrate to enable them to detect their prey.

7.1.4 Distribution

In Thailand, *A. gregoryi* has been recorded from the mountains of Chiang Mai Province (Hämäläinen and Pinratana 1999) and more recently from Loei and Phitsanulok provinces (Novelo-Gutiérrez and Sites 2019) (Fig. 7.8).

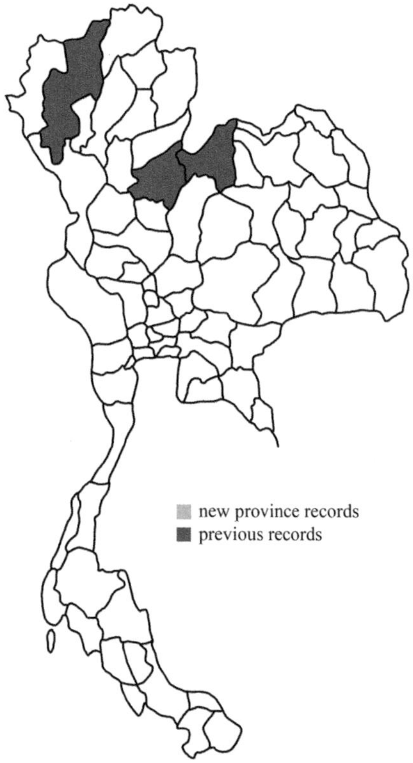

Fig. 7.8 Distribution of *Anotogaster*

References

Carle FL. A new Zoraena (Odonata: Cordulegastridae) from eastern North America, with a key to the adult Cordulegastridae of America. Ann Ent Soc Am. 1983;75(3):335–9.

Dijkstra K-DB, Bechly G, Bybee SM, et al. The classification and diversity of dragonflies and damselflies (Odonata). In: Zhang Z-Q (ed) Animal biodiversity: an outline of higher-level classification and survey of taxonomic richness (Addenda 2013). Zootaxa. 2013;3703:36–45.

Fleck G. Phylogenetic affinities of Petaluridae and basal Anisoptera families (Insecta: Odonata). Stutt Beit Nat A, NS. 2011;4:83–104.

Fraser FC. A revision of the Fissilabioidea (Cordulegasteridae, Petaliidae and Petaluridae) Part 1. Cordulegasteridae. Mem Indian Mus. 1929;9(3):69–167.

Fraser FC. The fauna of British India, including Ceylon and Burma. Odonata, vol. 3. London: Taylor & Francis; 1936.

Hämäläinen M, Pinratana BA. Atlas of the dragonflies of Thailand. Distribution maps by provinces. Bangkok: Brothers of Saint Gabriel in Thailand; 1999.

Kalkman V, Theischinger G, Richards SJ. Dragonflies and damselflies of the Muller range, Papua New Guinea. In: Richards SJ, Gamui BG, editors. Rapid biological assessments of the Nakanai Mountains and the upper Strickland Basin: surveying the biodiversity of Papua New Guinea's sublime karst environments, RAP Bull Biol Assess, vol. 60. Conservation International: Arlington; 2013. p. 175–81.

Lieftinck MA. Handlist of Malaysian Odonata. A catalogue of the dragonflies of the Malay Peninsula, Sumatra, Java and Borneo, including the adjacent small islands. Treubia. 1954;22(Suppl):1–202.

Lohmann H. Revision der Cordulegastridae. 1. Entwurf einer neuen Klassifizierung der Familie (Odonata: Anisoptera). Opusc Flum Zool. 1992;96:1–18.

Marinov M, Ashbee M. Dragonflies and damselflies of New Zealand. Auckland: Auckland University Press; 2019.

Needham JG, Westfall MJ Jr, May ML. Dragonflies of North America. The Odonata (Anisoptera) fauna of Canada, the continental United States, northern Mexico and the greater Antilles. Gainesville: Scientific Publishers; 2014.

Novelo-Gutiérrez R, Sites RW. The probable larva of *Anotogaster gregoryi* Fraser, 1923, with new distributional records of the genus from northern Thailand (Odonata: Cordulegastridae). Zootaxa. 2019;4565(1):138–44.

Orr AG. A guide to the dragonflies of Borneo. Their identification and biology. Kota Kinabalu, Sabah: Natural History Publications (Borneo); 2003.

Paulson DR. Dragonflies and damselflies of the west. Princeton University Press; 2009.

Paulson DR, Schorr M, Abbott J, et al. World Odonata list. OdonataCentral. University of Alabama; 2023. https://www.odonatacentral.org/app/#/wol/. Accessed 18 July 2023.

Rowe R. The dragonflies of New Zealand. Auckland: Auckland University Press; 1987.

Tennessen KJ. *Cordulegaster talaria*, n. sp. (Odonata: Cordulegastridae) from west Central Arkansas. Proc Ent Soc Washington. 2004;106(4):830–9.

Theischinger G, Hawking J. The complete field guide to dragonflies of Australia. Collinwood: CSIRO Publishing; 2006.

Macromiidae

8

Contents

8.1 Key to Genera of Macromiidae of Thailand 258
8.2 *Epophthalmia* Burmeister, 1839 259
8.2.1 General Appearance 259
8.2.2 Description 259
8.2.3 Remarks 262
8.2.4 Distribution 263
8.3 *Macromia* Rambur, 1842 264
8.3.1 General Appearance 264
8.3.2 Description 264
8.3.3 Remarks 267
8.3.4 Distribution 268
References 269

The Macromiidae is a relatively small family comprising 4 genera and 126 species (Paulson et al. 2023) of medium-sized to large dragonflies that are distributed throughout the world except the Neotropical region (Garrison et al. 2006). The macromiids have been traditionally considered by several authors as belonging to the family Corduliidae (e.g., Kirby 1890, Lieftinck 1971, Orr 2003) or even as a subfamily of Libellulidae (e.g., Needham and Westfall Jr 1955). Others have given them the rank of family (e.g., Fraser 1957, Gloyd 1959, Carle and Louton 1994, Michalski 2012). Recent phylogenetic studies based on morphological and molecular analyses consider macromiids to be monophyletic (Ware et al. 2007, Fleck et al. 2008) and the sister group of Corduliidae *sensu stricto* (Dijkstra et al. 2013).

Adult macromiids patrol the shorelines of lakes, ponds, and streams; they have a ground color of black, brown, or metallic-green marked with bright-yellow stripes on the thorax and yellow lines, spots, and dots on the abdomen (Orr 2003, Tze-wai

R. Novelo-Gutiérrez, R. W. Sites, *The Dragonfly Nymphs of Thailand (Odonata: Anisoptera)*, https://doi.org/10.1007/978-3-031-33712-3_8

et al. 2011). Nymphs live in various substrates (Fig. 8.1) partially hidden in soft beds of mud, silt, or sand or among root mats (Lieftinck 1950) or even “amidst curtains or masses of water-weed” (Fraser 1919).

Fig. 8.1 Stream at Tumbon Krung Ching, Klong Nob, Nakhon Si Thammarat Province

Two genera occur in Thailand: *Epophthalmia* Burmeister, 1839, with 2 species, and the more diverse *Macromia* Rambur, 1842, with 13 species known and 6 more yet to be described (Makbun 2022).

8.1 Key to Genera of Macromiidae of Thailand

1. Labial palp strongly developed, moderately reflexed, and lacking setae, distal margin with 6–7 strong, hook-like, asymmetrically developed teeth (Figs. 3.4b and 8.4); premental setae absent; ligula moderately well-developed, biconvex; ventral sulcus of prementum basally bifurcated . . . . ***Epophthalmia***

1’ Labial palp scoop-shaped, with 4–5 long setae, distal margin with 6–7 widely rounded crenations bearing strong setae of different sizes (Fig. 8.10); 5–7 long premental setae on each side of midline; ligula poorly to well-developed, straight, triangular, or convex; ventral sulcus of prementum not basally bifurcated . . . . ***Macromia***

8.2 *Epophthalmia* Burmeister, 1839

Fig. 8.2 Dorsal habitus of the nymph of *Epophthalmia*

8.2.1 General Appearance

The nymphs of *Epophthalmia* are characterized by their large size (30–35 mm), body robust and mostly lacking long setae, labial palps huge, legs very long, caudal appendages sharply pointed, and coloration reddish-brown to dark-brown (Fig. 8.2).

8.2.2 Description

Head. The head is small, roughly rectangular, twice as wide as long, and mostly bare (Fig. 8.3). The labrum has the anterior margin biconvex and widely concave at the middle; it is not flattened ventrally and has some scattered, stiff, short setae. The clypeus is mostly bare. The frons can be moderately produced dorsally in a semicircular, almost bare shelf, with its anterior surface flattened and the dorsal edge slightly thicker or triangular and bluntly pointed. The antennae are short and seven-segmented; the scape is barrel-shaped, the pedicel is cylindrical, the flagellomeres are slender, and the third and seventh antennomeres or only the third antennomere is the longest. The vertex is inflated with a median longitudinal groove, and three ocelli are present.

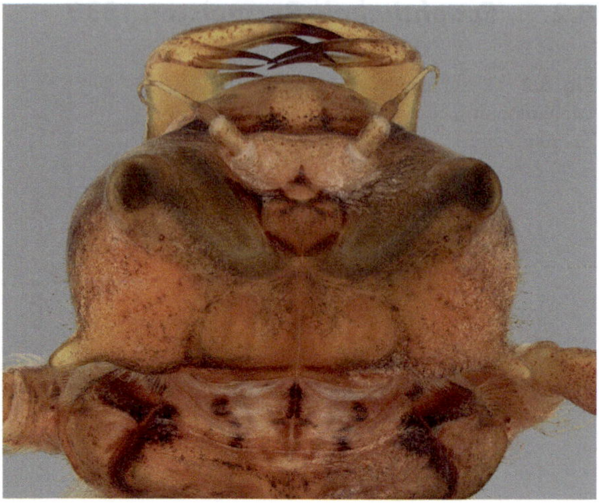

Fig. 8.3 The head and pronotum in dorsal view

The compound eyes are poorly developed, strongly protruding dorsally, and rounded (Figs. 8.3 and 8.4). The occiput is large and covered with scattered short, stiff setae and has straight lateral and posterior margins; on the posterolateral corner is a papillose tubercle of variable length (0.7–1.0 mm) directed laterally and slightly caudally (Fig. 8.3). Mandibles have the formula L 1234 0 ab/R 1234 y ab, with tooth $b > a$ in both mandibles. The maxilla's galeolacinia has seven teeth; the three dorsal teeth are slightly incurved and of similar length and robustness, and the four ventral teeth are of different sizes. The ventral pad of the hypopharynx is semicircular, with abundant long setae along the lateral margins and short, scattered setae on the ventral surface.

Labium. The prementum-postmentum articulation reaches the posterior margin of the mesocoxae. The prementum is subtriangular with the lateral margins smooth and subparallel along the narrow basal 0.30 and then widely divergent along the apical 0.70; a longitudinal, bare, central sulcus is in the basal half of the smooth ventral surface and is divided into two arms at the basal 0.25 (Fig. 8.4); premental setae are absent; the ligula is well-developed, biconvex, and finely serrulated and has a shallow V-shaped cleft between both convexities. The labial palps (Fig. 8.4) are strongly developed, are slightly reflexed, and cover the anterior part of the head as a mask; the mesal margins are divided into six very strong, hook-like, and asymmetrically developed teeth, separated from each other by deep, rounded spaces; the fifth tooth is the longest and widest at the base, and the basal tooth has the mesal margin finely serrulated; the apical 0.30–0.60 of each tooth is dark-brown; palpal setae are absent. The movable hook is relatively short, slender, sharply pointed, and smooth.

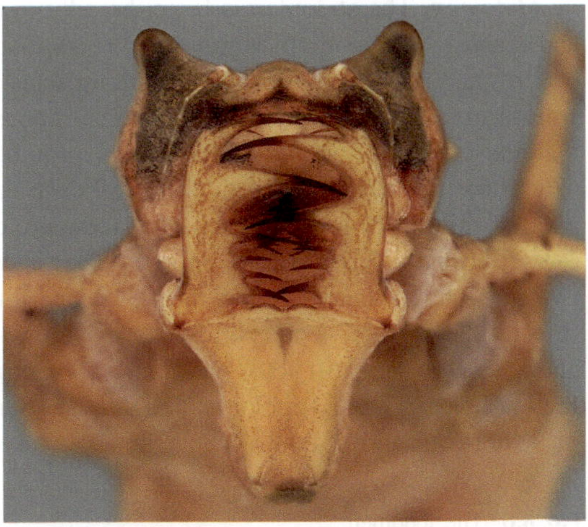

Fig. 8.4 The head in frontal view, showing the compound eyes strongly protruding dorsally, large teeth on the labial palps, and biconvex ligula

Thorax. The prothorax has a large, transverse, more or less flattened pronotal disc (Fig. 8.3); the anterior margin is sinuate, whereas the posterior margin is strongly convex at the middle; the sides are moderately inflated. The propleuron is extended laterally, and the ventral margin is setose. The legs are relatively long (Figs. 8.2 and 8.5) (i.e., when fully extended, the hind femur reaches the anterior margin of S8), the coxae are setose, and the middle and hind femora have three poorly defined, reddish-brown bands; the distal fourth of the fore tibia and the ventral surface of the fore tarsus have tridentate setae; the distal half of the middle tibia and the full length of the hind tibia have stout, short, spiniform setae; the pretarsal claws are simple and long (longer than the basal tarsomeres), and each has a pulvilliform empodium. The anterior and posterior wing sheaths are parallel (Fig. 8.2), spotted with dark-brown, and reach the basal half and posterior margin of S5, respectively.

Fig. 8.5 Ventral habitus of the nymph of *Epophthalmia*

Abdomen. The abdomen is large, broadly ovate, and slightly tapering caudally and reaches its maximum width at S6 (Figs. 8.2 and 8.5); it has a complex color pattern, and most of its surface lacks long setae but is covered with minute, dark, spiniform setae, which increase in size and robustness on the posterior and lateral margins of the terga. Spine-like dorsal protuberances are on S3–S9 or S4–S9 (Fig. 8.6). The lateral margins of S8–S9 each end in a large, acute, more or less parallel spine. S10 is very short. The sterna are orange-yellow and smooth; sterna 3–9 are each divided into three sternites. The caudal appendages are acutely pointed, twice as long as the dorsal length of S10, and form a thick, compact anal pyramid (Fig. 8.6). The epiproct and paraprocts are yellow, pyramidal, and of the same length and have abundant stiff setae on the mesal margins and spiniform setae on the surfaces. The cerci are triangular and slightly shorter than the epiproct and have abundant short, spiniform setae.

Fig. 8.6 The abdomen and caudal appendages in lateral view

Measurements (in mm). The total length including caudal appendages is 30–35; abdomen (ventral, including appendages) 21–26; maximum width of the abdomen 12.3–13.0; maximum width of the head 6.8–7.2; and length of hind femur (lateral) 12.0–14.8.

8.2.3 Remarks

Epophthalmia is a lacustrine genus that is widespread in the Oriental region (Orr 2003). In Thailand, two species have been recorded: *E. frontalis* Selys, 1871, and *E. vittigera* (Rambur, 1842) (Hämäläinen and Pinratana 1999). The nymph of *E. frontalis* is unknown (Bedjanič 2000), but the nymph of *E. vittigera* was described by Lieftinck (1931) by supposition. Nymphs are usually found in the substrates of ponds, lakes, and marshes and covered by mud or sand (Tze-wai et al. 2011).

8.2.4 Distribution

Epophthalmia has been recorded from the provinces of Bangkok, Chachoengsao, Chaiyaphum, Chiang Mai, Kanchanaburi, Lampang, Mae Hong Son, Nakhon Nayok, Phetchaburi, Ranong, Rayong, Sa Kaew, Saraburi, Tak, Trang (Hämäläinen and Pinratana 1999), Chiang Rai (Katatani et al. 2004), Chon Buri (Kosterin 2006), and Nan (Katatani et al. 2010) and more recently from Khon Kaen (Day et al. 2012). ***New province records:*** Kalasin, Loei, Nakhon Si Ayutthaya, Phetchabun, and Phitsanulok (Fig. 8.7).

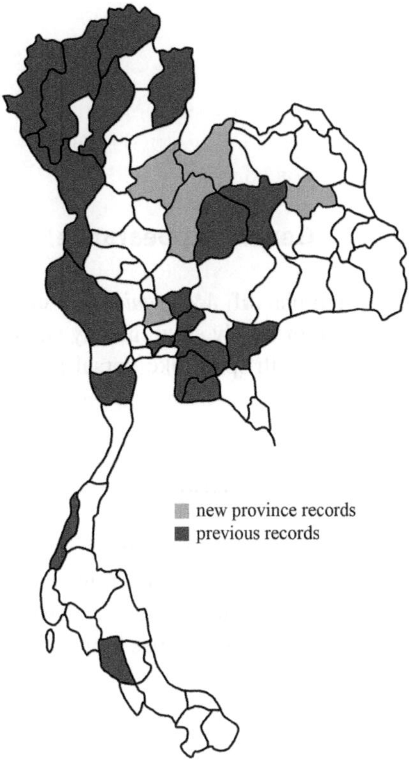

Fig. 8.7 Distribution of *Epophthalmia*

8.3 *Macromia* Rambur, 1842

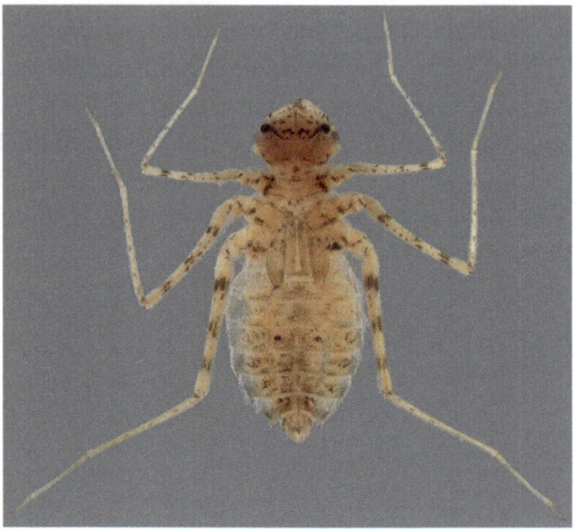

Fig. 8.8 Dorsal habitus of a young nymph of *Macromia*

8.3.1 General Appearance

The nymphs of *Macromia* are characterized by their medium to large size (16–29 mm); body stout, mostly smooth, cylindrical, or depressed; legs very long; abdomen with spine-like dorsal protuberances; and large lateral spines on S8–S9 (Fig. 8.8).

8.3.2 Description

Head. The head is almost twice as wide as long and with the lateral margins either straight and slightly converging posteriorly or strongly convex (Fig. 8.9). The labrum has a small, triangular convexity at the middle of the apical margin, is not flattened ventrally, and has long setae.

Fig. 8.9 The head in dorsal view

The clypeus is bare. The frons is highly variable in shape, with the anterior margin straight, acutely angulate, or with a distinct horn-like tubercle (Fig. 8.10); the entire margin is covered with a row of long, stout setae, and a tuft of short scale-like setae is on the highest part of the horn.

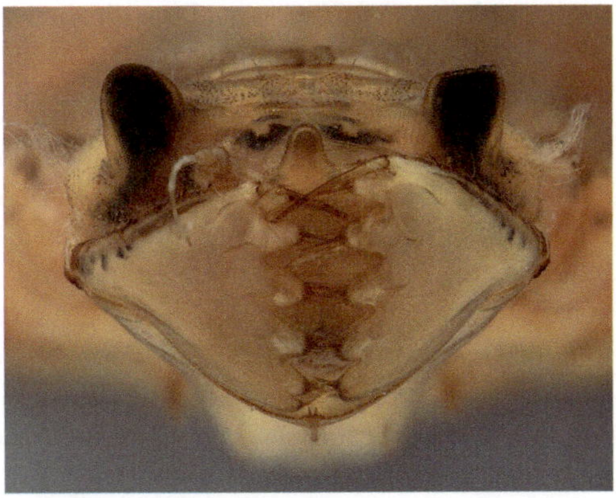

Fig. 8.10 The head in frontal view, showing the horn-like projection on the frons, compound eyes strongly protruding dorsally, and large crenations on the labial palps

The antennae can be short or long (i.e., the tip does not reach the anterior margin of the labrum or surpass it by much) and seven-segmented; the scape is cylindrical, the pedicel is subcylindrical or globose, the flagellomeres are slender, and the size proportions of the antennomeres are highly variable, but usually the third and seventh antennomeres are the longest.

The vertex is small, flat, or slightly raised, and there are three ocelli. The compound eyes are small and strongly protruding dorsally (Fig. 8.10). The occipital lobes are rounded (Fig. 8.9), covered with minute spiniform setae or sparse scale-like setae, and with or without a large dorsal tubercle on each side of the midline; a posterolateral papillose or horn-like tubercle is on each side; the posterior margin of the occiput is widely concave.

Fig. 8.11 The prementum in ventral view

Mandibles have the formula L 1234 0 abk/R 1234 y abk (*M. arachnomima*), or L 1234 0 ab/R 1234 y ab in other species, with tooth $b > a$ in both mandibles. The maxilla's galeolacinia has seven teeth; the three dorsal teeth are slightly incurved and of similar length and robustness, the four ventral teeth are of different sizes, the apical tooth is the largest, and the basal tooth is strongly reduced. The ventral pad of the hypopharynx is subrectangular with abundant long setae on all margins, and the posterior margin is sinuate.

Labium. The prementum-postmentum articulation reaches the middle of the mesosternum. The prementum is wider than long (Fig. 8.11), and the lateral margins either gradually converge toward the basal part and then suddenly narrow at the basal 0.40 or strongly converge from the apical to basal part; it is beset with setae, which become spiniform ventrolaterally; a ventral median sulcus occurs in the basal 0.40, with the surface on each side of the sulcus gently convex or with a strong longitudinal carina in the basal 0.60; 5–7 long setae are on each side of the prementum, with 2–6 small setae closer to the middle; the ligula is poorly or moderately developed, with the anterior margin almost straight, sinuate or widely convex, and finely serrulated or coarsely serrated, and with a dorsal subapical row of setae, which increase in size and robustness to the center. The labial palps are scoop-like and have four to five long setae and one setella close to the palp's articulation; the distal margin has six to seven widely rounded crenations (Fig. 8.10), which bear five strong setae of different sizes; the dorsal margin bears a row of delicate setae or long, stiff setae; the ventral margin is weakly serrulated. The movable hook is relatively short, slender, sharply pointed, and smooth.

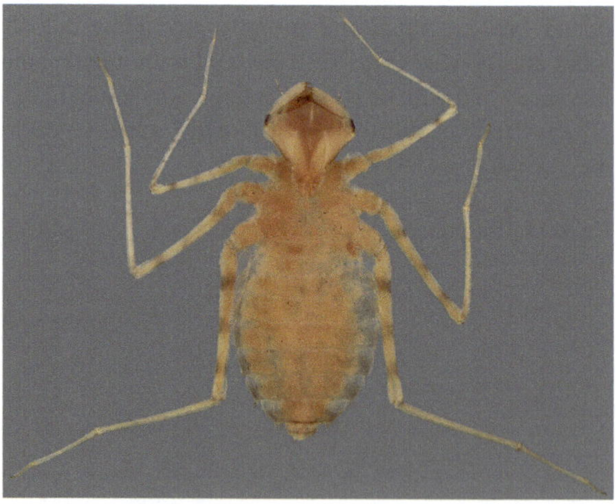

Fig. 8.12 Ventral habitus of a young nymph of *Macromia*

Thorax. The prothorax is narrower than the head. The pronotal disc is rectangular, with or without a low, blunt tubercle on each posterolateral corner. The proepisternum is projected anteriorly on the ventrolateral portion as a large, blunt cone. The meso- and metapleura are shallowly concave. The wing sheaths are parallel, mostly smooth, and the anterior wing sheath does not reach the posterior margins of S5 or S6, whereas the posterior wing sheath surpasses them. The legs are very long (Figs. 8.8 and 8.12) (i.e., when fully extended, the posterior end of the metafemur reaches at least to the posterior margin of S8); the femora and tibiae have either short, delicate setae and spiniform setae or only sparse, very long and stiff setae; these setae are mainly on the dorsal and ventral margins and are noticeably more abundant on the tibiae; the tarsi bear two ventral rows of intermingled hair-like setae and tridentate scale-like setae or only long, stiff setae. The pretarsal claws are very long, sharp, and simple, and each has a pulvilliform empodium.

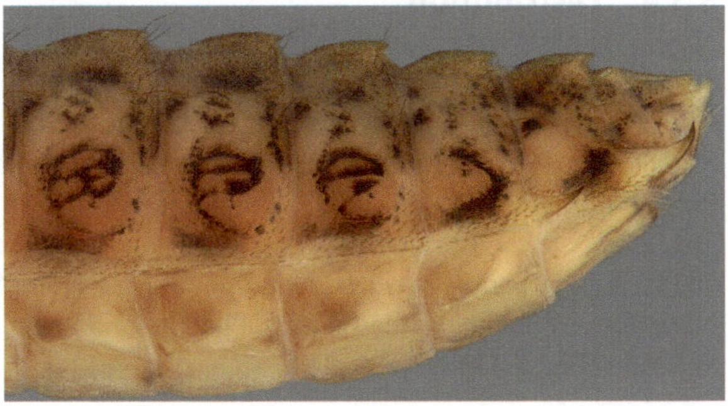

Fig. 8.13 The abdomen and caudal appendages in lateral view

Abdomen. The abdomen is large and either depressed or cylindrical. Dorsal protuberances are on S2–S10, S3–S9, or S5–S10; those on S2–S3 are almost vertical, and those on S3–S9 are variable and can be low and short, caudally directed spines that diminish in size posteriorly or strong and robust spines that increase in size and robustness posteriorly; dorsal protuberances are drastically reduced or lacking on S10. The lateral margins are setose and end in strong incurved spines on S8–S9 (Fig. 8.13). The posterior margins of terga 5–9 have a row of small, spiniform setae. The sterna are densely setose or quite smooth, and sterna 3–9 are each divided into three sternites, with the longitudinal sutures parallel or gradually convergent posteriorly. The female gonapophyses are vestigial and evident as tiny conical rudiments. The caudal appendages are acutely pointed and together are pyramidal or conical; the epiproct can be either wider basally than long and slightly keeled dorsally or longer than its basal width and rounded dorsally and with rows of long, clavate setae; the paraprocts are mostly setose; the cerci are triangular or conical. Usually, the epiproct is the longest of the caudal appendages.

Measurements (in mm). The total length including caudal appendages is 17–21; abdomen (ventral, including appendages) 9–13; maximum width of the abdomen 6–7; maximum width of the head 5.2–5.5; and length of hind femur (lateral) 8.3–9.4.

8.3.3 Remarks

Thirteen species of *Macromia* have been recorded in Thailand, and several undescribed species also have been reported (Hämäläinen and Pinratana 1999, Makbun 2022). The nymphs are mostly stream-dwellers (Lieftinck 1950), although some live in muddy pools (Michalski 2012) where they can be found sprawled on or concealed in the bottom, "among decaying vegetable matter," or "rotten leaves and debris accumulated between boulders in mid-stream" (Lieftinck 1954). *Macromia arachnomima* Lieftinck, 1953, nymphs live "in places where a sufficient current is maintained under all circumstances on exposed mats of stiff, very finely branched rootlets of trees that hung down freely into the water" (Lieftinck 1953). Other nymphs are mostly sand-dwellers.

8.3.4 Distribution

Macromia has been recorded from the provinces of Chaiyaphum, Chanthaburi, Chiang Mai, Chiang Rai, Kanchanaburi, Krabi, Lampang, Nakhon Nayok, Nakhon Ratchasima, Phrae, Ranong, Songkhla, Tak (Hämäläinen and Pinratana 1999), Phitsanulok (Ferro et al. 2009), Nan (Katatani et al. 2010), Loei (Kosterin et al. 2011), and Surat Thani (Day et al. 2012) and more recently from Trang (Muraki 2014). ***New province records:*** Chumphon, Mae Hong Son, Nakhon Si Thammarat, Narathiwat, Phangnga, Phatthalung, Phetchabun, Phetchaburi, Ratchaburi, Uttaradit, and Yala (Fig. 8.14).

Fig. 8.14 Distribution of *Macromia*

References

Bedjanič M. Description of the last larval instar of *Epophthalmia vittata cyanocephala* Hagen, 1867 (Anisoptera: Corduliidae). Odonatologica. 2000;29(1):57–61.

Carle FL, Louton JA. The larva of *Neopetalia punctata* and establishment of Austropetaliidae fam. nov. (Odonata). Proc Ent SocWashington. 1994;96(1):147–55.

Day L, Farrell D, Günther A, et al. New provincial records of Odonata from Thailand mostly based on photographs. Agrion. 2012;16(1):16–25.

Dijkstra K-DB, Bechly G, Bybee SM, et al. The classification and diversity of dragonflies and damselflies (Odonata). In: Zhang Z-Q (ed) Animal biodiversity: an outline of higher-level classification and survey of taxonomic richness (Addenda 2013). Zootaxa. 2013;3703:36–45.

Fleck GB, Ullrich M, Brenk C, et al. A phylogeny of anisopterous dragonflies (Insecta, Odonata) using mtRNA genes and mixed nucleotide/doublet models. J Zool Syst Evol Res. 2008;46(4):310–22.

Fraser FC. Descriptions of new Indian Odonata larvae and exuviae. Rec Indian Mus. 1919;16:459–60.

Fraser FC. A reclassification of the order Odonata. R Zool Soc New South Wales, Sydney; 1957, pp. 1–133.

Garrison RW, von Ellenrieder N, Louton JA. Dragonfly genera of the New World. An illustrated and annotated key to the Anisoptera. Baltimore: The Johns Hopkins University Press; 2006.

Gloyd LK. Elevation of the *Macromia* group to family status (Odonata). Ent News. 1959;70:197–205.

Hämäläinen M, Pinratana BA. Atlas of the dragonflies of Thailand. Distribution maps by provinces. Bangkok: Brothers of Saint Gabriel in Thailand; 1999.

Katatani N, Kitagawa K, Muraki A, Matuzawa T. Notes on Thai Odonata, part 3. The Odonata from Chiang Rai. Aeschna. 2004;41:59–70.

Katatani N, Kitagawa K, Muraki A, et al. Notes on Thai Odonata, part 5. The Odonata from Nan and Phrae. Aeschna. 2010;46:48–58.

Kirby WF. A synonymic catalogue of Neuroptera Odonata. London: Guerney and Jackson; 1890.

Kosterin OE. The journey of a dilettante. Agrion. 2006;10(1):5–7.

Lieftinck MA. A revision of the genus *Epophthalmia* Burm. (Odonata: Corduliinae), with notes on habits and larvae. Treubia. 1931;13(1):21–80.

Lieftinck MA. On south east Asiatic species of *Macromia*. Treubia. 1950;20:677–89.

Lieftinck MA. New dragonflies (Odonata) from Borneo, with notes on their habits and larvae. Treubia. 1953;22:381–406.

Lieftinck MA. Handlist of Malaysian Odonata. A catalogue of the dragonflies of the Malay peninsula, Sumatra, Java and Borneo, including the adjacent small islands. Treubia. 1954;22(1):1–202.

Lieftinck MA. Studies in oriental Corduliidae (Odonata) I. Tijdschr Ent. 1971;114(1):1–63.

Makbun N. *Macromia siamensis*, a new species from North Thailand (Odonata: Macromiidae). Zootaxa. 2022;5133(3):346–54.

Michalski J. A manual for the identification of the dragonflies and damselflies of New Guinea, Maluku, and the Solomon Islands. Kanduanum Books; 2012.

Muraki A. Taxonomic notes on the genus *Macromia* in Thailand (1). Aeschna. 2014;50:101–23.

Needham JG, Westfall MJ Jr. A manual of the dragonflies of North America. The Odonata (Anisoptera). Including the Greater Antilles and the provinces of the Mexican border. Berkeley: University of California Press; 1955.

Orr AG. A guide to the dragonflies of Borneo. Their identification and biology. Kota Kinabalu, Sabah: Natural History Publications (Borneo); 2003.

Paulson DR, Schorr M, Abbott, J et al. (Coordinators). World Odonata list, Odonata Central, University of Alabama; 2023. https://www.odonatacentral.org/app/#/wol/. Accessed 18 July 2023.

Tze-wai T, Ka-king L, Kwan BSP, et al. The dragonflies of Hong Kong. Hong Kong: Friends of the Country Parks, Cosmos Books Ltd; 2011.

Ware JL, May ML, Kjer KM. Phylogeny of the higher Libelluloidea (Anisoptera: Odonata): an exploration of the most speciose superfamily of dragonflies. Mol Phylogen Evol. 2007;45(1):289–310.

Corduliidae

9

Contents

9.1 Key to Genera of Corduliidae of Thailand ... 272
9.2 *Hemicordulia* Selys, 1870 ... 273
9.2.1 General Appearance ... 273
9.2.2 Description ... 273
9.2.3 Remarks ... 276
9.2.4 Distribution ... 277
9.3 *Procordulia* Martin, 1907 ... 277
9.3.1 General Appearance ... 277
9.3.2 Description ... 278
9.3.3 Remarks ... 279
9.3.4 Distribution ... 279
9.4 *Somatochlora* Selys, 1871 ... 280
9.4.1 General Appearance ... 280
9.4.2 Description ... 280
9.4.3 Remarks ... 283
9.4.4 Distribution ... 283
References ... 284

The Corduliidae (*sensu stricto*) is a family of small- to medium-sized dragonflies comprising 22 genera and 164 species (Paulson et al. 2023), which "consist of mostly Holarctic genera, a few Neotropical groups and the predominantly Australasian group around *Hemicordulia*" (Dijkstra et al. 2013). The corduliids have been considered by some authors as a subfamily of Libellulidae (e.g., Fraser 1936, Needham and Westfall Jr 1955, Garrison et al. 2006), whereas others have given them the rank of family (e.g., Lieftinck 1971, Walker and Corbet 1975, Michalski 2012). Recent phylogenetic studies based on morphological and molecular analyses consider Corduliidae as the sister group of Macromiidae (Dijkstra et al. 2013).

R. Novelo-Gutiérrez, R. W. Sites, *The Dragonfly Nymphs of Thailand (Odonata: Anisoptera)*, https://doi.org/10.1007/978-3-031-33712-3_9

Adult corduliids have emerald-green eyes and a ground color of black, brown, or metallic-green marked with pale-whitish or yellow on the thorax and abdomen. Males exhibit an acute anal angle on the hind wings, and small auricles are present on AS2 (Paulson 2009), except in *Hemicordulia* (Orr 2003). Corduliids are strong fliers and spend much of their time hovering, mostly inhabiting submontane torrential streams, but also can be found at lakes, ponds, and bogs. Nymphs live usually in mucky edges of primary forest streams and bogs (Fig. 9.1).

Fig. 9.1 Stream at Doi Suthep National Park, Huai Kaew, above Nam Tok Monthathan, in the Province of Chiang Mai

Three genera occur in Thailand: *Hemicordulia* Selys, 1870, with two species; *Procordulia artemis* Lieftinck, 1930; and *Somatochlora* Selys, 1871, with one undetermined species (Hämäläinen and Pinratana 1999, Hämäläinen 2002, Ferro et al. 2009).

9.1 Key to Genera of Corduliidae of Thailand

1 Long premental setae 6 on each side of midline; 6–7 palpal setae; abdominal surface smooth, almost glabrous . . . . ***Procordulia***

1’ Long premental setae 7–11 on each side of midline; 7–10 palpal setae; abdominal surface covered with minute spiniform setae 2

2(1’) Sternum 3 divided into five sternites; left mandible lacking tooth “y”; body mostly lacking long setae . . . . ***Hemicordulia***

2’ Sternum 3 divided into three sternites; left mandible with tooth “y”; body moderately to strongly setose . . . . ***Somatochlora***

9.2 *Hemicordulia* Selys, 1870

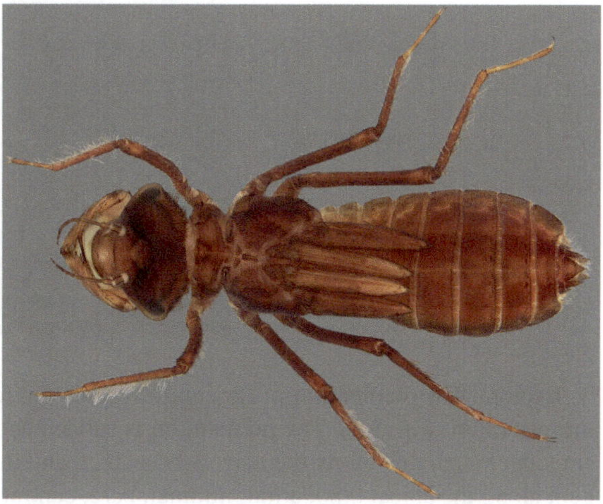

Fig. 9.2 Dorsal habitus of the nymph of *Hemicordulia*

9.2.1 General Appearance

The nymphs of *Hemicordulia* are characterized by their small to medium size (16–25 mm), body robust and mostly lacking long setae, antennae long, prementum with a basoventral sulcus, S10 short and sunken into S9, caudal appendages sharply pointed, and coloration brown to dark-brown (Fig. 9.2).

9.2.2 Description

Head. The head is small, subpentagonal, and twice as wide as long (Fig. 9.3). The labrum has the anterior margin shallowly concave at the middle, not flattened ventrally, and is beset with stiff, long, golden setae. The clypeus is mostly bare. The frons is almost straight along its anterior margin and is beset with long, stiff, yellow setae, and most of its dorsal surface has minute, spiniform setae. The antennae are long and seven-segmented (Fig. 9.3); the scape and pedicel are barrel-shaped, the flagellomeres are cylindrical and slender, and the third antennomere is the longest. The vertex is more or less flat with abundant small setae and three ocelli. The compound eyes are small, rounded, and not prominent. The occiput is large, covered with short, stiff setae, and with lateral margins straight and posterior margin widely concave. Mandibles have the formula L 1234 0 ab/R 1234 y abd, with tooth *d* the largest in both mandibles. The maxilla's galeolacinia has seven teeth; the three dorsal teeth are slightly incurved and of similar lengths and robustness, and the four ventral teeth are of different sizes, with the subbasal tooth directed ventrally. The ventral pad of the hypopharynx is wide and subrectangular and has abundant long setae on the apical half of its surface and along the lateral margins.

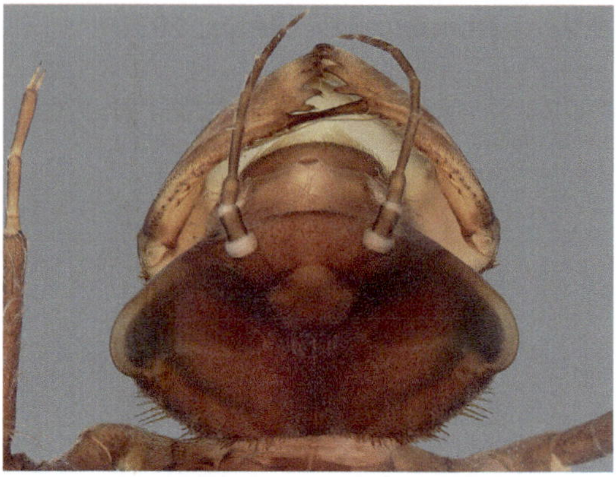

Fig. 9.3 The head in dorsal view. Notice the long antennae

Labium. The prementum-postmentum articulation reaches the middle part of the mesosternum (Fig. 9.4). The prementum is subpentagonal with the lateral margins smooth and parallel along the narrow basal 0.20 and then sinuate and widely divergent along the apical 0.80; a longitudinal, bare, central sulcus is in the basal half of the smooth ventral surface; the premental setae number is 7 + 4 or 7 + 5 on each side of the midline; the ligula is well-developed, triangular, and finely serrulated and has small spiniform setae. The labial palps are spoon-shaped, covering the anterior part of the head as a mask (Fig. 9.5); the apical margins are crenate, with 9–10 finely serrulated crenations; the crenations are separated from each other by deep spaces, and each bears a graded series of 4–6 elongate, spiniform setae; the dorsolateral margin of the palp has a row of short, stout, spiniform setae on the basal half,

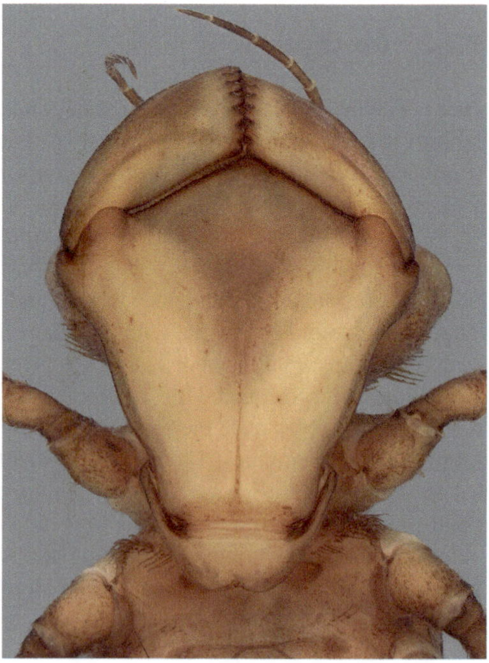

Fig. 9.4 The prementum in ventral view

followed by a row of short, hair-like, delicate setae; the ventral margin is finely serrulated and has a row of elongate spiniform setae; there are 7–9 palpal setae. The movable hook is relatively long, stout, sharply pointed, and smooth.

Fig. 9.5 The head in frontal view. Notice the large crenations on the labial palps

Thorax. The thorax is narrower than the head (Fig. 9.2). The prothorax has a large, transverse, subtrapezoidal pronotal disc, and the anterior and posterior margins of the pronotum are sinuate with the posterior margin strongly convex at the middle; the sides are straight and divergent posteriorly. The ventral margin of the proepisternum is dorsoventrally thick and covered with long, stout, and stiff setae. The legs are relatively short (i.e., when fully extended, the hind femora do not reach the posterior margin of S6) (Fig. 9.6); the femora are spiny on the dorsal and ventral margins; the tibiae are setose, and the distal 0.30 of the fore tibia and ventral surface of the tarsomere 2 of the fore tarsus have tridentate setae; the remainder of the legs have stout, short, and long spiniform setae; the pretarsal claws are simple, and each has a pulvilliform empodium. The anterior and posterior wing sheaths are parallel and reach the posterior margin of S5 and basal half of S6, respectively.

Fig. 9.6 Ventral habitus of the nymph of *Hemicordulia*

Abdomen. The abdomen is fusiform and truncate caudally (Figs. 9.2 and 9.6) and reaches its maximum width at S6. The surface is brown with a large, pale, smooth spot on each side of the midline, and it mostly lacks long setae but is covered with minute, dark, spiniform setae, including along the posterior margins of the terga. The middorsal protuberances can be tiny or well-developed on some segments or lacking (Fig. 9.7); small lateral spines are usually present on S8–S9 or they can be lacking. S10 is very short and sunken into S9. The sterna are light-brown and covered with minute, spiniform setae (Fig. 9.6); sterna 3–6 are each divided into five sternites and sterna 7–9 into three sternites, and sternum 10 forms a continuous ring with the respective tergum. The female gonapophyses are very reduced. The caudal appendages are acutely pointed and 5× the dorsal length of S10, forming a thick and compact anal pyramid. The epiproct and paraprocts are triangular (Figs. 9.2, 9.6, and 9.7) with elongate spiniform setae on the mesal borders and stiff setae on the lateral surfaces; the paraprocts are slightly longer than the epiproct. The cerci are digitiform, slightly shorter than the epiproct, and with some hair-like setae (Fig. 9.7).

Fig. 9.7 The abdomen and caudal appendages in lateral view

Measurements (in mm). The total length including caudal appendages is 20–25; abdomen (ventral, including appendages) 13–15; maximum width of the abdomen 7.6–8.4; maximum width of the head 6.5–6.7; and length of hind femur (lateral) 7.5–8.5.

9.2.3 Remarks

Hemicordulia breeds in mountain lakes and pools in mountain streams in the Oriental region (Fraser 1936), in "soaks and marshes in forest, and at forest edges" (Orr 2005), but also "in stream flowing slowly through low pH swamp forest … and even in quite highly disturbed forest" (Dow 2011). In Thailand, two species have been recorded: *H. tenera* Lieftinck, 1930, and *Hemicordulia* sp. (Hämäläinen and Pinratana 1999). The nymph of *H. tenera* is apparently unknown.

9.2.4 Distribution

Hemicordulia has been recorded from the provinces of Chiang Mai, Loei, and Songkhla (Hämäläinen and Pinratana 1999). ***New province record:*** Phitsanulok (Fig. 9.8).

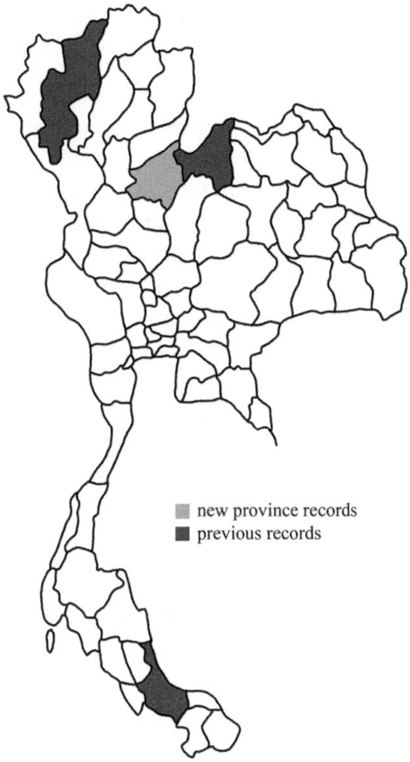

Fig. 9.8 Distribution of *Hemicordulia*

9.3 *Procordulia* Martin, 1907

In the absence of specimens to examine, the description presented here has been adapted from Lieftinck (1933).

9.3.1 General Appearance

The nymphs of *Procordulia* are characterized by their medium size (20 mm), body robust and mostly lacking long setae, antennae long, abdomen lacking middorsal protuberances and posterolateral spines, S10 short and sunken into S9, caudal appendages short and sharply pointed, and coloration dirty yellowish-brown with greenish intermingling.

9.3.2 Description

Head. The head is subpentagonal and about twice as wide as long. The frons has a row of long and very soft hairs on the middle of the anterior margin. The antennae are long and seven-segmented; the scape and pedicel are barrel-shaped, the flagellomeres are cylindrical and slender, and the third antennomere is the longest. The compound eyes are small, rounded, and prominent. The occiput is large and covered with short, stiff setae, the lateral margins and occipital angles are evenly rounded, and the posterior margin is wide and shallowly concave.

Labium. The prementum-postmentum articulation reaches the posterior margin of the mesosternum. The prementum is subpentagonal, slightly wider than long, and the lateral margins are sinuate and widely divergent distally; premental setae number is 6 + 3 or 6 + 4 on each side of the midline; a row of dorsolateral, thick, but very short spiniform setae extends parallel to the lateral margins of the prementum; the ligula is well-developed and broadly triangular and bears several short, irregularly distributed, spiniform setae. The labial palps are subtriangular; the apical margins have 7–8 finely serrulated crenations, which are separated from each other by a distance of twice the length of a crenation, and each bears a graded series of 2–4 elongate, spiniform setae; the dorsolateral margin has a row of about 15 short, stout, spiniform setae in the basal half; the ventral margin is finely serrulated with a row of elongate, spiniform setae; 6–7 (usually 7) palpal setae and a small group of very minute setae are in the middle at the base. The movable hook is relatively long, stout, and sharply pointed.

Thorax. The thorax is narrower than the head and smooth. The prothorax has a large, transverse, subtrapezoidal pronotal disc, with thin tufts of setae only on the dorsolateral margins, and the posterior margin is strongly convex at the middle. The legs are long and slender (i.e., when fully extended, the hind femur reaches the posterior margin of S8); the femora and tibiae are spiny on the dorsal and ventral margins. The anterior and posterior wing sheaths are parallel and reach the posterior margin of S5 and basal half of S6, respectively.

Abdomen. The abdomen is ovate, very broad, and truncate caudally and reaches its maximum width at S7. The surface is conspicuously blotched with pale-yellowish, ochreous, and dark-brown, especially on the lateral stripes of the terga; the middle portion of the abdominal segments is cinnamon-colored with paired basal blocks of reddish-brown on each side of the midline. The lateral stripes are much paler but have large crescent-shaped, black spots bordering the yellow spots, which themselves are double and rounded. Most of the dorsal surface is very smooth, almost glabrous, and lacks long setae but has minute, dark, spiniform setae on the posterior margins of the terga. Middorsal protuberances and posterolateral spines are entirely wanting. S10 is very short and sunken into S9. The sterna are uniformly yellowish-brown. The caudal appendages are short, acutely pointed, and form a thick and compact anal pyramid. The epiproct and paraprocts are triangular, and the paraprocts are slightly longer than the epiproct. The cerci are digitiform and shorter than the epiproct.

Measurements (in mm). The total length including caudal appendages is 20; abdomen (ventral, including appendages) 12.5; maximum width of the abdomen 8.4; maximum width of the head 6.0–6.5; and length of hind femur 7.0–7.3.

9.3.3 Remarks

Only one species of *Procordulia* has been recorded in Thailand: *P. artemis* Lieftinck, 1930 (Ferro et al. 2009). *P. artemis* is confined to montane forest above 1250 m and breeds in peaty bog holes (Orr 2005). Nymphs are bottom-dwellers in shallow crater lakes, with the bottom "very regular and almost flat, and evenly covered with a fairly deep layer of silt or mud" (Lieftinck 1933).

9.3.4 Distribution

Procordulia has been recorded only from Loei Province (Ferro et al. 2009) (Fig. 9.9).

Fig. 9.9 Distribution of *Procordulia*

9.4 *Somatochlora* Selys, 1871

Fig. 9.10 Dorsal habitus of the nymph of *Somatochlora uchidai* (some leg segments lost)

9.4.1 General Appearance

The nymphs of *Somatochlora* are characterized by their small to medium size (17–28 mm), body robust (Fig. 9.10) and moderately to very setose, abdominal dorsal protuberances present or lacking, lateral spines present on S8–S9 or S9 or absent, S10 short and sunken into S9, caudal appendages sharply pointed, and coloration yellowish-brown to dark-brown.

9.4.2 Description

Head. The head is small, subpentagonal, and almost twice as wide as long (Fig. 9.11). The labrum has the anterior border shallowly concave at the middle and not flattened ventrally and is beset with abundant stiff, long, golden setae. The clypeus is mostly bare. The frons is straight to shallowly convex on the anterior margin and beset with long, stiff, yellow setae, and most of its dorsal surface has minute, spiniform setae. The antennae are long (i.e., longer than head length) and seven-segmented; the scape is barrel-shaped, the pedicel is cylindrical, the flagellomeres are cylindrical and slender, and the third antennomere is the longest.

Fig. 9.11 The head and prothorax in dorsal view (antennae broken)

The vertex is more or less flat with abundant small setae and three ocelli. The compound eyes are small, rounded, and not very prominent. The occiput is large (Fig. 9.11), covered with short, stiff setae, and the lateral and posterior margins are straight. Mandibles have the formula L 122'34 y abb'/R 1234 y abd, with $a = b$ in the left mandible, and tooth *d* is the largest in right mandible. The maxilla's galeolacinia has seven teeth; the three dorsal teeth are slightly incurved and of similar lengths and robustness, and the four ventral teeth are of different sizes, with the subbasal tooth directed ventrally. The ventral pad of the hypopharynx is wide and subrectangular and has abundant long setae on the apical half of its surface and on the lateral margins, and the posterior margin is concave.

Labium. The prementum-postmentum articulation reaches the base or middle of the mesosternum. The prementum is subrhomboidal with the lateral margins smooth, sinuate, and widely divergent apically (Fig. 9.12); a longitudinal, bare, central sulcus is in the basal third of the smooth ventral surface; 8–11 long premental setae plus an additional 3–6 small setae are on each side of the midline; the ligula is well-developed, triangular, and finely serrulate and has small spiniform setae.

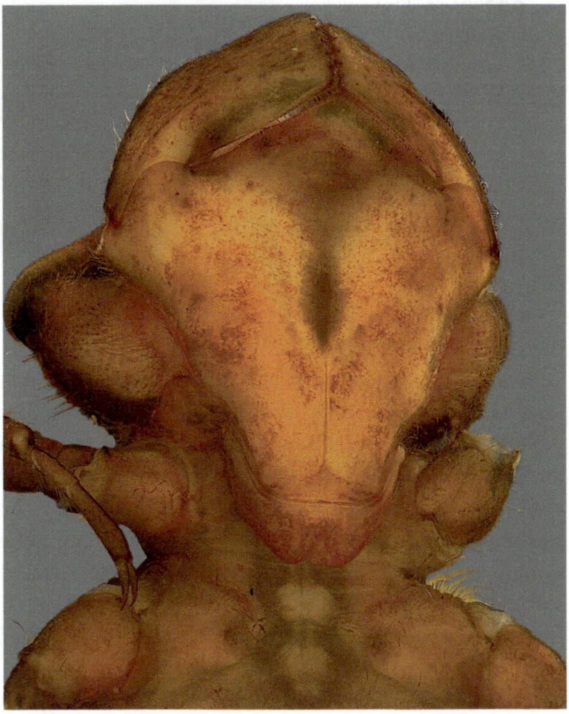

Fig. 9.12 The prementum in ventral view

The labial palps are spoon-shaped, covering the anterior part of the head as a mask; the apical margins are crenulate, with 8–11 finely serrulated crenations, the crenations are separated from each other by shallow spaces, and each bears a graded series of 4–6 elongate spiniform setae; the dorsolateral margin of the palp has a basal short row of stout, spiniform setae followed by a row of long, bristle-like, stiff setae; the ventral margin is finely serrulated with a row of elongate spiniform setae; there are 7–10 palpal setae. The movable hook is relatively short, stout, smooth, and sharply pointed.

Thorax. The thorax is narrower than the head (Fig. 9.10). The prothorax has a large, transverse, subtrapezoidal pronotal disc; the posterolateral corners are somewhat elevated and projecting (Fig. 9.11). The ventral margin of the proepisternum is thick and has long, stout, and stiff setae. The legs are setose with rows of spiniform setae on the femora and tibiae and are relatively long (i.e., when fully extended, the hind femur reaches the posterior margin of S8); the distal fourth of the fore tibia and the ventral surface of the middle tarsomere of the fore tarsus have tridentate setae, and the remainder of the legs have stout, short and long, spiniform setae, or all legs lack tridentate setae; the pretarsal claws are simple and each has a pulvilliform empodium. The anterior and posterior wing sheaths are parallel and usually extend beyond the posterior margin of S5 and almost to the posterior margin of S6, respectively (Fig. 9.10).

Fig. 9.13 Ventral habitus of the nymph of *Somatochlora uchidai* (some tarsi lost)

Abdomen. The abdomen is obovate, truncate caudally, and distinctly wider than the head and thorax, and its greatest width is at S6 (Fig. 9.13); its coloration is yellowish-brown to brown and nearly uniform; usually most of its surface lacks long setae but is covered with minute, dark, spiniform setae including along the posterior margins of the terga. The middorsal protuberances can be well-developed on S3–S9 or S4–S9, reduced to mere knobs, or lacking; lateral spines may be present on S8–S9, only on S9, or lacking. S10 is very short and sunken into S9. The sterna are light-brown, covered with minute and delicate setae on 1–7, and have minute, spiniform setae on 8–9; sterna 3, 7 and 9, or 7–9 are each divided into three sternites, sterna 4–6 and sometimes 8 are divided into five sternites, and sternum 10 forms a continuous ring with the respective tergum. The caudal appendages are acutely pointed, about as long as the dorsal length of S9 + S10 together, and form a thick and compact anal pyramid (Fig. 9.14). The epiproct and paraprocts are pyramidal and have stiff setae on the mesal borders and spiniform setae on the lateral and mesal surfaces; the paraprocts are slightly longer than the epiproct. The cerci are digitiform and slightly shorter than the epiproct (Fig. 9.14) and have some spiniform and hair-like setae.

Fig. 9.14 The abdomen and caudal appendages in lateral view

Measurements (in mm). The total length including caudal appendages is 17–28; abdomen (ventral, including appendages) 10–15; maximum width of the abdomen 6.5–9.1; maximum width of the head 5.6–7.4; and length of hind femur (lateral) 5.2–7.7.

9.4.3 Remarks

Somatochlora typically breeds in bogs, fens, and forest streams and to a lesser extent in lakes (Paulson 2009). In Thailand, only one undetermined species has been recorded (Hämäläinen and Pinratana 1999). Because we were unable to collect any nymphs, the nymphal description provided here is based on specimens of *Somatochlora uchidai* Förster, 1909; *S. clavata* Oguma, 1913; and *S. viridiaenea* (Uhler, 1858) from Japan. Measurements were obtained from Walker and Corbet (1975).

9.4.4 Distribution

Somatochlora has been recorded in Thailand only from Chiang Mai Province (Hämäläinen and Pinratana 1999) (Fig. 9.15).

Fig. 9.15 Distribution of *Somatochlora*

References

Dijkstra K-DB, Bechly G, Bybee SM, et al. The classification and diversity of dragonflies and damselflies (Odonata). In: Zhang Z-Q (ed) Animal biodiversity: an outline of higher-level classification and survey of taxonomic richness (Addenda 2013). Zootaxa. 2013;3703:36–45.

Dow RA. *Hemicordulia tenera*. The IUCN red list of threatened species 2011: e.T174541A7085109; 2011. doi: https://doi.org/10.2305/IUCN.UK.2011-2.RLTS.T174541A7085109.en. Downloaded on 11 May 2018.

Ferro ML, Sites RW, Vitheepradit A. Contributions to the faunistics of Odonata in Thailand. Insecta Mundi. 2009;104:1–24.

Fraser FC. The fauna of British India, including Ceylon and Burma. Odonata, vol. 3. London: Taylor & Francis; 1936.

Garrison RW, von Ellenrieder N, Louton JA. Dragonfly genera of the New World. An illustrated and annotated key to the Anisoptera. Baltimore: The Johns Hopkins University Press; 2006.

Hämäläinen M. The species list of Thai dragonflies increase steadily—an update. Malangpo. 2002;19:176–9.

Hämäläinen M, Pinratana BA. Atlas of the dragonflies of Thailand. Distribution maps by provinces. Bangkok: Brothers of Saint Gabriel in Thailand; 1999.

Lieftinck MA. The life-history of *Procordulia artemis* Lieft. (Odon., Cordul.), with comparative notes on the biology of *P. sumbawana* (Förster). Int Rev Hydrobiol. 1933;28:399–440.

Lieftinck MA. Studies in Oriental Corduliidae (Odonata) I. Tijdschr Ent. 1971;114(1):1–63.

Michalski J. A manual for the identification of the dragonflies and damselflies of New Guinea, Maluku, and the Solomon Islands. Kanduanum Books; 2012.

Needham JG, Westfall MJ Jr (1955) A manual of the dragonflies of North America. The Odonata (Anisoptera). Including the greater Antilles and the provinces of the Mexican Border. University of California Press, Berkeley.

Orr AG. A guide to the dragonflies of Borneo. Their identification and biology. Kota Kinabalu, Sabah: Natural History Publications (Borneo); 2003.

Orr AG. Dragonflies of peninsular Malaysia and Singapore. A pocket guide. Kota Kinabalu, Sabah: Natural History Publications (Borneo); 2005.

Paulson DR. Dragonflies and damselflies of the west. Princeton University Press; 2009.

Paulson DR, Schorr M, Abbott, J et al. (Coordinators). World Odonata list. Odonata Central, University of Alabama; 2023. https://www.odonatacentral.org/app/#/wol/. Accessed 18 July 2023.

Walker EM, Corbet PS. The Odonata of Canada and Alaska. Toronto: University of Toronto Press; 1975.

Libellulidae

10

Contents

10.1 Key to Genera of Libellulidae 292
10.2 *Acisoma* Rambur, 1842 303
10.2.1 General Appearance 303
10.2.2 Description 303
10.2.3 Remarks 306
10.2.4 Distribution 306
10.3 *Aethriamanta* Kirby, 1889 307
10.3.1 General Appearance 307
10.3.2 Description 307
10.3.3 Remarks 310
10.3.4 Distribution 310
10.4 *Agrionoptera* Brauer, 1864 311
10.4.1 General Appearance 311
10.4.2 Description 312
10.4.3 Remarks 314
10.4.4 Distribution 315
10.5 *Brachydiplax* Brauer, 1868 316
10.5.1 General Appearance 316
10.5.2 Description 316
10.5.3 Remarks 319
10.5.4 Distribution 319
10.6 *Brachythemis* Brauer, 1868 320
10.6.1 General Appearance 321
10.6.2 Description 321
10.6.3 Remarks 324
10.6.4 Distribution 324
10.7 *Bradinopyga* Kirby, 1893 325
10.7.1 General Appearance 325
10.7.2 Description 326
10.7.3 Remarks 327
10.7.4 Distribution 327
10.8 *Camacinia* Kirby, 1889 328
10.8.1 General Appearance 328

R. Novelo-Gutiérrez, R. W. Sites, *The Dragonfly Nymphs of Thailand (Odonata: Anisoptera)*, https://doi.org/10.1007/978-3-031-33712-3_10

10.8.2 Description 328
10.8.3 Remarks 330
10.8.4 Distribution 330
10.9 *Cratilla* Kirby, 1900 331
10.9.1 General Appearance 331
10.9.2 Description 332
10.9.3 Remarks 334
10.9.4 Distribution 335
10.10 *Crocothemis* Brauer, 1868 336
10.10.1 General Appearance 336
10.10.2 Description 336
10.10.3 Remarks 339
10.10.4 Distribution 340
10.11 *Diplacodes* Kirby, 1889 341
10.11.1 General Appearance 341
10.11.2 Description 341
10.11.3 Remarks 344
10.11.4 Distribution 345
10.12 *Hydrobasileus* Kirby, 1889 346
10.12.1 General Appearance 346
10.12.2 Description 346
10.12.3 Remarks 350
10.12.4 Distribution 350
10.13 *Hylaeothemis* Ris, 1909 351
10.13.1 General Appearance 351
10.13.2 Description 351
10.13.3 Remarks 354
10.13.4 Distribution 354
10.14 ?*Indothemis* Ris, 1909 355
10.14.1 General Appearance 356
10.14.2 Description 356
10.14.3 Remarks 359
10.14.4 Distribution 359
10.15 *Lathrecista* Kirby, 1889 360
10.15.1 General Appearance 360
10.15.2 Description 360
10.15.3 Remarks 361
10.15.4 Distribution 361
10.16 *Lyriothemis* Brauer, 1868 362
10.16.1 General Appearance 362
10.16.2 Description 362
10.16.3 Remarks 366
10.16.4 Distribution 366
10.17 *Macrodiplax* Brauer, 1868 367
10.17.1 General Appearance 367
10.17.2 Description 367
10.17.3 Remarks 371
10.17.4 Distribution 371
10.18 *Nannophya* Rambur, 1842 372
10.18.1 General Appearance 372
10.18.2 Description 372
10.18.3 Remarks 376
10.18.4 Distribution 376

10.19 *Nannophyopsis* Lieftinck, 1935 ... 377
10.19.1 General Appearance ... 377
10.19.2 Description ... 377
10.19.3 Remarks ... 378
10.19.4 Distribution ... 378
10.20 *Neurothemis* Brauer, 1867 ... 379
10.20.1 General Appearance ... 379
10.20.2 Description ... 380
10.20.3 Remarks ... 383
10.20.4 Distribution ... 384
10.21 *Onychothemis* Brauer, 1868 ... 385
10.21.1 General Appearance ... 385
10.21.2 Description ... 385
10.21.3 Remarks ... 389
10.21.4 Distribution ... 389
10.22 *Orchithemis* Brauer, 1878 ... 390
10.22.1 General Appearance ... 390
10.22.2 Description ... 390
10.22.3 Remarks ... 391
10.22.4 Distribution ... 392
10.23 *Orthetrum* Newman, 1833 ... 392
10.23.1 General Appearance ... 393
10.23.2 Description ... 393
10.23.3 Remarks ... 397
10.23.4 Distribution ... 397
10.24 *Palpopleura* Rambur, 1842 ... 398
10.24.1 General Appearance ... 398
10.24.2 Description ... 399
10.24.3 Remarks ... 402
10.24.4 Distribution ... 402
10.25 *Pantala* Hagen, 1861 ... 403
10.25.1 General Appearance ... 403
10.25.2 Description ... 403
10.25.3 Remarks ... 407
10.25.4 Distribution ... 407
10.26 *Potamarcha* Karsch, 1890 ... 408
10.26.1 General Appearance ... 408
10.26.2 Description ... 409
10.26.3 Remarks ... 413
10.26.4 Distribution ... 413
10.27 *Rhodothemis* Ris, 1909 ... 414
10.27.1 General Appearance ... 414
10.27.2 Description ... 414
10.27.3 Remarks ... 419
10.27.4 Distribution ... 419
10.28 *Rhyothemis* Hagen, 1867 ... 420
10.28.1 General Appearance ... 420
10.28.2 Description ... 420
10.28.3 Remarks ... 424
10.28.4 Distribution ... 424
10.29 *Tetrathemis* Brauer, 1868 ... 425
10.29.1 General Appearance ... 426
10.29.2 Description ... 426
10.29.3 Remarks ... 430
10.29.4 Distribution ... 430

10.30 *Tholymis* Hagen, 1867 431
10.30.1 General Appearance 431
10.30.2 Description 431
10.30.3 Remarks 435
10.30.4 Distribution 435
10.31 *Tramea* Hagen, 1861 436
10.31.1 General Appearance 436
10.31.2 Description 437
10.31.3 Remarks 441
10.31.4 Distribution 441
10.32 *Trithemis* Brauer, 1868 442
10.32.1 General Appearance 442
10.32.2 Description 442
10.32.3 Remarks 446
10.32.4 Distribution 446
10.33 *Urothemis* Brauer, 1868 447
10.33.1 General Appearance 447
10.33.2 Description 448
10.33.3 Remarks 452
10.33.4 Distribution 452
10.34 *Zygonyx* Hagen, 1867 453
10.34.1 General Appearance 453
10.34.2 Description 454
10.34.3 Remarks 458
10.34.4 Distribution 458
10.35 *Zyxomma* Rambur, 1842 459
10.35.1 General Appearance 459
10.35.2 Description 460
10.35.3 Remarks 463
10.35.4 Distribution 464
References 468

The Libellulidae is the largest family of Odonata with about 1034 species in 140 genera (Paulson et al. 2023). The libellulids are small- to medium-sized dragonflies and are distributed around the world. Phylogenetic studies based on morphological and molecular analyses consider Libellulidae to be monophyletic with Macromiidae + Corduliidae as the sister group (Letsch et al. 2009) or Corduliidae (*s. str.*) as the sister group (Ware et al. 2007, Fleck et al. 2008b, Dijkstra et al. 2013).

Libellulid adults are distinguished from other anisopterans by the large eyes meeting dorsally at the top of the head, as well as by the presence of a boot-shaped anal loop in the hind wing venation, although the latter feature is absent in some Oriental genera (i.e., *Nesoxenia*, *Risiophlebia*, *Tetrathemis*, and others). Most of the adult males occurring at lentic waters have brilliant colors, contrasting with those from lotic waters, which have brown or black bodies with pale stripes and spots. Also, many libellulids have the wings strikingly patterned with bands and spots or are fully colored. In addition, there can be a strong dimorphism in color between the sexes. Most of the species are perchers, although those of lotic waters behave mostly as fliers, and many of them exhibit strong territorial behavior. The nymphs are principally bottom-dwellers, inhabiting both lotic and lentic environments such as creeks, waterfalls (Fig. 10.1), streams (Fig. 10.2a), rivers, as well as ponds, swamps, bogs, lakes, and lagoons (Fig. 10.2b); some even inhabit phytotelmata. They are characterized by the spoon-shaped labial palps that cover the anterior part of the head as a mask, the absence of a longitudinal central sulcus on the ventral surface of the prementum, and the lack of male and female gonapophyses.

In total, 45 genera containing 90 species of libellulids have been recorded in Thailand (Plates 4, 5, and 6) (Hämäläinen and Pinratana 1999, Hämäläinen 2002, Sribal et al. 2018, Buppachat et al. 2020, Kompier et al. 2021).

Fig. 10.1 Typical habitat of *Zygonyx* at Sai Yoi Waterfall, Doi Suthep National Park, in Chiang Mai Province

Fig. 10.2 Typical habitats of Libellulidae. Stream at Surat Thani Province (**a**) and pond at Trang Province (**b**)

10.1 Key to Genera of Libellulidae

1 Abdomen lacking posterolateral spines 2

1' At least abdominal segment 9 with posterolateral spines 3

2(1) Long premental setae 7–8 plus 5–6 shorter setae on each side of the midline; 6–7 palpal setae; antennomere 7 longest; in dorsal view, lateral margins of cerci strongly concave; fully developed nymph 9.5–10.3 mm . . . . ***Acisoma***

2' Long premental setae absent, only a set of minute setae on each side of the midline (Fig. 10.3); palpal setae lacking; antennomere 3 longest; in dorsal view, lateral margins of cerci slightly concave at the apex; fully developed nymph 15 mm . . . . ***Hylaeothemis***

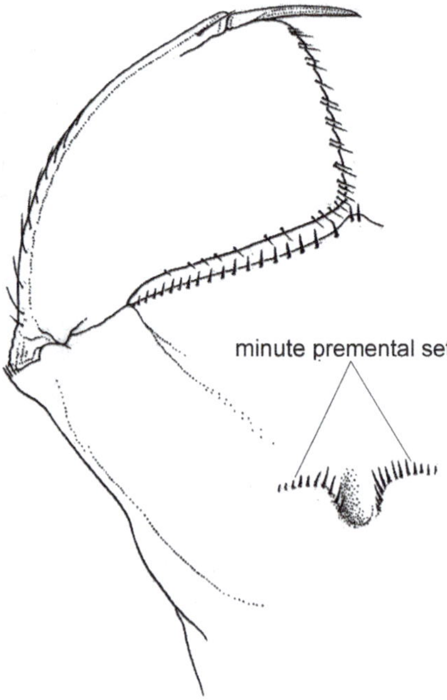

Fig. 10.3 Partial figure of prementum and left labial palp of *Hylaeothemis* (after Sasamoto and Kawashima 2009)

3(1') S8 lacking posterolateral spines, S9 with posterolateral spines short and deflexed; paraprocts hook-like, deflexed; cerci deflexed at the tip (Fig. 10.4) . . . . ***Rhodothemis***

3' S8 with minute to very large posterolateral spines, those on S9 variable; paraprocts variable but never hook-like nor deflexed; cerci variable 4

4(3') Middorsal protuberance on S2 5

4' Middorsal protuberance lacking on S2 8

5(4) Fully developed nymph less than 15 mm; body strongly setose, setae bristle-like, long and black (Fig. 10.19); antennomeres 3 and 5–7 approximately the same size; 7 palpal setae; epiproct 0.7× its basal width . . . . ***Aethriamanta***

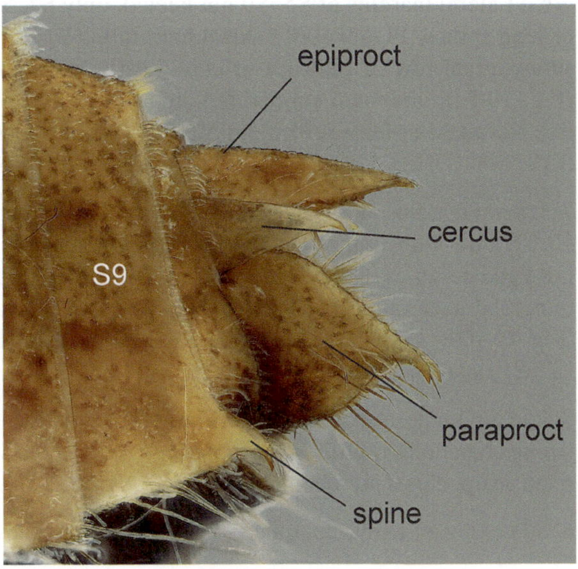

Fig. 10.4 Caudal appendages of *Rhodothemis* in lateral view

5' Fully developed nymph more than 20 mm; body mostly smooth or mainly with spiniform or claviform setae; antennomere 3 longest; ≥8 palpal setae; epiproct as long as or longer than its basal width 6

6(5') Middorsal protuberances small, only on S2–S4; occiput small; occipital lobes poorly developed, lacking tubercles (Fig. 10.126); distal margin of labial palp with 9 strong crenations (Fig. 10.128); 13–14 long palpal setae . . . . ***Pantala***

6' Middorsal protuberances on S2–S8 or S2–S10 of variable sizes; occiput large, occipital lobes well-developed, with a posterolateral tubercle; distal margin of labial palp with obsolete crenations or finely serrulated; 8–11 long palpal setae 7

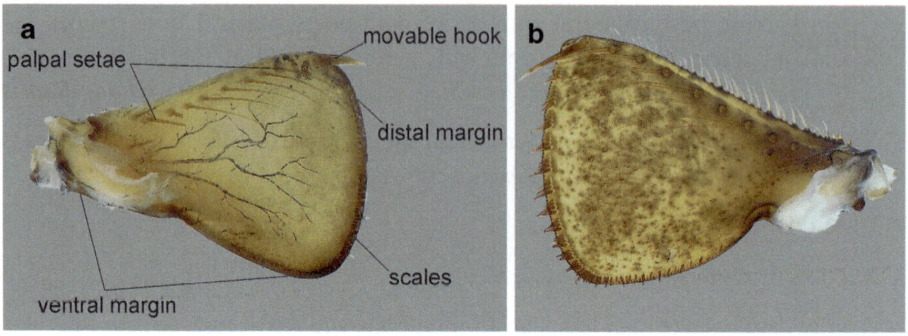

Fig. 10.5 Left labial palp of *Zygonyx* sp. (**a**) and right labial palp of *Onychothemis* sp. (**b**)

7(6') Lateral margins of S5–S9 not keeled, only S7–S9 or S8–S9 ending in short or long spines (Fig. 10.188); distal margin of labial palp finely serrulated with a submarginal row of piliform or claviform setae; movable hook short and thick (Fig. 10.5a); abdomen subparallel-sided (Fig. 10.191); sternum 5 divided into five sternites; at least left mandible with molar tooth "a"; S10 set off from S9 .... ***Zygonyx***

7' Lateral margins of S5–S9 keeled, each ending in a strong spine (Figs. 10.104 and 10.109); distal margin of labial palp with obsolete crenations, each crenation bearing a graded series of 4–5 long spiniform setae (Fig. 10.5b); movable hook long and thin; abdomen almost circular (Fig. 10.108), in lateral view highly arched (Fig. 10.109); sternum 5 divided into three sternites; mandibles lacking molar tooth "a"; S10 sunken into S9 .... ***Onychothemis***

8(4') S7 with posterolateral spines; head wide; distal margin of labial palp with six strong crenations, each crenation bearing one enlarged seta; ligula undulate, lacking setae on tip; dorsal protuberances on S4–S9 .... ***Nannophyopsis***

8' S7 lacking posterolateral spines; other features variable 9

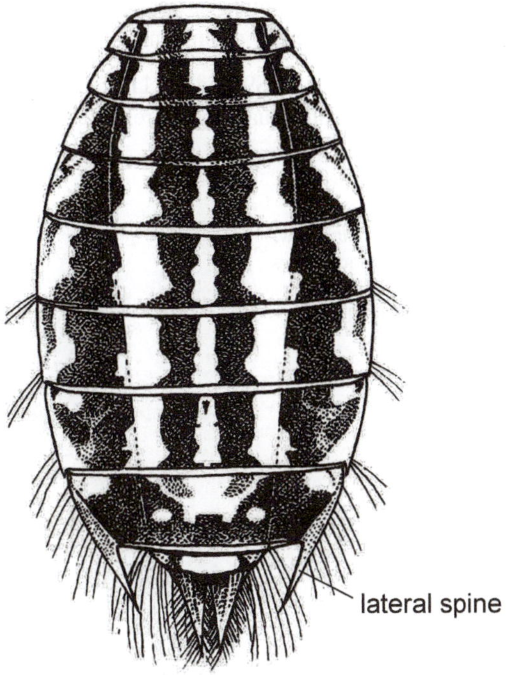

Fig. 10.6 The abdomen of *Camacinia* in ventral view (after Lieftinck published in van Tol [1992])

9(8') Ventral coloration of the abdomen with longitudinal dark and pale stripes (Fig. 10.6); posterolateral spines on S8 minute, those on S9 long and thin (Fig. 10.6); abdomen lacking dorsal protuberances . . . . ***Camacinia***

9' Ventral coloration of the abdomen variable, never as in Fig. 10.6; other combination of features 10

10(9') Anterior margin of ligula crenate (Fig. 10.7); compound eyes small, dorsally prominent (Figs. 10.8, 10.115, and 10.135) 11

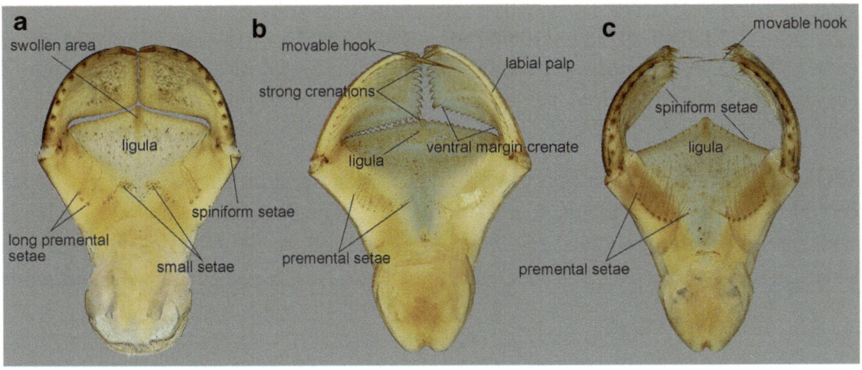

Fig. 10.7 Prementa showing crenate ligulae of *Orthetrum* (**a**), *Cratilla* (**b**), and *Potamarcha* (**c**). All in dorsal view

10’ Anterior margin of ligula variable but never crenate; compound eyes variable in size, laterally or anterolaterally prominent 13

11(10) S10 set off from S9 (Fig. 10.112); epiproct twice as long as its basal width; distal margin of labial palp usually with crenations obsolete (Fig. 10.8), 5–9 palpal setae . . . . ***Orthetrum***

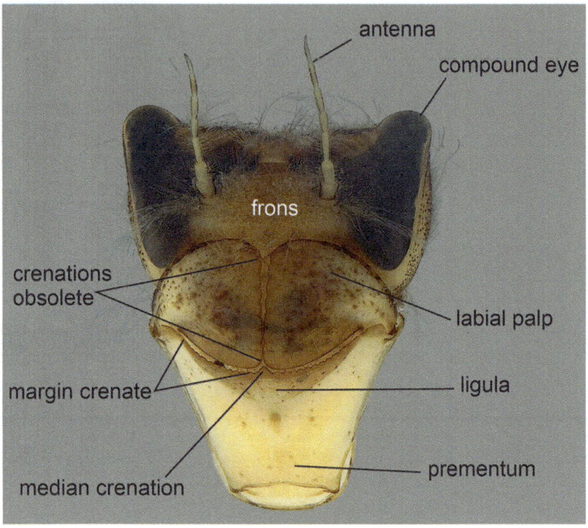

Fig. 10.8 The head of *Orthetrum* in frontal view. Notice the compound eyes produced dorsally and crenations of the labial palp obsolete

11’ S10 sunken into S9 (Figs. 10.45 and 10.132); epiproct 0.30× longer than its basal width; distal margin of labial palp with 10–12 well-developed crenations (Figs. 10.7b and 10.135); number of palpal setae variable 12

12(11’) Ventral margin of labial palp strongly crenate, with a stout spiniform seta at the tip of each crenation (Fig. 10.7b); distal margin of labial palp with 11–12 well-developed crenations; 12–13 palpal setae; anterior margin of ligula strongly crenate (Fig. 10.7b) . . . . ***Cratilla***

12’ Ventral margin of labial palp not crenate, with a row of spiniform setae (Fig. 10.7c); distal margin of labial palp with ten well-developed crenations (Fig. 10.135); eight palpal setae; anterior margin of ligula moderately crenate (Fig. 10.7c) . . . . ***Potamarcha***

13(10’) Long premental setae absent, only 15 small setae on each side of a scooped central area arranged in an inverted “U” (Fig. 10.9); labial palp with 6 short setae; ligula with anterior margin undulate; middorsal protuberances on S4–S9, densely beset with setae. . . . . ***Orchithemis***

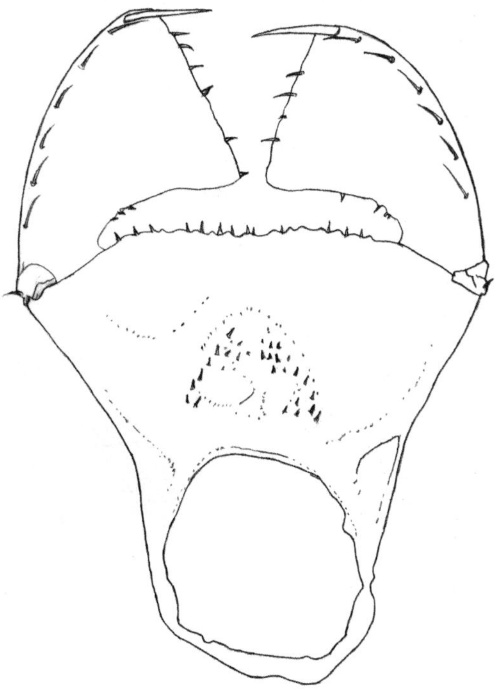

Fig. 10.9 The prementum of *Orchithemis pulcherrima* in dorsal view (redrawn from Butler 2013)

13' At least one long premental seta present on each side of the midline, usually ≥5; labial palp with 5 or more long setae; ligula triangular or widely convex; middorsal protuberances variable. 14

14(13') Compound eyes large with posterolateral corners acute (Fig. 10.10); tip of ligula widely convex; middorsal protuberances well-developed on S3–S8, spine-like and beset with spiniform setae, that on S8 usually surpassing posterior margin of S9 or more; posterolateral spines on S8–S9 large 15

14' Compound eyes variable but posterolateral corners never acute (sometimes laterally pointed in *Macrodiplax* [as in Fig. 10.85]); other features variable 16

15(14) Sternum 9 divided into three sternites; distal margin of labial palp with 9–10 well-developed crenations (Figs. 10.182 and 10.184); anterior margin of metasternellum narrowly rounded, 0.16× the midlength of metasternellum; posterolateral spine lengths on S8–S9 0.6 and 1.0 mm, respectively (Figs. 10.181 and 10.186); epiproct as long as its basal width; cerci 0.85× epiproct length . . . . ***Urothemis***

15' Sternum 9 divided into five sternites; distal margin of labial palp with 9–10 weakly developed crenations (Figs. 10.10 and 10.64); anterior margin of metasternellum straight, 0.50× the midlength of metasternellum; posterolateral spine lengths on S8–S9

3.4 and 3.7 mm, respectively (Figs. 10.61 and 10.63); epiproct 0.60× longer than its basal width; cerci 0.14× epiproct length . . . . ***Hydrobasileus***

Fig. 10.10 The head of *Hydrobasileus croceus* in dorsal view, showing large compound eyes with acute posterolateral corners

16(14') Distal margin of labial palp with crenations well-developed (Fig. 10.156) 17

16' Distal margin of labial palp with crenations obsolete (Fig. 10.11) or weakly developed (Fig. 10.177) 19

17(16) S6–S10 each with a grayish, large, middorsal spot (Fig. 10.72); the tip of ligula rounded (Fig. 10.74); 12 premental setae (7 + 5 or 8 + 4); palpal setae 8–9; distal margin of labial palp with 10–11 small crenations, each crenation bearing only 1 large spiniform setae . . . . ***?Indothemis***

17' S6–S10 midline without grayish spots; the tip of ligula triangular; premental setae 12–14; palpal setae 8–11; distal margin of labial palp crenations each with 2–4 spiniform setae 18

18(17') Middorsal protuberances lacking, but if present, minute and only on S4–S6; 11 palpal setae; 14 premental setae on each side of the midline; crenations on distal margin of labial palp each bearing a pair of spiniform setae; fully developed nymph 16.3–19.0 mm . . . . ***Lathrecista***

18' Middorsal protuberances spine-like on S4–S9 (Fig. 10.158); 8 palpal setae plus 1 setella; 12 premental setae on each side of the midline; 7–8 crenations on distal margin of labial palp each bearing a graded series of 4 spiniform setae; fully developed nymph 12 mm . . . . ***Tetrathemis***

19(16') Epiproct with a dorsal full-length longitudinal carina (as in Fig. 10.160). 20

19' Epiproct in dorsal view without a full-length longitudinal carina 23

20(19) Occiput widely rounded posteriorly (Fig. 10.175); compound eyes large (Fig. 10.175); metasternellum with anterior margin half its midlength; sterna 6–9 each divided into three sternites, sterna 3–4 or 3–5 divided into five sternites; S10 set off from S9 (Fig. 10.174) . . . . ***Trithemis***

20' Occiput more or less angulate posteriorly (Figs. 10.161 and 10.195); compound eyes rather small (Figs. 10.35, 10.161, and 10.195); metasternellum with anterior margin as long as or longer than its midlength; sterna 6–7 each divided into three sternites, sterna 3–5 and 8–9 divided into five sternites; S10 sunken or set off from S9 21

21(20') Middorsal protuberances on S4–S9 well-developed, spine-like (Fig. 10.38); antennomere 6 longest; 11–12 premental setae; 6 palpal setae; posterolateral spines on S8–S9 short, 0.3 and 0.4 mm, respectively (Fig. 10.38); fully developed nymph 13.5–15.3 mm . . . . ***Brachythemis***

21' Middorsal protuberances on S3–S9 or S3–S10, spine-like but low and reduced on S10 (Figs. 10.165 and 10.199); antennomere 3 longest; premental setae variable; palpal setae variable; posterolateral spines on S8–S9 longer than above (Figs. 10.160 and 10.194); fully developed nymph 17.5–20.5 mm 22

22(21') Premental setae 11–13 (8 + 3 to 10 + 3); 8–10 palpal setae; posterolateral spine lengths on S8–S9 large, 0.5–0.6 and 1.0–1.1 mm, respectively (Fig. 10.199); S10 sunken into S9 (Figs. 10.194 and 10.199); epiproct 0.2× longer than its basal width . . . . ***Zyxomma***

22' Premental setae 7–9 (5 + 2 to 7 + 2); five palpal setae; posterolateral spine lengths on S8–S9 short, 0.4 and 0.5 mm, respectively (Fig. 10.165); S10 set off from S9 (Figs. 10.160 and 10.165); epiproct twice as long as its basal width . . . . ***Tholymis***

23(19') Middorsal protuberances present on some abdominal segments 24

23' Middorsal protuberances absent on all abdominal segments 28

24(23) Long premental setae 1–2 plus 3–4 small setae; compound eyes small (Fig. 10.147); middorsal protuberances on S4–S9 short, spine-like; 5–6 palpal setae; sterna 3–5 and 9 each divided into five sternites, sterna 6–8 divided into three sternites; lateral margins of S9 strongly convergent caudally (Fig. 10.146); epiproct as long as its basal width . . . . ***Rhyothemis***

24' Long premental setae 7 or more; compound eyes usually prominent (e.g., Fig. 10.30); middorsal protuberances variable; seven or more palpal setae; sternum 9 divided into three or five sternites; lateral margins of S9 not as above; epiproct longer than its basal width 25

25(24') Middorsal protuberances on S4–S8 small, low (Fig. 10.27); posterolateral spines on S8–S9 short (Fig. 10.24); ligula triangular; 10–11 palpal setae plus 1 setella . . . . ***Agrionoptera***

25' Middorsal protuberances on S3–S8, S4–S8, or S4–S9 well-developed (Figs. 10.32, 10.82, and 10.88); posterolateral spines on S8–S9 or S9 large (Figs. 10.29, 10.78, and 10.84); ligula rounded; 7–12 palpal setae 26

26(25') Sterna S8–S9 each divided into five sternites; paraprocts almost twice as long as epiproct (Fig. 10.32); metasternellum with anterior margin narrow, rounded, ≤0.20× its midlength (Fig. 10.31); 9–12 palpal setae; occiput small; eyes strongly protuberant laterally (Fig. 10.30); antennomere 6 longest . . . . ***Brachydiplax***

26' Sterna S8–S9 each divided into three sternites; paraprocts equal to or slightly longer than epiproct (Figs. 10.82 and 10.88); metasternellum with anterior margin straight, 0.50× its midlength (Fig. 10.87); 7–10 palpal setae; occiput large; eyes variable; antennomeres 6 and 7 longest 27

27(26') Sternum 3 divided into three sternites; middorsal protuberances on S4–S9, semierect, spine-like (Fig. 10.82); metasternellum with anterior margin as long as its midlength; posterolateral spines on S8–S9 incurved (Fig. 10.78), 0.5 and 0.8 mm in length, respectively; palpal setae 7–10 plus one setella .. . . ***Lyriothemis***

27' Sternum 3 divided into five sternites; middorsal protuberances on S3–S8, spine-like, diminishing in height caudally (Fig. 10.88); metasternellum with anterior margin half as long as its midlength; posterolateral spines on S8–S9 straight (Fig. 10.84), 0.6 and 1.1 mm in length, respectively; palpal setae 8 plus one setella . . . . ***Macrodiplax***

28(23') Premental setae 19–21 on each side of midline; 14–16 palpal setae, no setella; S10 set off from S9 . . . . ***Bradinopyga***

28' Premental setae 15 or fewer on each side of midline; 11 or fewer palpal setae with or without 1 setella; S10 sunken into S9 29

29(28') Ligula anterior margin widely convex (Fig. 10.11); posterolateral spines on S8–S9 very large, twice and three times longer the middorsal length of their respective segments, those of S9 reaching the tips of paraprocts (Fig. 10.167); epiproct 0.32× longer than its basal width; fully developed nymph 26.3–27.5 mm . . . . ***Tramea***

29' Ligula anterior margin with sides straight and convergent to median apex (e.g., Fig. 10.121); posterolateral spines on S8–S9 short; epiproct variable; fully developed nymph less than 20 mm 30

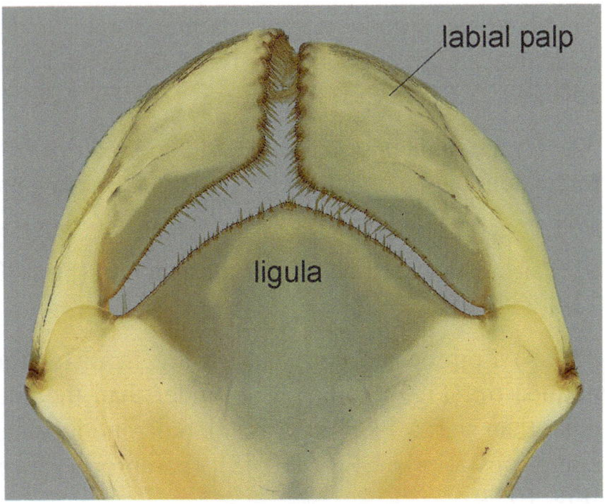

Fig. 10.11 The ligula and palps of *Tramea* in ventral view

30(29') Premental setae 9–10 (6 + 3 or 6 + 4); six palpal setae; sternum 3 divided into three sternites; epiproct as long as its basal width (Fig. 10.90); fully developed nymph 7.5–11.0 mm . . . . ***Nannophya***

30' Premental setae 12–15 (8 + 4 to 10 + 5); 9–11 palpal setae; sternum 3 divided into five sternites; epiproct longer than its basal width (e.g., Fig. 10.50); fully developed nymph 11.7–18.0 mm 31

31(29') Mandibles with incisor teeth largely fused, usually rounded on distal margin (Fig. 10.12a), with a large, mound-like, subapical swelling on lateral surface (Fig. 10.12b); abdomen dorsally concolorous (Fig. 10.50); premental setae 14–15 (10 + 4 or 10 + 5); sternum 9 divided into five sternites (anterolateral sternite very small) . . . . ***Crocothemis***

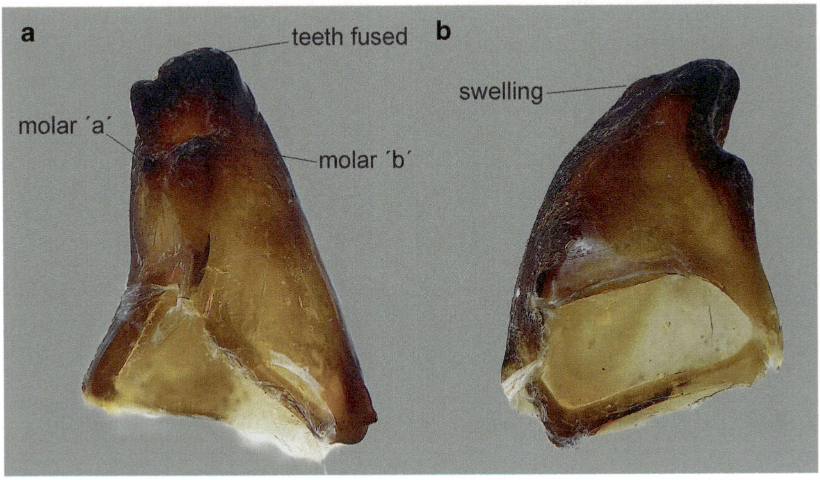

Fig. 10.12 Left mandible of *Crocothemis servilia*: (**a**) mesal (internal) view and (**b**) dorso-basal view

31’ Mandibles with incisor teeth acute, distinct, and not fused; abdomen with a pale middorsal stripe or spots (Figs. 10.55, 10.98, and 10.119); other characteristics variable 32

32(31’) A continuous pale, wide, middorsal stripe from anterior margin of pronotum throughout the tip of epiproct (Fig. 10.119); premental setae 9 + 3; 9–10 palpal setae . . . . ***Palpopleura***

32’ Pale middorsal stripe only on the abdomen (Figs. 10.55 and 10.98); premental setae 12–13; 9–11 palpal setae 33

33(32’) Epiproct abruptly narrowing in distal fourth to form a long, sharply pointed tip (Fig. 10.13a); head wider than the abdomen; abdominal sterna 1–6 paler than sterna 7–9 (Fig. 10.57); pale middorsal spots on S6–S9 gradually widening posteriorly (Fig. 10.55); paraprocts subequal in length to epiproct (Fig. 10.59) . . . . ***Diplacodes***

33’ Epiproct gradually narrowing to a sharply pointed tip (Fig. 10.13b); head narrower than the abdomen; abdominal sterna with a complex color pattern, alternating transverse pale and brown stripes (Fig. 10.100); a pale middorsal, parallel-sided stripe on S6–S10 (Fig. 10.98); paraprocts longer than epiproct (Fig. 10.102) . . . . ***Neurothemis***

Genera not included: *Amphithemis*, *Atratothemis*, *Nesoxenia*, *Phyllothemis*, *Pseudothemis*, *Pseudotramea*, *Risiophlebia*, *Sympetrum*, *Tyriobapta*

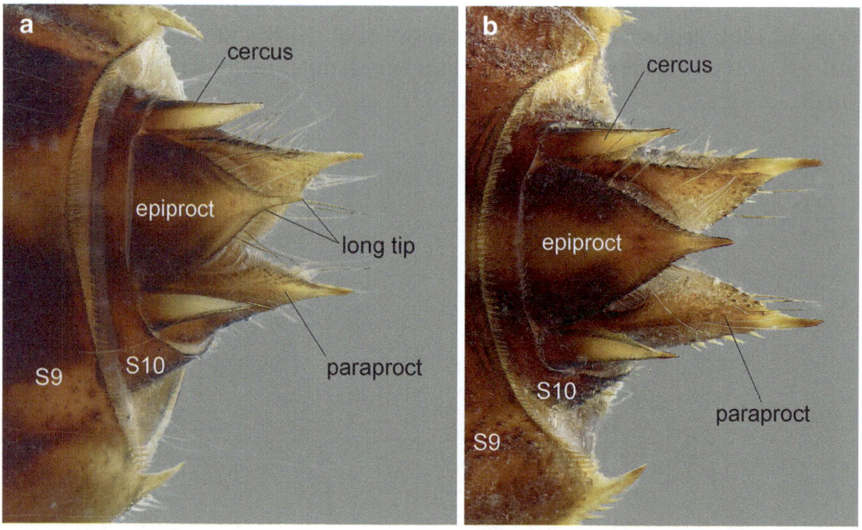

Fig. 10.13 The caudal appendages in dorsal view of *Diplacodes* (**a**) and *Neurothemis* (**b**)

10.2 *Acisoma* Rambur, 1842

Fig. 10.14 Dorsal habitus of the nymph of *Acisoma*

10.2.1 General Appearance

The nymphs of *Acisoma* are characterized by their very small size (9.5–10.3 mm), with the body robust, compound eyes set well forward on the head, abdomen lacking dorsal protuberances and posterolateral spines, S10 short and sunken into S9, and caudal appendages with the tips sharply pointed and decurved (Fig. 10.14).

10.2.2 Description

Head. The head is large, roughly subtrapezoidal, wider than long, and dorsally granulose (Fig. 10.15). The labrum anterior margin is biconcave with abundant stiff, long, golden setae and not flattened ventrally. The clypeus is mostly bare. The frons anterior surface is vertical, the dorsal surface is flat, and the anterior margin is straight to shallowly convex and has long, stiff, yellow setae. The vertex is more or less flat and granulose and has short stiff setae and three ocelli. The antennae are seven-segmented; the scape and pedicel are brown and barrel-shaped; the flagellomeres are yellow, cylindrical, and short; the basal third of antennomere 6 and the apical third of antennomere 7 are brown; and antennomere 7 is the longest. The compound eyes are rounded, prominent, and set well forward on the head and have seven longitudinal, black stripes each (Fig. 10.15). The occiput is large, granulose, and covered with short, stiff setae, except on nine oval, bare areas; the lateral margins are shallowly convex and convergent posteriorly; the posterior corners are rounded and have short, stiff setae. The posterior margin of the head is shallowly concave and has long, stiff setae (Fig. 10.15).

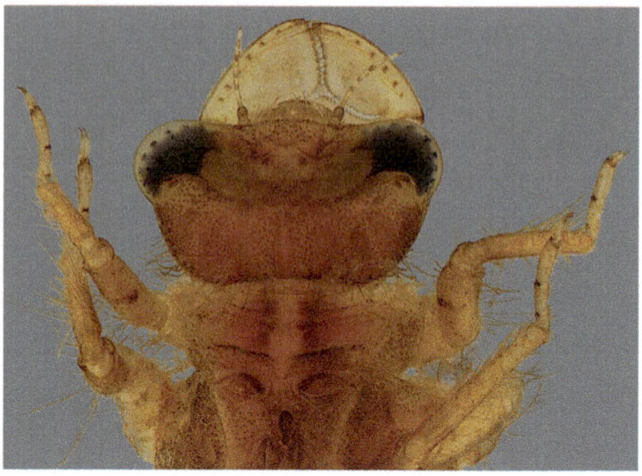

Fig. 10.15 The head and pro- and mesothorax in dorsal view. Notice the large mesospiracles

Mandibles have the formula L 1234 0 abd/R 1234 y abd, with d > a > b in the left mandible and d > b > a in the right mandible. The maxilla's galeolacinia has six teeth; two of the dorsal teeth are slightly incurved and of similar lengths and robustness, and the basalmost dorsal tooth is slender and shorter; the three ventral teeth are of different sizes and robustness, the basal tooth is minute, and the apical tooth is the largest. The ventral pad of the hypopharynx is wide and more or less kidney-shaped, the abundant long setae are on the apical half of its surface and lateral margins, and the posterior margin is strongly concave.

Labium. The prementum-postmentum articulation reaches the anterior margin of the mesosternum. The prementum is light-brown and subrhomboidal, and the lateral margins are setose, straight, and widely divergent to the apex (Fig. 10.16); 7 + 5 to 8 + 6 premental setae are on each side of the midline. The ligula is well-developed, triangular (Fig. 10.16), and finely serrulated on the distal margin and has a submarginal row of regularly spaced, spiniform setae and four spiniform setae at the tip. The labial palp is creamy-pale, mottled on the anterior surface, and the apical margin is finely serrulated and has indistinct crenations, each bearing a group of spiniform setae arranged in a graded series of three; the anterodorsal margin has a row of short setae; the ventral margin is finely serrulated and has a submarginal row of elongate, spiniform setae; and there are 6–7 palpal setae. The movable hook is relatively short, slender, sharply pointed, and smooth. A row of small setae is at the palp's articulation.

Fig. 10.16 Ventral habitus of the nymph of *Acisoma*

Thorax. The prothorax is narrower than the head. The pronotal disc is light-brown, transverse, and subrectangular and has a middorsal, longitudinal, creamy-pale stripe bearing two long, stiff, erect setae in the middle (Fig. 10.15); the anterior margin is widely concave; the sides are rounded with long, stiff setae; and the posterior margin is widely convex, granulose, and beset with long, stiff setae and has a very thick ridge in the middle. The proepisternum anterior margin is strongly convex and hirsute. The pleuron is granulose; the flat dorsal part of the mesepisternum is hirsute; the mesospiracles are large (Fig. 10.15) and the body's highest point in lateral view (Fig. 10.17). The legs are moderately hairy; the fore and middle legs are relatively short and the hind legs are relatively long (i.e., when fully extended, they far surpass S10) (Fig. 10.14); the femora and tibiae have a row of long, stout, stiff setae mainly on the dorsal margins; the distal fourth of the ventral surface of the fore tibia has tetradentate setae, whereas the remainder of the legs have short, spiniform setae and long, delicate setae; the pretarsal claws are simple and each has a pulvilliform empodium. The anterior wing sheaths are parallel and the posterior ones slightly divergent (Fig. 10.14), and they reach the basal half of S6 and surpass the anterior margin of S7, respectively. The anterior margin of the mesolaterosternite is straight and fringed with long, abundant stiff setae. The anterior margin of the metasternellum is straight and 0.75× as wide as the metasternellum midline length.

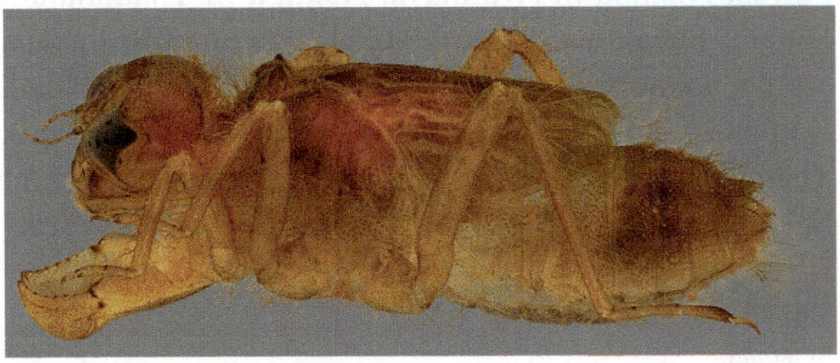

Fig. 10.17 The nymph of *Acisoma* in lateral view

Abdomen. The abdomen is abruptly truncate posteriorly and widest at S6 (Figs. 10.14 and 10.16). The posterior margins of the terga are mostly fringed with spiniform setae and have tufts of long, bristle-like setae on the midline of S5–S9 (Fig. 10.14). Middorsal protuberances and lateral spines are lacking. S10 is very short and sunken into S9 (Figs. 10.14 and 10.17). The lateral margins of 3–9 have long setae. The sternum is brown and granulose (Fig. 10.16); the posterior margins of sterna 5–9 have a row of small, spiniform setae, and the posterior margin of sternum 9 also has a fringe of long, stiff setae. Sterna 2–9 have several white, subquadrate spots on the laterosternites (Fig. 10.16) and several white dots on the mediosternites. Sterna 3 and 6–9 are each divided into three sternites and sterna 4–5 into five sternites. The caudal appendages are short and thick, and the tips are decurved (Fig. 10.17). The epiproct and paraprocts are triangular and beset with spiniform setae on the surfaces and margins. The epiproct is as long as its basal width (Fig. 10.14). The paraprocts are slightly longer than the epiproct, and their dorsal and ventral margins are slanting and converge to small, sharply pointed, decurved tips (Fig. 10.17). The cerci are digitiform and 0.75× as long as the epiproct, and their tips are strongly divergent.

Measurements (in mm). The total length including caudal appendages is 9.5–10.3; abdomen (ventral, including appendages) 5.3–5.9; maximum width of the abdomen 3.7–4.3; maximum width of the head 3.3–3.5; and length of hind femur 2.7–3.0.

10.2.3 Remarks

Acisoma is found in "open, often disturbed habitats including grassy swamps, drains and overgrown padi, with nymphs dwelling amongst shallow vegetation" (Orr 2005). In Thailand, only *Acisoma panorpoides* Rambur, 1842, has been recorded (Hämäläinen and Pinratana 1999).

10.2.4 Distribution

Acisoma is widespread and common and has been recorded from the provinces of Ang Thong, Bangkok, Chai Nat, Chaiyaphum, Chanthaburi, Chiang Mai, Chon Buri, Chumphon, Kanchanaburi, Krabi, Loei, Lop Buri, Nakhon Pathom, Nakhon Sawan, Nakhon Si Ayutthaya, Nakhon Si Thammarat, Narathiwat, Pathum Thani, Pattani, Phangnga, Phrae, Phuket, Prachuap Khiri Khan, Rayong, Sakon Nakhon, Samut Sakhon, Sing Buri, Udon Thani, Yala (Hämäläinen and Pinratana 1999), Kamphaeng Phet, Krabi, Phatthalung, Ratchaburi, Songkhla, Suphan Buri, Surat Thani, Tak (Ferro et al. 2009), Khon Kaen, Nan, Phitsanulok, Prachin Buri, and Sukhothai (Day et al. 2012) and more recently from Trat (Kosterin 2015) (Fig. 10.18).

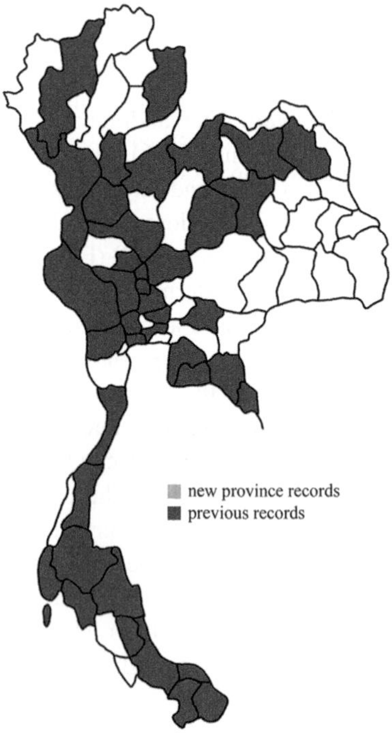

Fig. 10.18 Distribution of *Acisoma*

10.3 *Aethriamanta* Kirby, 1889

Fig. 10.19 Dorsal habitus of the nymph of *Aethriamanta*

10.3.1 General Appearance

The nymphs of *Aethriamanta* are characterized by their small size (10.5–12.5 mm), with the body brown and covered abundantly by long black, bristle-like setae, hind legs long, abdomen with well-developed dorsal protuberances on S2–S8 and posterolateral spines on S8–S9, and S10 short and sunken into S9 (Fig. 10.19).

10.3.2 Description

Head. The head is large and almost twice wider than long and has a strong, mottled pattern (Fig. 10.20). The labrum anterior border is biconcave with abundant stiff, long, golden setae and is not flattened ventrally. The clypeus is bare. The frons anterior margin is straight, black, and beset with long, stiff, light-brown setae, and its dorsal surface is shallowly convex. The antennae are seven-segmented, the scape is ring-shaped, the pedicel is barrel-shaped, and the flagellomeres are cylindrical and short; antennomeres 3 and 5–7 are of the same length and the largest; antennomeres 1, 2, and 6 are darker. The vertex is shallowly convex and has three ocelli.

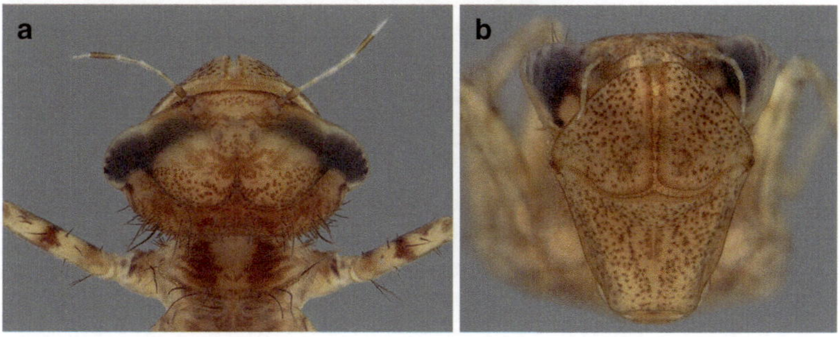

Fig. 10.20 (**a**) The head in dorsal view. (**b**) The head in frontal view showing the prementum and labial palps densely mottled

The compound eyes are rounded and laterally prominent (Fig. 10.20a). The occiput is large and covered with short, stout, stiff setae as well as long, bristle-like setae, which are mainly on the posterior and lateral margins (Fig. 10.20a); the lateral margins are more or less straight and convergent posteriorly; the posterior corners are widely rounded. The posterior margin of the head is sinuate. Mandibles have the formula L 1234 0 ab'b/R 1234 y abcd, with b > a in the left mandible, and tooth *d* is the largest in the right mandible. The maxilla's galeolacinia has six teeth; the two dorsal teeth are slightly incurved and of similar lengths and robustness, and the basalmost dorsal tooth is slender and shorter; the three ventral teeth are of different sizes and robustness, the basal tooth is minute, and the apical tooth is the largest. The ventral pad of the hypopharynx is wide and more or less kidney-shaped and has abundant long setae on the apical half of its surface and lateral margins; the posterior margin is strongly concave.

Labium. The prementum-postmentum articulation reaches the middle of the mesosternum. The prementum is subpentagonal and densely mottled on the ventral surface; the lateral margins are smooth, sinuate, and widely divergent to the apex (Fig. 10.20b); and 5 + 5 to 7 + 6 premental setae are on each side of the midline. The ligula is well-developed, triangular, and finely serrulated on the distal margin. The labial palp lateral surface is densely mottled (Fig. 10.20b); the apical margin has indistinct crenations, each bearing a group of two or three spiniform setae, one large and one or two smaller; the dorsolateral margin is mostly smooth; the ventral margin is finely serrulated and bears a row of elongate, spiniform setae; there are seven palpal setae. The movable hook is relatively short, sharply pointed, and smooth. Three short, stout, spiniform setae are at the palp's articulation.

Fig. 10.21 Ventral habitus of the nymph of *Aethriamanta*

Thorax. The prothorax is narrower than the head. The pronotal disc is transverse and semicircular; the anterior margin is widely concave; the posterior margin is widely convex, and it has a row of long, black, bristle-like setae directed posteriorly, the middle third is thick and raised, and the sides are rounded. The proepisternum anterior margin is setose with long setae. The legs are setose and pale and have dark rings on the femora and tibiae; the fore and middle legs are relatively short and the hind legs are relatively long (i.e., when fully extended, they far surpass S10) (Fig. 10.19); the femora and tibiae have rows of long, stout, stiff, bristle-like setae mainly on the dorsal margins; the distal fourth of the ventral surfaces of the fore and middle tibiae have tetradentate setae, whereas the remainder of the legs have simple, short and long setae; the pretarsal claws are simple, and each has a pulvilliform empodium. The anterior and posterior wing sheaths are parallel and have a row of black, stiff, bristle-like setae on the costal margins, and they reach the basal half and posterior margin of S7, respectively. The anterior margin of the mesolaterosternite is shallowly convex and fringed with abundant long, stiff setae. The width of the anterior margin of the metasternellum is half the metasternellum midline length.

Abdomen. The abdomen is abruptly truncate and widest at S5–S7 (Figs. 10.19 and 10.21). The dorsal surface and posterior margins of S5–S9 are covered abundantly with small spiniform setae, including a row of very long, black, bristle-like setae on the posterior margins of S4–S9 (Fig. 10.22).

Fig. 10.22 The abdomen in lateral view of a young nymph. Notice the large dorsal protuberances and the long, stiff, bristle-like setae

The posterolateral spines on S8–S9 are sharply pointed, incurved, and 0.50× and 0.75× as long as the dorsal length of their respective segments (Fig. 10.22). S10 is very short and sunken into S9. The sternum is pale and has abundant brown dots and small, spiniform setae (Fig. 10.21); the posterior margins of sterna 2–4 have only long, reddish-brown, bristle-like setae; the posterior margins of sterna 5–10 have both bristle-like and spiniform setae; sterna 8–9 also have a transverse row of very long, stiff, bristle-like setae. Sterna 3–5 are each divided into five sternites and sterna 6–9 into three sternites. The caudal appendages are short and thick, their tips are small and decurved, and they form a thick, compact anal pyramid. The epiproct and paraprocts are triangular (Fig. 10.22). The epiproct is one third shorter than its basal width and ends in a minute tip. The paraprocts are longer than the epiproct, and their dorsal and ventral margins are shallowly convex. The cerci are as long as the epiproct, slightly incurved, and sharply pointed.

Measurements (in mm). The total length including caudal appendages is 10.5–12.5; abdomen (ventral, including appendages) 5.2–6.6; maximum width of the abdomen 3.7–4.5; maximum width of the head 3.8–4.1; and length of hind femur 3.3–3.5.

10.3.3 Remarks

Aethriamanta is found in "weedy ponds and lowland marshes, with nymphs among floating vegetation" (Orr 2005), in "mare envahie de végétation" [ponds invaded by vegetation] (Fleck and Legrand 2005), and in riverine lagoons and swamps (Theischinger and Hawking 2006). In Thailand, three species have been recorded: *A. aethra* Ris, 1912, *A. brevipennis* (Rambur, 1842), and *A. gracilis* (Brauer, 1878) (Hämäläinen and Pinratana 1999).

10.3.4 Distribution

Aethriamanta is widely distributed, but not common, and has been recorded from the provinces of Bangkok, Chachoengsao, Chaiyaphum, Chiang Mai, Chiang Rai, Chon Buri, Kanchanaburi, Lampang, Lop Buri, Nakhon Nayok, Nakhon Pathom, Nakhon Sawan, Nakhon Si Ayutthaya, Nonthaburi, Pattani, Phangnga, Phayao, Phrae, Prachuap Khiri Khan, Rayong, Sakon Nakhon, Saraburi, Udon Thani (Hämäläinen and Pinratana 1999), Chanthaburi, Trang (Kitagawa and Katatani 2002, 2003), and Kalasin (Ferro et al. 2009) and more recently from Khon Kaen, Samut Sakhon, and Surat Thani (Day et al. 2012). ***New province records:*** Loei and Ubon Ratchathani (Fig. 10.23).

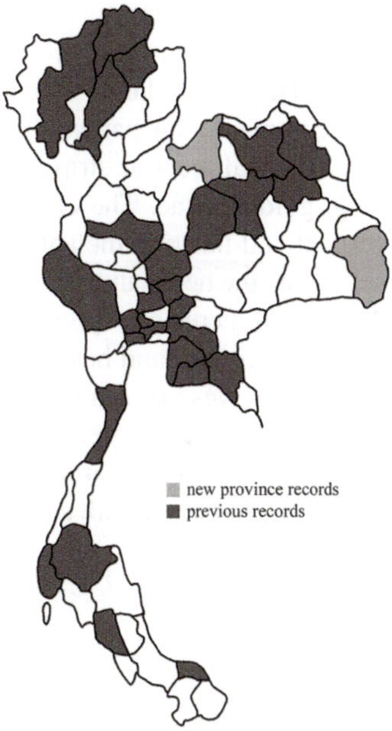

Fig. 10.23 Distribution of *Aethriamanta*

10.4 *Agrionoptera* Brauer, 1864

Fig. 10.24 Dorsal habitus of the nymph of *Agrionoptera insignis*

10.4.1 General Appearance

The nymphs of *Agrionoptera* are characterized by their small size (14–15 mm), with the body dark-brown; head mottled and striped; hind legs long; abdomen with or without small, low, blunt dorsal protuberances on S4–S8 and short posterolateral spines on S8–S9; and S10 short and sunken into S9 (Fig. 10.24).

10.4.2 Description

Head. The head is large, twice wider than long, and spotted and striped with brown in a complex color pattern (Fig. 10.25). The labrum is mostly light-brown with a pale, vertical, central stripe, and the anterior margin is biconcave with abundant long, golden setae. The clypeus is brown and bare. The frons anterior surface is vertical and flat, and the anterior margin is straight. The antennae are long, almost as long as the head, and seven-segmented. The scape and pedicel are barrel-shaped, the flagellomeres are cylindrical, and antennomere 6 is the longest. The vertex is shallowly convex and has minute spiniform setae, and three ocelli are evident. The compound eyes are rounded and laterally prominent (Fig. 10.25). The occiput is relatively large and covered with short setae, its lateral margins are convergent posteriorly, and the hind corners are widely rounded. The posterior margin of the head is straight and beset with stout setae (Fig. 10.25).

Fig. 10.25 The head in dorsal view

Mandibles have the formula L 1234 0 ab/R 1234 y abcd, with b > a in the left mandible and b > d > a > c in the right mandible. The maxilla's galeolacinia has seven teeth; the three dorsal teeth are slightly incurved, the two apicalmost teeth are subequal in length and robustness, and the basal tooth is shorter and slender; the four ventral teeth are of different sizes and robustness, the basal tooth is minute, and the apical tooth is the largest. The ventral pad of the hypopharynx is wide and more or less kidney-shaped, the posterior margin is widely concave, and it has large setae on its surface and margins; the insertions of these setae are protuberant and darker.

Labium. The prementum-postmentum articulation reaches the posterior margin of the mesosternum. The prementum is subpentagonal; the lateral margins are smooth along the basal half, but fringed with short setae along the distal half, and are sinuate and widely divergent to the apex (Fig. 10.26); the ventral surface is sparsely mottled; 8 + 4 to 9 + 5 premental setae are on each side of the midline.

The ligula is well-developed, triangular, and finely serrulated on the distal margin and has a submarginal row of spiniform setae of different sizes on each side of the midline and two setae at the tip. The labial palp is abundantly mottled on the lateral surface; the apical margin is crenate and has 12–13 weak crenations, each bearing a graded series of 4 spiniform setae, one of which is larger; the dorsolateral margin has a row of 12–14 short, spiniform setae along the basal half, but the distal half is fringed with small, delicate setae; the ventral margin is finely serrulated and has a tight row of spiniform setae; 10–11 palpal setae, 1 basal setella, and a group of 8 min, spiniform setae are on the basomesal surface. The movable hook is sharply pointed and smooth. A row of four small, spiniform setae is at the palp's articulation.

Fig. 10.26 Ventral habitus of the nymph of *Agrionoptera insignis*

Thorax. The prothorax is narrower than the head (Fig. 10.24). The pronotal disc is transverse, the anterior margin is almost straight, the sides are rounded with tufts of stout setae, the posterolateral margins are sinuate, and the posterior margin is short and straight. The proepisternum is hirsute with long setae. The legs are setose, spiny, and pale with two wide, dark bands on the femora (Fig. 10.24); the tibiae are darker on both proximal and distal ends and have a dark band in the middle; the hind legs are relatively long (i.e., when fully extended, they far surpass S10) (Figs. 10.24 and 10.26); the femora and tibiae have rows of long, stout, stiff, bristle-like setae mainly on the dorsal margins; the distal ends of the ventral surfaces of the fore and middle tibiae have numerous tridentate setae, whereas the hind tibia has only stout, spiniform setae; the tarsi are yellow and have tridentate setae on the ventrolateral row; the pretarsal claws are simple and each has a pulvilliform empodium. The anterior and posterior wing sheaths are parallel and reach the posterior margin and basal half of S5 and S6, respectively. The anterior margin of the mesolaterosternite is convex and beset with stiff, stout setae. The anterior margin of the metasternellum is straight, and its width is half the metasternellum midline length (Fig. 10.26).

Abdomen. The abdomen is abruptly truncate and widest at S6 (Figs. 10.24 and 10.26). The tergum is creamy-pale in the middle third on S1–S5 and light-brown laterally, whereas it is brown with dark-reddish-brown spots in a complex pattern on S6–S10; the lateral and posterior margins are covered abundantly with small spiniform setae. The middorsal protuberances on S4–S8 are short, low, and blunt (Fig. 10.27); the posterolateral spines on S8–S9 are short and sharply pointed (0.3 mm each) (Fig. 10.26), one third and one half the middorsal length of their respective segments. S10 is very short and sunken into S9 (Figs. 10.24 and 10.27). The sternum is light-brown on S1–S6, reddish-brown on S7–S10 (Fig. 10.26), and mostly smooth, but with a row of small, spiniform setae on the posterior margins of S8–S9 and tufts of very long, golden setae on the distal half of S9 (Fig. 10.27). Sterna 3–5 are each divided into five sternites and sterna 6–9 into three sternites. The caudal appendages are light-brown, pyramidal, sharply pointed (Fig. 10.27), setose, and spiny on the lateral margins and mesal surfaces. The epiproct is slightly longer than its basal width, and its dorsal margin is straight in lateral view. The paraprocts are slightly longer than the epiproct, the dorsal margins are concave in lateral view, and the tips are parallel. The cerci are digitiform and half the length of the epiproct (Fig. 10.27).

Fig. 10.27 The abdomen and caudal appendages in lateral view

Measurements (in mm). The total length including caudal appendages is 14–15; abdomen (ventral, including appendages) 8–9; maximum width of the abdomen 5.9–6.2; maximum width of the head 4.8–5.1; and length of hind femur 4.7–5.0.

10.4.3 Remarks

Agrionoptera is found "in lowland forest marshes, sluggish streams and shady drains in disturbed country; also in landward margin of mangroves" (Orr 2005). In Thailand, only two species have been recorded: *A. insignis* (Rambur, 1842)

(Hämäläinen and Pinratana 1999) and *A. sexlineata* Selys, 1879. Lieftinck (1962) mentioned that he described the nymph of *Agrionoptera insignis similis* Selys, 1878, under *Diplacina phoebe* Ris, 1919, with some uncertainty (Lieftinck 1930). In both of these papers, Lieftinck did not mention the presence of abdominal dorsal protuberances in the nymph of *Agrionoptera*, although he recognized this genus as heterogeneous (Lieftinck 1962). Thus, we think the presence/absence of abdominal dorsal protuberances in the nymphs of *Agrionoptera* could be geographic variation among populations.

10.4.4 Distribution

Agrionoptera has been recorded from the provinces of Kanchanaburi, Nakhon Nayok, Ranong, Songkhla, Yala (Hämäläinen and Pinratana 1999), Nakhon Ratchasima (Hämäläinen and Pinratana 2000), and Phangnga (Kosterin et al. 2010). ***New province records:*** Krabi and Trang (Fig. 10.28).

Fig. 10.28 Distribution of *Agrionoptera*

10.5 *Brachydiplax* Brauer, 1868

Fig. 10.29 Dorsal habitus of the nymph of *Brachydiplax*

10.5.1 General Appearance

The nymphs of *Brachydiplax* are small- to medium-sized (15–20 mm), with the body light-brown with a complex pattern of dark spots, hind legs short, abdomen with well-developed dorsal protuberances on S3–S8 and large posterolateral spines on S8–S9, and S10 annular and sunken into S9 (Fig. 10.29).

10.5.2 Description

Head. The head is transverse, twice wider than long, and light-yellowish-brown (Fig. 10.30). The labrum is entirely brown, and the anterior margin is biconcave and slightly flattened ventromedially and has abundant short setae. The clypeus is bare; the anteclypeus is creamy-pale with a large, brown, oval, basal spot on each side of the midline; the postclypeus is short and yellowish-brown. The frons anterior surface is vertical, flat, and creamy-pale; its anterior margin is rounded and is beset with short, stiff, reddish-brown setae; and its dorsal surface is shallowly convex. The antennae are seven-segmented, the scape and pedicel are barrel-shaped, the flagellomeres are cylindrical and subequal in length, and antennomere 6 is the longest; antennomeres 1–2 are light-brown, 3–5 and 7 are creamy-pale, and 6 is mostly dark-brown (Fig. 10.30). The vertex is large and shallowly convex, and three ocelli are evident.

Fig. 10.30 The head in dorsal view

The compound eyes are rounded and strongly prominent laterally (Fig. 10.30) and have five longitudinal, black, regularly spaced stripes. The occiput is short and covered with short, stout, stiff setae and long bristle-like setae mainly on the posterior and lateral margins; the lateral margins are very short, more or less straight, and convergent posteriorly; the posterior corners are widely rounded. The posterior margin of the head is sinuate. Mandibles have the formula L 1234 y ab'b/R 123 + 4 y ab'bd, with a > b in the left mandible and b > d > a in the right mandible. The maxilla's galeolacinia has seven teeth; the three dorsal teeth are slightly incurved, the two apicalmost teeth are subequal in length and robustness, and the basal tooth is shorter and slender; the four ventral teeth are of different sizes and robustness, the basal tooth is minute, and the apical tooth is the largest. The ventral pad of the hypopharynx is wide and more or less kidney-shaped and has abundant long setae on the apical half of the surface and the lateral margins, and the posterior margin is strongly concave in the middle.

Fig. 10.31 Ventral habitus of the nymph of *Brachydiplax*

Labium. The prementum-postmentum articulation surpasses the anterior margin of the mesosternum (Fig. 10.31). The prementum is subpentagonal; the ventral surface is sparsely mottled, creamy-pale (Fig. 10.31), and translucent in the middle; and the lateral margins are smooth, widely concave, and divergent to the apex (Fig. 10.31); 7 + 4 to 8 + 6 premental setae are on each side of the midline; the ligula is well-developed, triangular, and finely serrulated on the distal margin, and it has a submarginal row of spiniform setae on each side of the midline. The labial palp lateral surface is mostly translucent and mottled; the apical margin has weakly developed, almost indistinct crenations, each bearing a group of three spiniform setae, one larger and two smallcr; thc dorsolateral margin of the palp is smooth; the ventral margin is finely serrulated and has a row of seven, elongate, regularly spaced, spiniform setae intermingled with minute spiniform setae; and there are 9–12 palpal setae. The movable hook is relatively long, sharply pointed, and smooth. A row of 4–5 short, stout, spiniform setae is at the palp's articulation.

Thorax. The prothorax is narrower than the head (Figs. 10.29 and 10.30). The pronotal disc is brown with a longitudinal, middorsal, creamy-pale stripe and a pale spot on the lateral margins; the anterior margin is shallowly concave and has a row of short, brown setae; the posterior margin is widely convex; the sides are rounded and beset with rows of short, stout, spiniform setae and a row of long, stiff setae. The proepisternum anterior margin is hirsute. The pterothorax is pale with brown, irregular stripes and has a granular appearance. The legs are pale with dark rings on the femora and tibiae (Fig. 10.29), setose, and relatively short (i.e., when fully extended, the hind femur does not reach the posterior margin of S6) (Figs. 10.29 and 10.31); the tibiae have rows of long, stout, stiff, reddish-brown, bristle-like setae mainly on the dorsal margins; tridentate setae are on the distal fourth of the ventral (internal) surface of the fore tibia, on the distal margin of the middle tibia, and on the ventrolateral row of the fore tarsus, whereas the remainder of the legs have simple, short and long setae; the pretarsal claws are simple and each has a pulvilliform empodium. The wing sheaths are parallel and reddish-brown with pale venation (Fig. 10.29); a row of spiniform setae is on the costal and anal margins; the anterior and posterior wing sheaths reach the anterior margin and basal half of S6, respectively. The anterior margin of the mesolaterosternite is convex and beset with stiff, stout setae. The anterior margin of the metasternellum is very short and narrowly rounded, and its width is 0.20× the metasternellum midline length.

Abdomen. The abdomen is widest at S6–S7 and tapers slightly and gradually posteriorly (Figs. 10.29 and 10.31). The dorsal surface, posterior margins of S2–S10, and lateral margins of S5–S10 are covered abundantly with small, spiniform setae. The terga are creamy-pale on S2–S3 and the middle third of S4–S5, and the remainder are light-brown with a middorsal, longitudinal, pale stripe on S6–S9 and a large, anterolateral pale spot on S5–S9. The middorsal protuberances are well-developed and spine-like on S3–S8 or S4–S8 (Fig. 10.32). The posterolateral spines are large and sharply pointed on S8–S9 (Fig. 10.32); the spines of S8 are divergent, 0.4 mm long, and half the dorsal length of the segment; the spines of S9 are parallel, 0.8 mm long, and one third longer than the dorsal length of the segment (Fig. 10.32). S10 is very short and sunken into S9 (Fig. 10.29). The sterna are creamy-pale, but the distal two thirds of sterna 8–9 are mainly brown (Fig. 10.31); sterna 2–9 have three transverse, full-length rows of dark dots; laterosternites 4–7 have irregular, light-brown spots, and laterosternites 5–7 each has a large black dot. Sterna 2–6 are smooth; sterna 7–9 have minute, golden, spiniform setae; and sterna 7–10 posterior margins have a row of spiniform setae. Sterna 3–5 and 8–9 are each divided into five sternites and sterna 6–7 into three sternites. The caudal appendages are long and thick and have sharply pointed tips (Fig. 10.32). The epiproct is mostly yellow and triangular; it has a longitudinal, almost full-length, middorsal carina; the tip is slightly decurved; and it is 0.17× longer than its basal width. The paraprocts are mostly light-brown, pyramidal, and almost twice longer than the epiproct (Fig. 10.32); the dorsal and ventral margins are beset with rows of stout, spiniform setae and bristle-like setae; and the tips are long, slender, and straight. The cerci are mostly yellow but with brown at the base, more or less digitiform, the shortest of the caudal appendages, and half the length of the epiproct (Fig. 10.32), and the tips are slightly decurved.

Fig. 10.32 The abdomen and caudal appendages in lateral view

Measurements (in mm). The total length including caudal appendages is 15–20; abdomen (ventral, including appendages) 11–13; maximum width of the abdomen 5.3–6.0; maximum width of the head 4.6–5.2; and length of hind femur 4.0–4.2.

10.5.3 Remarks

Brachydiplax is found in forested swamps (Butler 2012) and also in "ponds, drains and mangrove swamps in exposed, disturbed habitats" (Orr 2005). In Thailand, four species have been recorded: *B. chalybea* Brauer, 1868; *B. farinosa* Krüger, 1902; *B. sobrina* (Rambur, 1842); and *Brachydiplax* sp. (Hämäläinen and Pinratana 1999).

10.5.4 Distribution

Brachydiplax is widely distributed and has been recorded from the provinces of Ang Thong, Bangkok, Chachoengsao, Chaiyaphum, Chanthaburi, Chiang Mai, Chiang Rai, Chon Buri, Kalasin, Kanchanaburi, Krabi, Lampang, Loei, Nakhon Nayok, Nakhon Pathom, Nakhon Sawan, Nakhon Si Ayutthaya, Narathiwat, Nonthaburi, Pathum Thani, Pattani, Phangnga, Phetchabun, Phuket, Prachuap Khiri Khan, Ranong, Ratchaburi, Rayong, Sakon Nakhon, Saraburi, Satun, Songkhla, Suphan Buri, Surat Thani, Tak, Trat, Ubon Ratchathani, Yala (Hämäläinen and Pinratana 1999), and Phetchaburi (Ferro et al. 2009) and more recently from Khon Kaen,

Samut Sakhon, and Sukhothai (Day et al. 2012). ***New province records:*** Phitsanulok, Phatthalung, and Trang (Fig. 10.33).

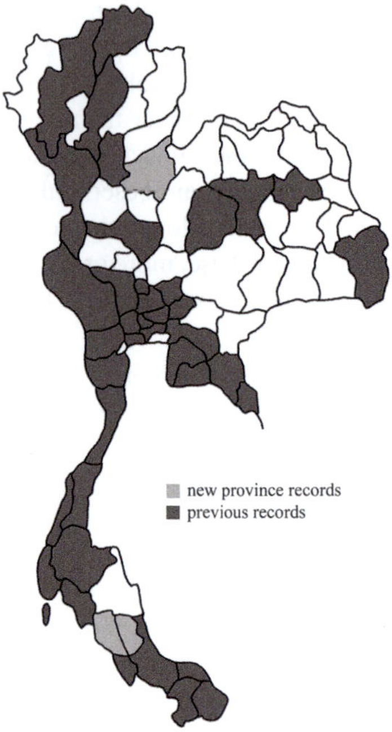

Fig. 10.33 Distribution of *Brachydiplax*

10.6 *Brachythemis* Brauer, 1868

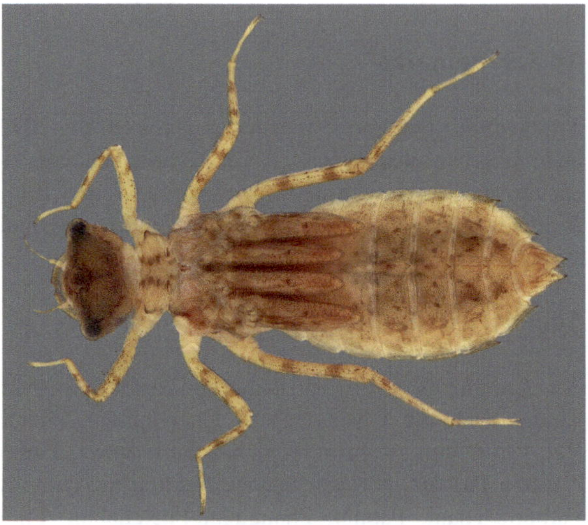

Fig. 10.34 Dorsal habitus of the nymph of *Brachythemis contaminata*

10.6.1 General Appearance

The nymphs of *Brachythemis* are small-sized (13–15 mm), with the body light-brown with a complex pattern of dark spots and dots, hind legs short, abdomen with dorsal protuberances on S4–S9 and short posterolateral spines on S8–S9, and S10 annular (Fig. 10.34).

10.6.2 Description

Head. The head is small, subpentagonal, wider than long, and light-brown (Fig. 10.35). The labrum is pale and has three brown spots, one at the middle and one on each side, its anterior margin is biconcave ventrally, and it has abundant short setae. The anteclypeus is bare and creamy-pale with a large, diffuse, brown, oval, basal spot on each side of the midline, and the postclypeus is light-brown basally and yellow apically with a group of minute setae in the middle. The frons anterior surface is vertical, flat, and creamy-pale; the dorsal margin is convex in frontal view and beset with long, stiff, yellow setae; and the dorsal surface is convex in the middle and concave laterally.

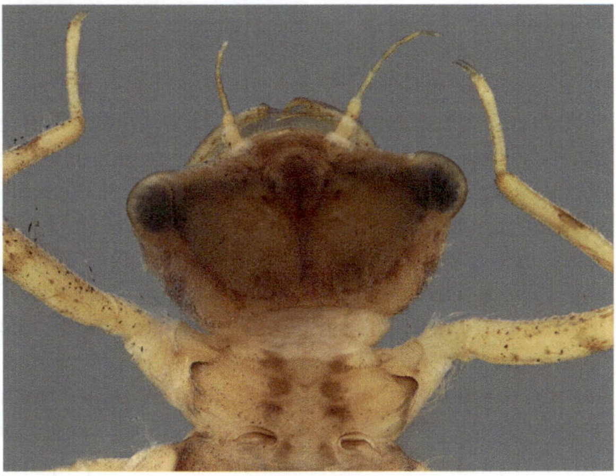

Fig. 10.35 The head in dorsal view. Notice the small compound eyes and long antennae

The antennae are relatively long and seven-segmented (Fig. 10.35), the scape and pedicel are light-brown and barrel-shaped, and the flagellomeres are cylindrical and subequal in length; antennomere 6 is the longest, antennomere 3 is mostly pale, and antennomeres 4–7 have the basal half light-brown and the apical half yellow. The vertex is large, convex, and reddish-brown and has three ocelli. The compound eyes are relatively small, rounded, and prominent laterally (Fig. 10.35). The occiput is large, reddish-brown, granulose, and covered with small spiniform setae on the posterior margin and has stiff setae on the posterior and lateral margins; an anterior, central, shallow concavity is at the union with the vertex; the lateral margins are straight and convergent posteriorly; the hind corners are slightly angulate. The posterior margin of the head is almost straight. Mandibles have the formula L 123'34 0 abd/R 1234 y- abd, with b > d > a in both mandibles. The maxilla's galeolacinia is as in *Brachydiplax*. The ventral pad of the hypopharynx is wide and more or less kidney-shaped and has abundant long setae on its surface and lateral margins, and the posterior margin is strongly concave.

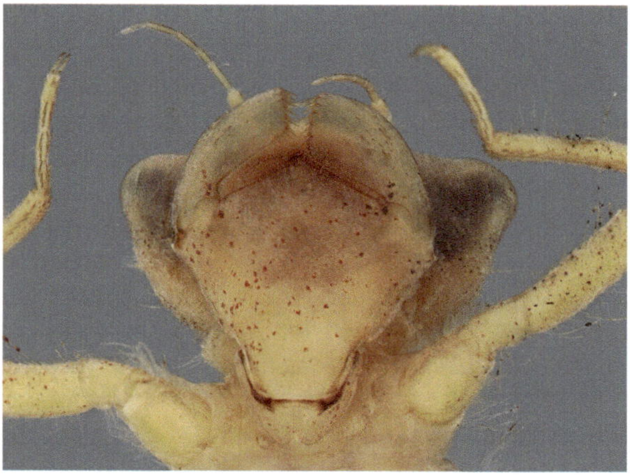

Fig. 10.36 The prementum in ventral view

Labium. The prementum-postmentum articulation reaches the anterior margin of the mesosternum. The prementum is subpentagonal, sparsely mottled on the ventral surface, creamy-pale laterally, and translucent medially, and the lateral margins are smooth, sinuate, and divergent to the apex (Fig. 10.36); 8 + 3 to 9 + 3 premental setae are on each side of the midline arranged in a semicircle; a group of 16–18 small spiniform setae is basal to the premental setae. The ligula is well-developed, triangular (Fig. 10.36), and finely serrulated on the distal margin and has a row of submarginal spiniform setae on each side of the midline. The labial palp lateral surface is mottled and has a large, central, light-brown spot; the apical margin has weakly developed, finely serrated crenations, each bearing a group of three graded spiniform setae; the dorsolateral margin of the palp has spiniform setae; the ventral margin is rather smooth but has a row of submarginal, elongate, regularly spaced, spiniform setae intermingled with minute, spiniform setae; six palpal setae plus six minute setae are at the basomesal surface close to the palp's articulation. The movable hook is relatively long, sharply pointed, and smooth. A row of four short, stout, spiniform setae is on the palp's articulation.

Fig. 10.37 Ventral habitus of the nymph of *Brachythemis contaminata*

Thorax. The prothorax is narrower than the head (Fig. 10.34). The pronotal disc is pale with a longitudinal brown stripe on each side of the creamy-pale middorsal stripe (Fig. 10.35); the anterior margin is sinuate; the posterior margin is widely convex; the sides are roundly angulate, slightly raised, and beset with rows of black, short, spiniform setae. The proepisternum is hirsute and strongly protrudes ventrally. The pterothorax is light-brown and has a granular appearance. The legs are pale with dark rings on the femora and tibiae, setose, and relatively short (Fig. 10.34) (i.e., when fully extended, the hind femur almost reaches the basal half of S6); the distal half of the ventral (internal) surface of the fore tibia and the ventrolateral margins of all tarsi are beset with tridentate setae, whereas the remainder of the legs have simple, short, spiniform setae and long setae; the pretarsal claws are simple and each has a pulvilliform empodium. The anterior and posterior wing sheaths are parallel (Fig. 10.34) and reddish-brown and almost reach the anterior margin of S5 and basal half of S6, respectively. The anterior margin of the mesolaterosternite is convex and is beset with stiff, stout setae. The anterior margin of the metasternellum is straight and as wide as the metasternellum midline length (Fig. 10.37).

Fig. 10.38 The abdomen and caudal appendages in lateral view

Abdomen. The abdomen is ovate (Figs. 10.34 and 10.37) and widest at S6 and tapers slightly and gradually posteriorly. The terga are creamy-pale on S2–S4 and the middle third of S5, and the remainder are light-brown with a complex pattern of dark spots and dots, which are bolder on S8–S9; the dorsal surface and posterior margins of S6–S9 and lateral margins of S2–S10 are covered abundantly with small, spiniform setae. Middorsal protuberances on S4–S9 are well-developed and spine-like, those on S6–S9 have the dorsal margins straight and parallel to the longitudinal axis of the abdomen (Fig. 10.38), and that on S9 reaches the posterior margin of S10. The posterolateral spines on S8–S9 are short, sharply pointed, and with thc tips slightly convergent (Figs. 10.34 and 10.37); the spine lengths are 0.3 and 0.4 mm and 0.50× and 0.75× as long as the middorsal lengths of their respective segments. S10 is very short and sunken into S9. Sterna 1–5 are light-yellow, sterna 6–9 are light-yellowish-brown, and sternum 10 is light-brown; all have minute, delicate setae, and the posterior margins of sterna 9–10 have a row of small, spiniform setae. Sterna 3–5 and 8–9 are each divided into five sternites and sterna 6–7 into three sternites. The caudal appendages are thick, sharply pointed, and longer than S9 + S10 (Fig. 10.38). The epi- and paraprocts are pyramidal and light-brown. The epiproct is 0.25× longer than its maximum basal width, and it has a longitudinal, full-length, middorsal carina, the tip is relatively long and slightly decurved, and the lateral margins are beset with a row of long, stiff setae. The paraprocts are as long as the epiproct and have a row of long, stiff setae on the dorsal and ventral margins and a central row of spiniform setae on the lateral surfaces; the tips are straight. The cerci are digitiform, the shortest of the caudal appendages, and 0.66× as long as the epiproct, the basal third is brown, the apical two thirds is creamy-pale, and the tips are slightly divergent.

Measurements (in mm). The total length including caudal appendages is 13.5–15.3; abdomen (ventral, including appendages) 8.5–9.9; maximum width of the abdomen 5.3–5.6; maximum width of the head 3.7–4.0; and length of hind femur 3.6–4.0.

10.6.3 Remarks

Brachythemis is found "around margins of ponds, lakes, drains and sluggish rivers and tolerates polluted water" (Orr 2005) and also in rice fields (Thipaksorn et al. 2003). In Thailand, only one species has been recorded: *B. contaminata* (Fabricius, 1793) (Hämäläinen and Pinratana 1999).

10.6.4 Distribution

Brachythemis is widely distributed and has been recorded from the provinces of Ang Thong, Bangkok, Chai Nat, Chaiyaphum, Chanthaburi, Chiang Mai, Chiang Rai, Chon Buri, Kalasin, Kanchanaburi, Khon Kaen, Krabi, Lampang, Loei, Lop Buri, Mae Hong Son, Maha Sarakham, Nakhon Nayok, Nakhon Pathom, Nakhon Ratchasima, Nakhon Sawan, Nakhon Si Ayutthaya, Nakhon Si Thammarat, Nong Khai, Nonthaburi, Pathum Thani, Phayao, Phetchaburi, Phrae, Phuket, Prachuap Khiri

Khan, Sakon Nakhon, Samut Prakan, Samut Sakhon, Saraburi, Sing Buri, Si Sa Ket, Songkhla, Surat Thani, Tak, Ubon Ratchathani, Udon Thani, Uttaradit, Yasothon (Hämäläinen and Pinratana 1999); Pattani (Thipaksorn et al. 2003), Rayong (Kosterin 2006), Chumphon, Kamphaeng Phet, Nan, Phangnga, Phetchabun, Ranong, Ratchaburi, Suphan Buri, Trang, Uthai Thani (Ferro et al. 2009), Chachoengsao, Prachin Buri, and Sukhothai (Day et al. 2012) and more recently from Trat (Kosterin 2015). ***New province records:*** Phatthalung and Phitsanulok (Fig. 10.39).

Fig. 10.39 Distribution of *Brachythemis*

10.7 *Bradinopyga* Kirby, 1893

In the absence of specimens to examine, the description presented here has been adapted from Sangal and Kumar (1970).

10.7.1 General Appearance

The nymphs of *Bradinopyga* are medium-sized (20.1–23.2 mm), with the body grayish and mottled with black, legs pale and banded, abdomen lacking dorsal protuberances, short posterolateral spines on S8–S9, S10 annular and set off from S9, and caudal appendages sharply pointed and slightly decurved.

10.7.2 Description

Head. The head is roughly subtrapezoidal, wider than long and wider than the thorax, and grayish with black spots. The antennae are short, thin, and seven-segmented; the scape is barrel-shaped, the pedicel and flagellomeres are cylindrical, and antennomere 3 is the longest. The vertex is wedge-shaped and has scattered setae, and three ocelli are evident. The compound eyes are moderately developed, rounded, and bulging anterolaterally. The occiput is moderately sized, and the lateral margins are more or less straight, not bulging, and convergent posteriorly. The posterior margin of the head is more or less straight.

Labium. The prementum is subpentagonal with the lateral margins widely divergent to the apex; 19–21 premental setae are on each side of the midline. The ligula is well-developed, triangular, finely serrulated, and with a submarginal row of small, spiniform setae on the distal margin; its dorsal surface is covered with abundant regularly spaced, minute, spiniform setae. The labial palp distal margin has indistinct crenations, each bearing a single, large, spiniform seta; the dorsolateral margin is smooth; the ventral margin is finely serrulated and has a submarginal row of regularly spaced, elongate, spiniform setae; 14–16 palpal setae and a row of 3–5 min setae are at the basomesal surface of the palp. The movable hook is relatively short (i.e., as long as or shorter than the preceding palpal seta), sharply pointed, and smooth.

Thorax. The prothorax is narrower than the head. The pronotal disc posterior margin is sinuate, and the lateral margins are widely rounded. The thoracic pleuron is mostly pale with two narrow, dark stripes on each side. The legs are pale with black bands on the femora and tibiae and dark spots on the tarsi, and all have scattered setae; the hind legs are relatively short (i.e., when fully extended, they barely reach the tips of the caudal appendages); the pretarsal claws are simple. The anterior and posterior wing sheaths are parallel and pale and have transverse dark bands, and they reach the posterior margin of S5 and basal half of S6, respectively.

Abdomen. The abdomen is ovate, widest at S6–S7, truncate posteriorly, elevated along the middorsal line, and more or less flattened ventrally and appears triangular in cross section. The terga have a complex color pattern of dots and spots on a paler background; the surfaces and posterior margins of S1–S9 are covered with small, spiniform setae. Middorsal protuberances are lacking; a row of spiniform setae is on the lateral margins of S5–S9; short, slightly incurved, sharply pointed spines are on S8–S9, and those on S9 reach the basal half of S10. S10 is very short and set off from S9. The sterna are light-gray. The caudal appendages are long, sharply pointed, as long as the dorsal length of S8 + S9, and with the tips long and slightly decurved. The epiproct and paraprocts are pyramidal, and their surfaces are abundantly covered with short, spiniform setae. The epiproct is one third longer than its maximum basal width; the male's epiproct has a large tubercle on the basal third. The paraprocts are the longest of the caudal appendages and appear moderately incurved in dorsal view. The cerci are digitiform and 0.57× as long as the epiproct.

Measurements (in mm). The total length including caudal appendages is 20.1–23.2; abdomen (excluding appendages) 13.6; maximum width of the abdomen 6.2; maximum width of the head 5.0; length of hind femur 5.3; epiproct 1.62; paraproct 1.96; and cercus 0.93.

10.7.3 Remarks

Bradinopyga is found in pools, irrigation channels, wells, and containers with standing water. Of the four known species of *Bradinopyga* (Paulson et al. 2022), only *B. geminata* (Rambur, 1842) has been recorded in Thailand (Hämäläinen and Pinratana 1999).

10.7.4 Distribution

In Thailand, *Bradinopyga* is known only from Chiang Mai Province (Hämäläinen and Pinratana 1999) (Fig. 10.40).

Fig. 10.40 Distribution of *Bradinopyga*

10.8 *Camacinia* Kirby, 1889

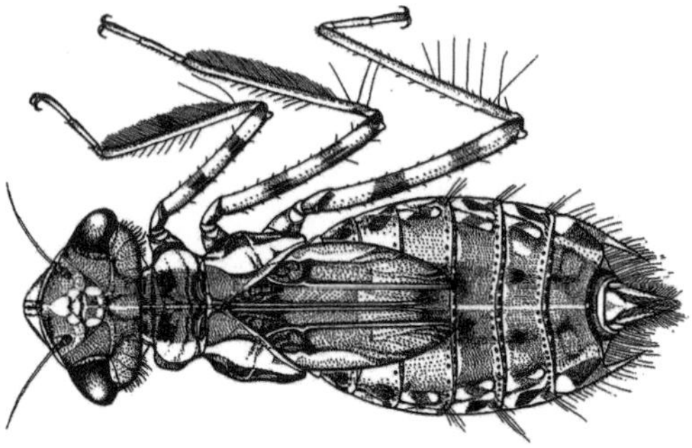

Fig. 10.41 Dorsal habitus of the nymph of *Camacinia gigantea* (after Lieftinck modified from van Tol [1992])

The nymph of *Camacinia* remains undescribed; thus, the description presented here is based on Lieftinck's drawing of *C. gigantea* (Brauer, 1867) (Fig. 10.41), published in van Tol (1992).

10.8.1 General Appearance

The nymphs of *Camacinia* are characterized by their body robust, long legs, abdomen lacking dorsal protuberances, posterolateral spines on S8–S9 with those on S9 long and thin, S10 annular and sunken into S9, and caudal appendages long and sharply pointed (Fig. 10.41).

10.8.2 Description

Head. The head is much wider than long. The dorsal surface of the frons has three pale spots grouped in front of the median ocellus. The vertex has three ocelli. The antennae are as long as the head and seven-segmented, the scape and pedicel are barrel-shaped, the flagellomeres are cylindrical, and apparently antennomeres 3, 6, and 7 are subequal in length and the longest. The compound eyes are large, rounded, and bulging laterally (Fig. 10.41). The occiput is short; the cephalic lobes are poorly developed and beset with large, strong, spiniform setae and have sparse, bristle-like setae. The posterior margin of the head is straight.

Thorax. The thorax is narrower than the head (Fig. 10.41). The pronotal disc is pale and has a large, wide, longitudinal, dark stripe that extends posteriorly onto the pterothorax; its anterior margin is more or less straight, the posterior margin is sinuate, and the lateral margins are largely rounded. The pterothorax is pale dorsolaterally and dark ventrolaterally. The legs are pale and spiny and have three well-defined dark bands on the femora; the hind legs are long (i.e., when fully extended, the tip of the hind tibia far surpasses the tip of the paraproct); the fore and middle tibiae are fringed with a dense brush of long setae on the dorsal (external) margins (Fig. 10.41); probably, the distal

Fig. 10.42 The abdomen and caudal appendages in lateral view (after Lieftinck modified from van Tol [1992])

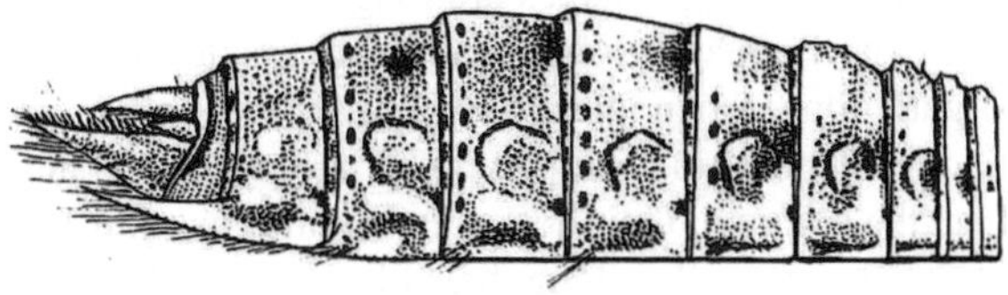

ends of the tibiae have tridentate setae, and the remainder of the legs have stout, short, spiniform setae. The anterior and posterior wing sheaths are parallel, pale ventrally, and darker dorsally and reach the posterior margin of S6 (Fig. 10.41).

Abdomen. The abdomen is elongate, spindle-shaped, and widest at S6–S7. The terga have a complex color pattern of dark and pale spots and blotches on the sides of S4–S6 and over the entire surfaces of S7–S9 and have a pale, longitudinal, narrow stripe on S7–S9 (Fig. 10.41). The posterior margins of S4–S8 have small, spiniform setae, and the posterodorsal corners of S6–S8 have tufts of long setae. Middorsal protuberances are lacking (Fig. 10.42). The posterolateral spines on S8–S9 are sharply pointed; those of S8 are minute, and those of S9 are very long and narrow, converge with each other posteriorly (Fig. 10.43), and are approximately 1.2× the middorsal length of S9 and reach the tip of the epiproct. S10 is annular and sunken into S9. The sterna are pale, but with four dark, wide, longitudinal stripes along sterna 2–8 as follows (Fig. 10.43): one sublateral stripe on each side increasing irregularly in width posteriorly and one submedian stripe on each side of the midline; sternum 9 has two large, dark spots, which converge posteromedially.

Fig. 10.43 The abdomen in ventral view (after Lieftinck modified from van Tol [1992])

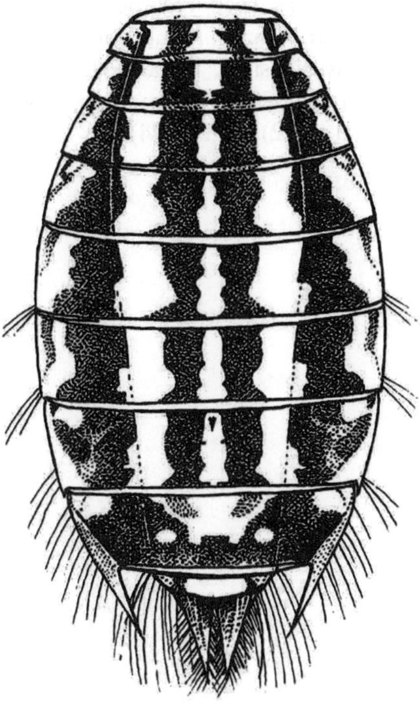

The caudal appendages are long, slender distally, sharply pointed, and strongly spiny on the margins and have long, bristle-like setae. The epiproct is triangular, slightly longer than its basal width, and strongly narrow at midlength, thus forming a long, slender tip. The paraprocts are pyramidal and approximately 0.35× longer than the epiproct; their dorsal and ventral margins have rows of strong, long, spiniform setae as well as long, hair-like setae; and their tips are long, slender, straight, and parallel (Fig. 10.43). The cerci are probably digitiform, the shortest of the caudal appendages, and 0.45× as long as the epiproct.

Measurements (in mm). Unknown, but judging by the adult size, nymphs probably range as follows: total length including caudal appendages 27–30; abdomen (ventral, including caudal appendages) 17–18; maximum width of the abdomen 9–10; maximum width of the head 8.5–9.0; length of hind femur 8.5–9.0.

10.8.3 Remarks

Camacinia is "restricted to low country and found in weedy ponds or boggy situations in marshy land" (Lieftinck 1934). The nymphs are probably semi-pelagic (Orr 2005). Only *C. gigantea* (Brauer, 1867) and *C. harterti* Karsch, 1890, have been recorded from Thailand (Hämäläinen and Pinratana 1999, Katatani et al. 2004).

10.8.4 Distribution

Camacinia has been recorded from the provinces of Bangkok, Chaiyaphum, Chanthaburi, Chiang Mai, Krabi, Kamphaeng Phet, Lampang, Mae Hong Son, Nakhon Ratchasima, Phrae (Hämäläinen and Pinratana 1999), and Chiang Rai (Katatani et al. 2004) and more recently from Phetchaburi, Surat Thani, and Trang (Day et al. 2012) (Fig. 10.44).

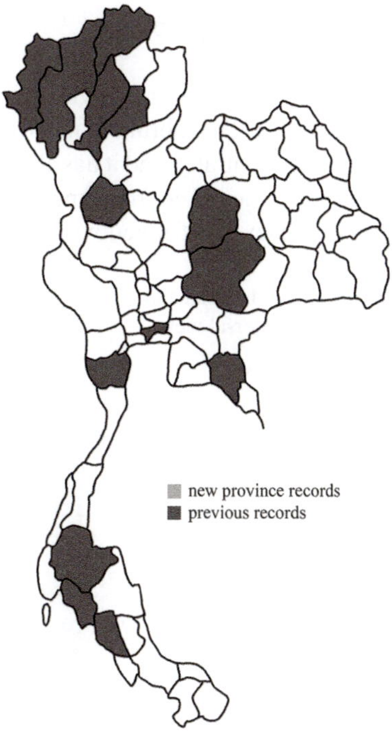

Fig. 10.44 Distribution of *Camacinia*

10.9 *Cratilla* Kirby, 1900

Fig. 10.45 Dorsal habitus of the nymph of *Cratilla*

10.9.1 General Appearance

The nymphs of *Cratilla* are characterized by their medium size (20.0–21.5 mm), with the body yellow to light-brown (Fig. 10.45), smooth, or with minute spiniform setae; compound eyes small; hind legs long; abdomen with small, low, blunt dorsal protuberances on S4–S5, but are vestigial or lacking on S6, with small posterolateral spines on S8–S9; and S10 short and sunken into S9 (Fig. 10.45).

10.9.2 Description

Head. The head is large, twice wider than long, and yellow without dark markings (Fig. 10.46). The labrum is mostly light-brown with the anterior margin brown and biconcave ventrally, and it has abundant long, golden setae. The clypeus is yellowish-brown and bare. The frons anterior surface is vertical and flat, and the anterior margin is convex and fringed with long, red-golden setae.

Fig. 10.46 The head and prothorax in dorsal view

The antennae are long, almost as long as the midlength of the head, and seven-segmented; the scape and pedicel are barrel-shaped, the flagellomeres are cylindrical, and antennomeres 3, 6, and 7 are the longest. The vertex is shallowly convex and has a pair of submedian tufts of setae and three ocelli. The compound eyes are small, anterolateral, and dorsally prominent (Fig. 10.46). The occiput is relatively large and covered with small, brown, spiniform setae; the lateral margins are furnished with stout spiniform setae and converge posteriorly; the cephalic lobes are widely rounded. The posterior margin of the head is straight and beset with short, spiniform setae. Mandibles have the formula L 12'234 0 ab/R 123 + 4 y- abd, with tooth b huge and >a in the left mandible and b > a > d in the right mandible. The maxilla's galeolacinia has seven teeth; the three dorsal teeth are slightly incurved, the two apicalmost teeth are of similar lengths and robustness, and the basal tooth is shorter and slender; the four ventral teeth are of different sizes and robustness, the basal tooth is small, and the subapical tooth is largely fused with the largest apical tooth. The ventral pad of the hypopharynx is wide, is more or less kidney-shaped, and has large setae on its surface and margins; the posterior margin is shallowly concave.

Labium. The prementum-postmentum articulation reaches the anterior margin of the mesosternum. The prementum is subrhomboidal and yellow, and the lateral margins are beset with minute, spiniform setae on the basal half, whereas the distal half is smooth, sinuate, and widely divergent to the apex; 6 + 5 or 7 + 4 premental setae are on each side of the midline. The ligula is widely convex and strongly crenate on the distal margin (Fig. 10.7b) and has a small, spiniform seta between each crenation, with the largest crenation at the tip. The labial palp is yellow, and its apical margin is strongly crenate with 11–12 large crenations, each bearing 1 seta or a graded series of 3 spiniform setae; the dorsolateral margin of the palp has a row of minute spiniform setae; the ventral margin is strongly crenate (Fig. 10.7b) and has a strong spiniform seta at the tip of each crenation; 12–13 palpal setae plus a row of 6–8 min spiniform setae are at the basomesal surface. The movable hook is long, slender, sharply pointed, and smooth. A tight row of 8–9 small, spiniform setae is at the palp's articulation.

Thorax. The prothorax is narrower than the head (Fig. 10.46). The pronotal disc anterior margin is sinuate, the sides are straight, the posterior corners are rounded, the posterolateral margins are straight, the posterior margin is short and straight, and all margins are beset with small, reddish-brown setae and long, bristle-like setae. The ventral margin of the pleuron is strongly setose with the setae long and delicate; the proepisternum is very short and strongly convex ventrally. The pterothorax is yellow with the dorsal margins brown, and it has abundant small, reddish-brown, spiniform setae. The legs are setose, spiny, and pale with a diffuse dark band on the femora (Fig. 10.45); the hind legs are relatively long (i.e., when fully extended, they far surpass S10); the ventral (internal) surfaces of the tibiae have numerous tridentate setae; the tarsi are yellow and have tridentate setae on the ventrolateral row; the pretarsal claws are simple and each has a pulvilliform empodium. The anterior and posterior wing sheaths are parallel and reach the basal half and posterior margin of S5, respectively. The anterior margin of the mesolaterosternite is convex and beset with stiff, stout setae. The anterior margin of the metasternellum is narrowly convex, and its width is 0.17× the metasternellum midline length.

Fig. 10.47 Ventral habitus of the nymph of *Cratilla*

Abdomen. The abdomen is elongate, fusiform, and widest at S6 (Fig. 10.47). The terga are creamy-pale to light-brown with three diffuse or well-marked spots along the midline of S6–S9 (Fig. 10.45). Most of the surfaces of S2–S10, including the lateral and posterior margins, are abundantly covered with small, reddish-brown, spiniform setae. Middorsal protuberances are small, low, and blunt on S4–S5 and vestigial on S6. The posterolateral spines are small and sharply pointed on S8–S9 (0.2 and 0.3 mm, respectively) and 0.27 and 0.50 as long as the middorsal length of their respective segments. S10 is very short and sunken into S9. The sterna are creamy-pale and glabrous (Fig. 10.47) but have a row of small, spiniform setae on the posterior margins of 7–10 and tufts of very long, golden setae on the distal half of 9. Sterna 3–5 are each divided into five sternites and sterna 6–9 into three sternites.

Fig. 10.48 The abdomen and caudal appendages in lateral view

The caudal appendages are creamy-pale, pyramidal, sharply pointed, setose, and spiny on the lateral margins and mesal surfaces (Fig. 10.48). The epiproct is 0.30× longer than its basal width, and its dorsal margin is straight in lateral view. The paraprocts are as long as the epiproct or slightly longer and have the dorsal and ventral margins shallowly concave in lateral view, and their tips are parallel. The cerci are digitiform and 0.45–0.50× as long as the epiproct.

Measurements (in mm). The total length including caudal appendages is 20.0–21.5; abdomen (ventral, including appendages) 13.0–13.2; maximum width of the abdomen 6.4–7.0; maximum width of the head 5.2–5.5; and length of hind femur 4.3–5.6.

10.9.3 Remarks

Cratilla is found in "closed forest and forested swamps from 0–1100 m" (Orr 2005); it "breeds in slowly running water or in spring water in jungle" (Lien and Matsuki 1988) and "in a wider range of standing water forest habitats … and sometimes in large phytotelmata" (Orr 2005). In Thailand, only two species have been recorded: *C. lineata* (Brauer, 1878) and *C. metallica* (Brauer, 1878) (Hämäläinen and Pinratana 1999).

10.9.4 Distribution

Cratilla has been recorded from the provinces of Chaiyaphum, Chanthaburi, Chiang Mai, Chon Buri, Chumphon, Kanchanaburi, Krabi, Lampang, Loei, Nakhon Nayok, Nakhon Ratchasima, Nakhon Si Thammarat, Nonthaburi, Phangnga, Prachuap Khiri Khan, Ranong, Sakon Nakhon, Songkhla, Suphan Buri, Tak, Trang, Yala (Hämäläinen and Pinratana 1999), Chiang Rai (Katatani et al. 2004), Nan, and Phrae (Katatani et al. 2010) and more recently from Chachoengsao, Khon Kaen, Phuket, Sa Kaew, Surat Thani, and Trat (Day et al. 2012) (Fig. 10.49).

Fig. 10.49 Distribution of *Cratilla*

10.10 *Crocothemis* Brauer, 1868

Fig. 10.50 Dorsal habitus of the nymph of *Crocothemis servilia*

10.10.1 General Appearance

The nymphs of *Crocothemis* are small-sized (14–18 mm), with the body reddish-brown; head large with a dorsal, transverse pale stripe; legs banded and short; abdomen lacking dorsal protuberances but with moderate posterolateral spines on S8–S9; S10 annular and sunken into S9; and caudal appendages long and sharply pointed (Fig. 10.50).

10.10.2 Description

Head. The head is large, roughly subtrapezoidal, wider than long, and wider than the thorax (Fig. 10.51). The labrum is pale but has three brown spots with one in the middle and one on each side; its anterior margin is biconcave and slightly flattened ventrally and has abundant stiff, long, golden setae. The clypeus is mostly bare and light-brown; the anteclypeus is weakly striated, and the postclypeus anterior margin has minute setae. The frons anterior surface is vertical, the anterior margin is straight, and the dorsal surface is shallowly convex and beset with short, yellow setae. The antennae are seven-segmented, the scape is annular, the pedicel is barrel-shaped, and both are light-brown; the flagellomeres are cylindrical and yellow, and antennomere 3 is the longest. The vertex is creamy-pale with brown spots, bare, shallowly convex, and transversely crossed by a wide, pale band that reaches the mesal margins of the compound eyes, and three ocelli are evident.

Fig. 10.51 The head in dorsal view

The compound eyes are rounded and prominent anterolaterally and have longitudinal black stripes (Fig. 10.51). The occiput is large, reddish-brown, and covered with short, spiniform setae mainly on the posterior margin (Fig. 10.51), and it has seven oval bare areas; the lateral margins are straight and convergent posteriorly; the hind angles are rounded and have sparse, short, spiniform setae. The posterior margin of the head is straight. The mandibles are peculiar because the incisive teeth 2–4 are fused and largely rounded at the apex and have a large, mound-like, subapical swelling on the external (lateral) surface; they have the formula L 1 2 + 3 + 4 0 ab'b/R 1 2 + 3 + 4 0 abd, with $a < b$ in both mandibles and $b > d$ in the right mandible. The maxilla's galeolacinia has seven teeth; the three dorsal teeth are slightly incurved, the two apicalmost teeth are of similar lengths and robustness, and the basal tooth is shorter and slender; the four ventral teeth are of different sizes and robustness, the basal tooth is minute, and the apical tooth is the largest. The ventral pad of the hypopharynx is wide and more or less kidney-shaped and has abundant long setae on most of its surface and anterior and lateral margins; the posterior margin is strongly concave.

Fig. 10.52 Ventral habitus of the nymph of *Crocothemis servilia*

Labium. The prementum-postmentum articulation almost reaches the middle of the mesosternum. The prementum is subrhomboidal and yellowish-brown with a short, longitudinal, white stripe on the midline, its ventral surface is sparsely mottled, and the lateral margins are mostly smooth, sinuate, and widely divergent to the apex (Fig. 10.52); 10 + 4 to 10 + 5 premental setae are on each side of the midline. The ligula is well-developed, triangular (Fig. 10.52), and finely serrulated on the distal margin, and a row of eight submarginal spiniform setae is on each side of the midline, with one seta at the tip; its dorsal surface is covered with abundant regularly spaced, small, spiniform setae. The labial palp is creamy-pale and densely mottled on the lateral surface; its apical margin is finely serrulated and has indistinct crenations, each bearing a single, large, spiniform seta; the dorsolateral margin is smooth; the ventral margin is finely serrulated and bears a row of elongate spiniform setae; 10–11 palpal setae plus a group of 8–10 min setae are at the basomesal surface of the palp. The movable hook is relatively short, sharply pointed, and smooth. A row of three short, stout, spiniform setae is on the palp's articulation.

Thorax. The prothorax is narrower than the head (Fig. 10.51). The pronotal disc is mostly brown with a longitudinal, pale stripe on the midline and an oval, pale spot on the posterolateral margins; the anterior margin is shallowly concave, the posterior margin is widely convex, and the sides are rounded and beset with short, stout, spiniform setae that become long, stiff setae on the posterolateral margins. The proepisternum ventral margin is straight and setose. The pterothorax is brown and smooth, and the sides are almost vertical. The legs are moderately setose and pale; the fore and middle legs are relatively short, and the hind legs are relatively long (Fig. 10.50) (i.e., when fully extended, they far surpass S10); the femora have three dark bands and two dorsal rows of large spiniform setae; the tibiae have rows of long, stout, brown spiniform setae mainly on the dorsal (external) margins; most of the mesal surfaces of the fore and middle tibiae bear tetradentate setae, and the remainder of the fore and middle tibiae have simple, short and long setae; the ventral surface of the hind tibia has short, stout, spiniform setae; all tarsi have tridentate setae in the ventromesal rows; the pretarsal claws are simple and each has a pulvilliform empodium. The anterior and posterior wing sheaths are parallel (Fig. 10.50) and reach the posterior margin of S6 and basal half of S7, respectively. The anterior margin of the mesolaterosternite is convex and beset with stiff, stout setae. The anterior margin of the metasternellum is straight and 0.66× as wide as the metasternellum midline length (Fig. 10.52).

Abdomen. The abdomen is ovate, widest at S6, and abruptly truncate posteriorly (Figs. 10.50 and 10.52); S1–S5 are mostly creamy-pale, S6–S8 are brown laterally and light-yellowish-brown in the middle third, and S9–S10 are brown; most of the posterior margins of the terga are fringed with spiniform setae, and S3–S9 have a few groups of small, spiniform setae.

Fig. 10.53 The abdomen and caudal appendages in lateral view

Middorsal protuberances are lacking (Fig. 10.53). A row of spiniform setae on the lateral margins of S6–S9 end in pale, short, sharply pointed posterolateral spines on S8–S9 (Figs. 10.50, 10.52, and 10.53), which are 0.43× and 0.67× as long as the middorsal length of their respective segments. S10 is very short and sunken into S9. Sterna 1–3 and mediosternites 4–7 are light-yellowish-brown, laterosternites 4–7 and sterna 8–10 are reddish-brown, laterosternites 4–8 have a pale oval spot, and mediosternites 2–3 and 9 have a pair of pale spots (Fig. 10.52). Sterna 1–4 are mostly smooth, and sterna 5–9 have minute spiniform setae on most of their surfaces, which increase in size and robustness on the posterior margins. Sterna 3–5 and 9 are each divided into five sternites and sterna 6–8 into three sternites. The caudal appendages are long, sharply pointed (Fig. 10.53), and as long as the dorsal length of S8 + S9, and the tips are slightly decurved. The epiproct and paraprocts are pyramidal, and their mesal surfaces are abundantly covered by short, stout spiniform setae; the lateral margins of the epiproct and dorsal and ventral margins of the paraprocts each bear a row of spiniform setae intermingled with long, stiff setae. The paraprocts have a row of four robust spiniform setae on the lateral surfaces. The epiproct is longer than its maximum basal width. The paraprocts are as long as their basal width. The cerci are digitiform and 0.60× as long as the epiproct.

Measurements (in mm). The total length including caudal appendages is 14–18; abdomen (ventral, including appendages) 10.3–10.6; maximum width of the abdomen 6.0–6.5; maximum width of the head 5.7; and length of hind femur 4.3–4.5.

10.10.3 Remarks

Crocothemis is found in "disturbed open habitats" (Orr 2005) as well as in ponds, marshes, tanks, sluggish sections of streams, and especially rice fields (Tze-wai et al. 2011). In Thailand, only *Crocothemis servilia* (Drury, 1770) has been recorded (Hämäläinen and Pinratana 1999).

10.10.4 Distribution

Crocothemis is widespread and very common, especially in lowlands, and has been recorded from the provinces of Ang Thong, Bangkok, Chachoengsao, Chai Nat, Chaiyaphum, Chanthaburi, Chiang Mai, Chiang Rai, Chon Buri, Chumphon, Kalasin, Kanchanaburi, Krabi, Lampang, Loei, Lop Buri, Mae Hong Son, Maha Sarakham, Nakhon Nayok, Nakhon Pathom, Nakhon Ratchasima, Nakhon Sawan, Nakhon Si Thammarat, Narathiwat, Nong Khai, Nonthaburi, Pathum Thani, Phangnga, Phetchaburi, Phrae, Phuket, Prachuap Khiri Khan, Ranong, Rayong, Roi Et, Sa Kaew, Sakon Nakhon, Samut Prakan, Samut Sakhon, Saraburi, Satun, Sing Buri, Si Sa Ket, Songkhla, Suphan Buri, Surin, Tak, Trang, Ubon Ratchathani, Udon Thani, Uttaradit, Yala, Yasothon (Hämäläinen and Pinratana 1999), Phatthalung, Prachin Buri (Thipaksorn et al. 2003), Kamphaeng Phet, Ratchaburi, Surat Thani, and Uthai Thani (Ferro et al. 2009) and more recently from Khon Kaen, Phitsanulok, Sukhothai, and Trat (Day et al. 2012) (Fig. 10.54).

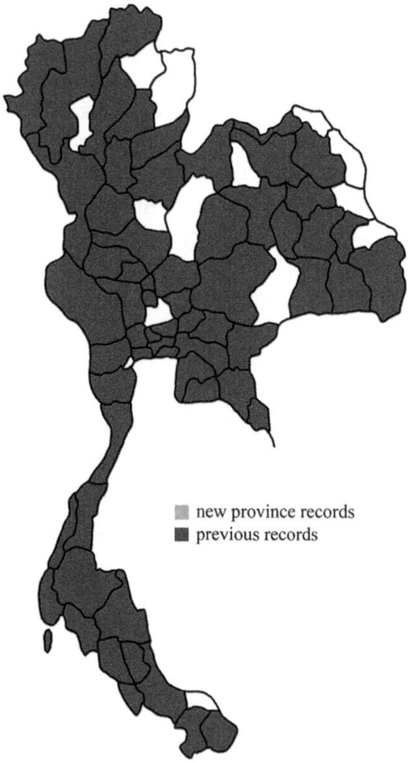

Fig. 10.54 Distribution of *Crocothemis*

10.11 *Diplacodes* Kirby, 1889

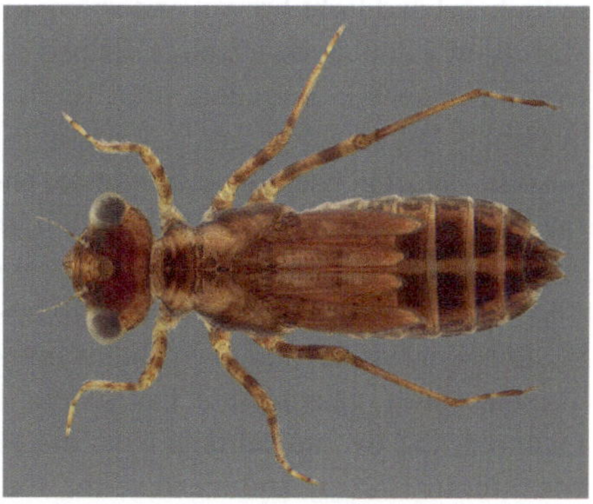

Fig. 10.55 Dorsal habitus of the nymph of *Diplacodes*

10.11.1 General Appearance

The nymphs of *Diplacodes* are small-sized (11.7–14.0 mm), with the body yellowish-brown, head wide, legs short and banded, abdomen lacking dorsal protuberances but with small posterolateral spines on S8–S9, S10 annular and sunken into S9, and caudal appendages sharply pointed (Fig. 10.55).

10.11.2 Description

Head. The head is roughly subtrapezoidal, wider than long, wider than the thorax, and with a complex color pattern (Fig. 10.56). The labrum is yellow with a transverse brown stripe on the distal half and one brown spot in the middle of the basal margin; the anterior margin is biconcave and slightly flattened ventrally and has short, golden setae. The clypeus is mostly bare and light-brown. The frons is almost flat; beset with short, yellow setae; and straight along the anterior margin.

Fig. 10.56 The head and prothorax in dorsal view

The antennae are short, thin, and seven-segmented (Fig. 10.56); the scape is light-brown, the pedicel is brown, and both are barrel-shaped; the flagellomeres are cylindrical and light-brown, and antennomere 3 is the longest. The vertex is yellow with brown dots, brown wedge-shaped twin spots, and three ocelli. The compound eyes are rounded, prominent (Fig. 10.56) anterolaterally, and diffusely striped. The occiput is large, yellow, and covered with short and long, stout, spiniform setae; the posterior margin has bristle-like setae, and three oval, bare areas are in the middle; the lateral margins are rounded, not bulging, and convergent posteriorly. The posterior margin of the head is shallowly convex in the middle. Mandibles have the formula L 1234 y- ab/R 1234 y abd, with b > a in both mandibles and b > d in the right mandible; a rugose, dull, oval area is immediately below the basalmost molar tooth. The maxilla's galeolacinia has seven teeth; the three dorsal teeth are slightly incurved, the two apicalmost teeth are of similar lengths and robustness, and the basal tooth is short and slender; the four ventral teeth are of different sizes and robustness, the basal tooth is minute, and the apical tooth is the largest. The ventral pad of the hypopharynx is wide and more or less kidney-shaped and has abundant long, anteriorly directed setae on most of its surface and margins; the posterior margin is strongly concave.

Fig. 10.57 Ventral habitus of the nymph of *Diplacodes*

Labium. The prementum-postmentum articulation reaches the anterior margin of the mesosternum (Fig. 10.57). The prementum is subrhomboidal and mostly creamy-pale and sparsely mottled on the ventral surface, and the lateral margins are smooth, sinuate, and widely divergent to the apex (Fig. 10.57); 8 + 4 to 8 + 5 premental setae are on each side of the midline. The ligula is well-developed, translucent, triangular, and finely serrulated on the distal margin and has a marginal row of 8–10 min, spiniform setae on each side of the midline and 2 small, submarginal setae at the tip; its dorsal surface is covered with abundant regularly spaced, minute spiniform setae.

Fig. 10.58 The head and prementum in frontal view

The labial palp is creamy-pale and sparsely mottled on the lateral surface; the apical margin is finely serrulated and has indistinct crenations (Fig. 10.58), each bearing a single, large, spiniform seta; the dorsolateral margin is smooth; the ventral margin is finely serrulated and bears a row of regularly spaced, elongate, spiniform setae; 10 palpal setae plus 1 setella and a group of 8–10 min setae are at the basomesal surface of the palp. The movable hook is relatively short (i.e., as long as the preceding palpal seta), sharply pointed, and smooth. A dorsal row of three, small, spiniform setae is at the palp's articulation.

Thorax. The prothorax is narrower than the head. The pronotal disc is mostly pale with a wide, longitudinal, brown stripe on the midline (Fig. 10.56); the posterior margin is widely convex, the posterolateral margins are rounded and beset with a group of large and short spiniform setae intermingled with bristle-like setae, and a short, stout, erect, spiniform seta is on each side of the midline. The proepisternum anterior margin has a row of four large, stout, spiniform setae followed by long setae. The pterothorax is mostly brown. The legs are moderately setose and pale, and the hind legs are relatively long (Fig. 10.57) (i.e., when fully extended, they far surpass S10); the femora have two dark rings and two dorsal rows of short spiniform setae; the fore and middle tibiae have three dark rings and rows of long, stout, stiff setae mainly on the dorsal (external) and ventral (internal) margins; the distal fourth of the mesal surfaces of the fore and middle tibiae bear tridentate setae, and the remainder of the fore and middle tibiae have simple, short and long setae; the ventral surface of the hind tibia has short, stout, spiniform setae; all tarsi have tridentate setae in the ventromesal rows; the pretarsal claws are simple and each has a pulvilliform empodium. The anterior and posterior wing sheaths are parallel and light-brown and reach the posterior margin of S5 and basal half of S6, respectively. The mesolaterosternite anterior margin is more or less straight and fringed with short, stout setae. The anterior margin of the metasternellum is straight and 0.66× as wide as the metasternellum midline length.

Fig. 10.59 The abdomen and caudal appendages in lateral view

Abdomen. The abdomen is ovate, widest at S6, and truncate posteriorly (Figs. 10.55 and 10.57). S1–S5 are mostly creamy-pale, S6 is light-yellowish-brown, S7–S10 are reddish-brown, and S6–S9 each have a longitudinal, wide, V-shaped pale spot. Terga 1–9 posterior margins are fringed with spiniform setae; the lateral thirds of terga 1–4 and all of terga 5–9 have minute spiniform setae as well as long, delicate, whitish setae. Middorsal protuberances are lacking (Fig. 10.59). A row of spiniform setae is on the lateral margins of S8–S9 and ends in pale, small, sharply pointed posterolateral spines, which are 0.35× and 0.47× as long as the dorsal length of their respective segments. S10 is very short and sunken into S9. Sterna 1–6 are light-yellowish-brown, sterna 7–10 are reddish-brown, and sterna 2–9 have a pair of dark spots in the middle (Fig. 10.57). The posterior margins of sterna 1–6 and 10 are smooth, and sterna 7–9 have a row of spiniform setae. Sterna 3–5 are each divided into five sternites and sterna 6–9 into three sternites. The caudal appendages are short, sharply pointed, and shorter than the dorsal length of S8 + S9, and their tips are slightly decurved (Fig. 10.59). The epiproct and paraprocts are pyramidal, and the mesal surfaces are abundantly covered by short, stout, spiniform setae. The epiproct is longer than its maximum basal width. The paraprocts have a row of three robust spiniform setae on the apical half of the lateral surfaces and are as long as their basal widths. The cerci are digitiform and 0.75× as long as the epiproct.

Measurements (in mm). The total length including caudal appendages is 11.7–14.0; abdomen (ventral, including appendages) 7.0–7.5; maximum width of the abdomen 3.7–4.0; maximum width of the head 3.8–4.0; and length of hind femur 3.0–3.3.

10.11.3 Remarks

Diplacodes is found in “marshes and shallow drains and ponds in open and disturbed habitats” (Orr 2005). In Thailand, only two species have been recorded: *D. nebulosa* (Fabricius, 1793) and *D. trivialis* (Rambur, 1842) (Hämäläinen and Pinratana 1999).

10.11.4 Distribution

Diplacodes is widespread and very common, especially in lowlands, and has been recorded from the provinces of Bangkok, Chai Nat, Chaiyaphum, Chanthaburi, Chiang Mai, Chiang Rai, Chon Buri, Chumphon, Kalasin, Kanchanaburi, Krabi, Lampang, Loei, Lop Buri, Mae Hong Son, Maha Sarakham, Nakhon Nayok, Nakhon Pathom, Nakhon Ratchasima, Nakhon Sawan, Nakhon Si Ayutthaya, Nakhon Si Thammarat, Narathiwat, Pathum Thani, Pattani, Phangnga, Phetchaburi, Phrae, Phuket, Prachuap Khiri Khan, Ranong, Rayong, Sa Kaew, Sakon Nakhon, Saraburi, Satun, Sing Buri, Si Sa Ket, Songkhla, Suphan Buri, Tak, Trang, Ubon Ratchathani, Uttaradit, Yala (Hämäläinen and Pinratana 1999), Buri Ram, Khon Kaen, Phitsanulok, Phrae, Roi Et, Surat Thani (Thipaksorn et al. 2003), Kamphaeng Phet, Phatthalung, Phayao, Ratchaburi, and Uthai Thani (Ferro et al. 2009) and more recently from Nan, Prachin Buri, and Trat (Day et al. 2012) (Fig. 10.60).

Fig. 10.60 Distribution of *Diplacodes*

10.12 *Hydrobasileus* Kirby, 1889

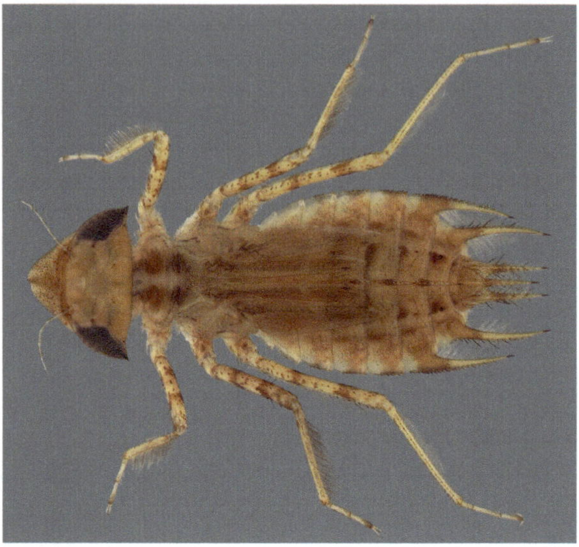

Fig. 10.61 Dorsal habitus of a fully developed nymph of *Hydrobasileus croceus*

10.12.1 General Appearance

The nymphs of *Hydrobasileus* are medium-sized (22–24 mm), with the body light-brown, hind legs long, abdomen with dorsal protuberances on S3–S8 and very large posterolateral spines on S8–S9, S10 annular and sunken into S9, and caudal appendages long and sharply pointed (Fig. 10.61).

10.12.2 Description

Head. The head is wide, almost twice wider than long, and light-yellowish-brown (Fig. 10.62). The labrum is yellow to mostly reddish-brown, and the anterior margin is biconcave and slightly flattened ventrally at the middle and has abundant long setae. The clypeus is light-yellowish-brown and bare. The frons anterior surface is vertical and flat; the anterior margin is rounded and beset with long, bristle-like, yellow setae; and the dorsal surface is shallowly convex. The antennae are seven-segmented (Fig. 10.62) and yellowish-brown, the scape and pedicel are barrel-shaped, the flagellomeres are cylindrical and subequal in length, and antennomere 7 is the longest. The vertex is bare and shallowly convex and has three ocelli.

Fig. 10.62 The head in dorsal view showing the acutely pointed compound eyes

The compound eyes are large and with the posterolateral corners strongly prominent laterally and acute (Fig. 10.62). The occiput is short and mostly bare dorsally, and a small group of short, stiff, dark setae is on each side of the midline; the posterior lobes are practically lacking. The posterior margin of the head is straight and beset with strong, spiniform setae of different sizes intermingled with long, bristle-like setae. Mandibles have the formula L 1234 0 abb'/R 1234 y abb'cd, with b > a in the left mandible and d > b > a > c in the right mandible (Fig. 2.5a, b); sometimes a tooth "3" is present in the left mandible. The maxilla's galeolacinia has seven teeth; the three dorsal teeth are slightly incurved, the two apicalmost teeth are of similar lengths and robustness, and the basal tooth is shorter and slender; the four ventral teeth are of different sizes and robustness, the basal tooth is minute, and the apical tooth is the largest. The ventral pad of the hypopharynx is wide, more or less kidney-shaped, and beset with large setae on the surface and margins; the lateral margins have long, stout spiniform setae, and the posterior margin is strongly concave.

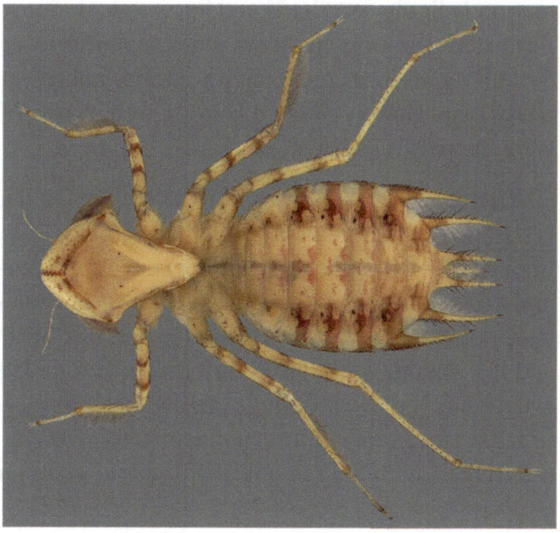

Fig. 10.63 Ventral habitus of the nymph of *Hydrobasileus croceus*

Labium. The prementum-postmentum articulation reaches the anterior margin of the metasternum. The prementum is subrhomboidal (Fig. 10.63), sparsely mottled on the ventral surface, creamy-pale laterally, and translucent in the middle; the lateral margins are smooth, sinuate, and divergent to the apex; 8 + 3, 8 + 4, or 9 + 3 premental setae are on each side of the midline. The ligula is well-developed and triangular, and the tip is widely rounded and finely serrulated on the apical margin; a submarginal row of 8–9 stout, spiniform setae is on each side of the midline.

Fig. 10.64 The prementum in frontal view showing the mottled labial palps

The labial palp lateral surface is abundantly mottled (Fig. 10.64); the apical margin is finely serrulated and has 9–10 weakly developed crenations, each bearing a graded series of 3–4 spiniform setae; the dorsolateral margin is smooth; the ventral margin is finely serrulated and has a row of elongate, stout, regularly spaced, spiniform setae intermingled with small, spiniform setae; 8–9 palpal setae and a row of 5–6 min spiniform setae are on the basomesal surface. The movable hook is relatively long, sharply pointed, and smooth. A row of four short, stout, spiniform setae is on the palp's articulation.

Thorax. The prothorax is narrower than the head (Fig. 10.61). The pronotal disc is pale with a longitudinal, irregular, reddish-brown stripe on each side of the midline; its posterior margin is widely convex and mostly smooth; the sides are rounded and beset with a row of short, stout, spiniform setae. The proepisternum ventral margin bears a few spiniform setae and tuft of long setae. The pterothorax is pale and has abundant minute spiniform setae. The legs are relatively long (Figs. 10.61 and 10.63) (i.e., when fully extended, the hind legs far surpass the posterior margin of S10), pale, spiny, setose, and with dark rings on the femora; the femora and tibiae are fringed with long, stout, stiff, yellow setae on the ventral (internal) and dorsal (external) margins, respectively; the distal fourth of the ventral (internal) surface of the fore and middle tibiae and the ventrolateral row of the fore tarsus bear tridentate setae; the remainder of the legs have stout, short and long spiniform setae; the pretarsal claws are simple and each has a pulvilliform empodium. The anterior and posterior wing sheaths are parallel (Fig. 10.61) and light-brown, have a row of spiniform setae on the distal half of the costal and anal margins, and reach the basal halves of S6 and S7, respectively. The anterior margin of the mesolaterosternite is shallowly convex and has short, stout, spiniform setae. The anterior margin of the metasternellum is straight and half as wide as the metasternellum midline length.

Fig. 10.65 The abdomen in lateral view showing the large dorsal and lateral spines

Abdomen. The abdomen is widest at S6–S7 and suddenly narrows at S9 (Figs. 10.61 and 10.63). The terga are light-yellowish-brown with a complex pattern of brown spots and dots, mainly on S5–S9; the dorsal surface and posterior margins of S2–S10 and lateral margins of S2–S9 are covered with small, spiniform setae, which increase in number and robustness posteriorly. The middorsal protuberances on S3–S8 are well-developed and spine-like and increase in size and robustness posteriorly (Fig. 10.65); the protuberance on S8 reaches the basal 0.10 of the epiproct, and those on S4–S8 have a longitudinal, dorsal row of spiniform setae. The posterolateral spines on S8–S9 are very large, 3.4 and 3.7 mm in length, respectively, and are sharply pointed and parallel and have a fringe of abundant long, yellow setae on the mesal margins (Figs. 10.61 and 10.63); those of S9 almost reach the tips of the paraprocts. S10 is short and sunken into S9. The sterna (Fig. 10.63) are mostly yellow with a complex pattern of large dark spots on laterosternites 4–9; sterna 2–9 have minute, spiniform setae on the posterior margins. Sterna 3–5 and 9 are each divided into five sternites and sterna 6–8 into three sternites. The caudal appendages are long, straight, sharply pointed, and strongly spiny on the margins and have long, reddish-brown, bristle-like setae (Figs. 10.61, 10.63, and 10.65). The epiproct is yellowish-brown and almost two thirds longer than its basal width. The paraprocts are yellowish-brown, pyramidal, and longer than the epiproct, their dorsal and ventral margins have rows of stout spiniform and bristle-like setae, and their tips are long, slender, and straight. The cerci are yellow, more or less digitiform, the shortest of the caudal appendages, and almost one seventh as long as the epiproct and have short tips.

Measurements (in mm). The total length including caudal appendages is 22.7–24.0; abdomen (ventral, including appendages) 14.3–14.8; maximum width of the abdomen 8.3–8.5; maximum width of the head 7.1–7.5; and length of hind femur 7.2–7.3.

10.12.3 Remarks

Hydrobasileus is found in "lowland lakes, swamp forest margins and larger, well vegetated ponds and drains in open country," and the nymphs are "pelagic among subsurface water weed" (Orr 2005). Of the three known species of *Hydrobasileus*, only *H. croceus* (Brauer, 1867) has been found in Thailand (Hämäläinen and Pinratana 1999).

10.12.4 Distribution

Hydrobasileus is widely distributed in Thailand and has been recorded from the provinces of Chachoengsao, Chaiyaphum, Chiang Mai, Chon Buri, Chumphon, Kanchanaburi, Krabi, Nakhon Nayok, Nakhon Pathom, Phrae, Ranong, Sakon Nakhon, Songkhla (Hämäläinen and Pinratana 1999), Chiang Rai (Katatani et al. 2004), Chanthaburi (Kosterin 2006), Prachuap Khiri Khan (Ferro et al. 2009), Nakhon Ratchasima, Phangnga, and Surat Thani (Day et al. 2012) and more recently from Trat (Kosterin 2015). ***New province records:*** Kalasin, Loei, Nakhon Si Thammarat, Narathiwat, Phatthalung, Phetchabun, Phuket, Satun, and Trang (Fig. 10.66).

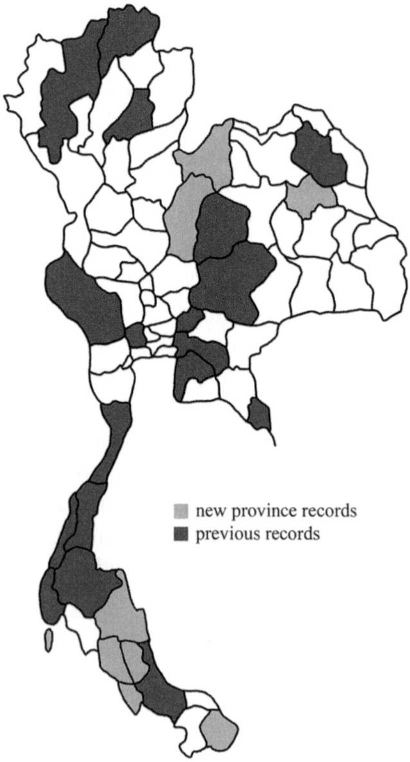

Fig. 10.66 Distribution of *Hydrobasileus*

10.13 *Hylaeothemis* Ris, 1909

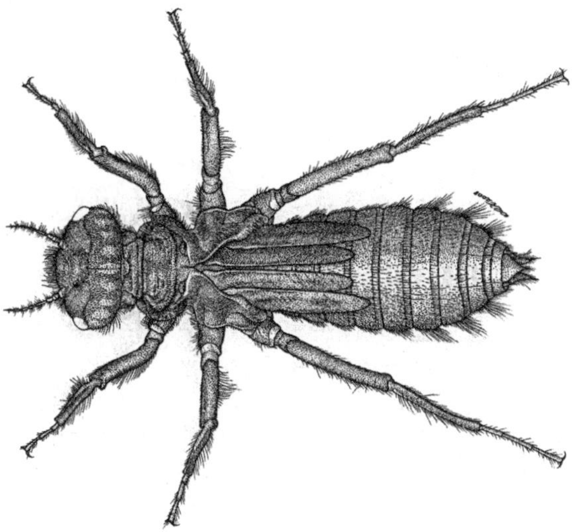

Fig. 10.67 Exuvia of the last instar of *Hylaeothemis clementia* (after Sasamoto and Kawashima [2009])

In the absence of specimens to examine, the description presented here has been adapted from Sasamoto and Kawashima (2009).

10.13.1 General Appearance

The nymphs of *Hylaeothemis* are small-sized (15.5 mm), with the body creamy-pale, slender, smooth, and shiny, compound eyes small, occiput large, hind legs moderately long, abdomen lacking dorsal protuberances and posterolateral spines, S10 short and extending beyond S9 to form a smooth outline with the anal pyramid, and caudal appendages short, thick, and sharply pointed (Fig. 10.67).

10.13.2 Description

Head. The head is buff, large, subrectangular, and wider than long (Fig. 10.68). The labrum is short, and its anterior margin is weakly arcuate and very shallowly concave in the middle. The anteclypeus is as long as the labrum, glabrous, and smooth. The antennae are almost as long as the head capsule and seven-segmented; the scape and pedicel are barrel-shaped, the flagellomeres are cylindrical and short, antennomere 7 is fusiform and acuminate, and antennomere 3 is the longest.

The compound eyes are small, rounded, and slightly prominent laterally. The occiput is large and covered with scattered spines and short setae on the dorsal

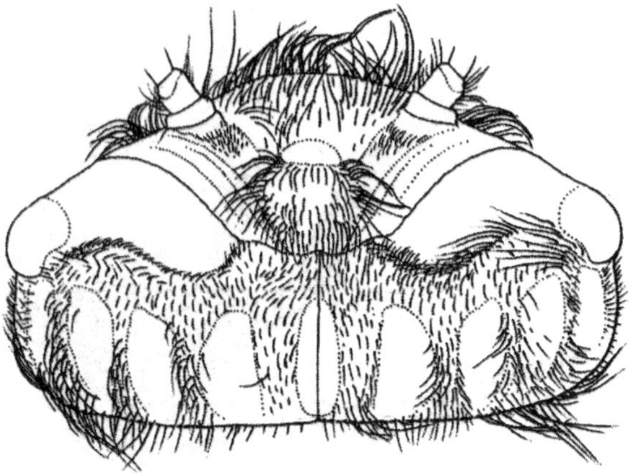

Fig. 10.68 The head in dorsal view (after Sasamoto and Kawashima [2009])

surface, except on the posterior, oval, bare areas; the lateral margins are shallowly convex and converge posteriorly; the posterior corners are rounded and bear long setae. The posterior margin of the head is rather straight.

Labium. The prementum-postmentum articulation reaches the posterior margin of the prosternum. The prementum is subpentagonal, and the lateral margins are sinuate and widely divergent to the apex; long premental setae are absent, although a set of minute setae is on each side of the midline (Fig. 10.69). The ligula is well-developed and triangular, and a row of submarginal setae are regularly spaced, with two setae at the tip. The labial palp apical margin is almost straight and has nine shallow, flattened serrations, each serration bearing 1–3 short setae; the dorsolateral margin of the palp has a row of setae along the basal half, and the remainder is smooth except for a relatively long seta close to the base of the movable hook; the ventral margin has a submarginal row of setae; palpal setae are absent. The movable hook is relatively short, slender, sharply pointed, and smooth. Three small, spiniform setae are at the palp's articulation.

Thorax. The prothorax is narrower than the head (Fig. 10.67). The pronotal disc is transverse, fusiform, and widely convex but straight at the middle, the sides are

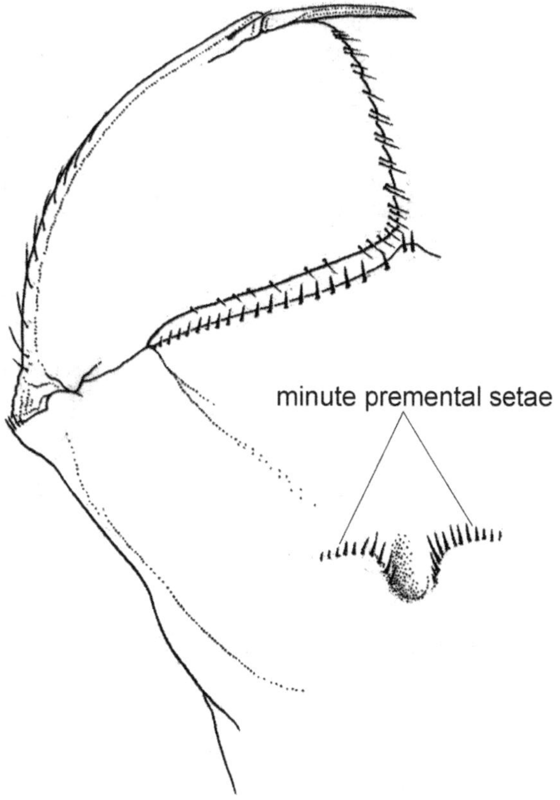

Fig. 10.69 Distal half of the prementum showing minute premental setae, left labial palp, and ligula (after Sasamoto and Kawashima [2009])

rounded and have stiff bristle-like setae, the posterior margin is thick, and the dorsal surface is widely glabrous. The pterothorax is large, wider than the head, and almost smooth and has a few long setae. The legs are moderately hairy and pale and lack dark bands; the fore and middle legs are relatively short, and the hind legs are relatively long (Fig. 10.67) (i.e., when fully extended, they far surpass S10); the pretarsal claws are small and simple. The anterior and posterior wing sheaths are parallel and reach the posterior margins of S4 and S5 (Fig. 10.67), respectively; the posterior wing sheaths are 2.5× as long as wide.

Abdomen. The abdomen is elongate, slender, fusiform, and widest at S6 (Fig. 10.67). The terga are mostly covered with minute spiniform setae and short and longer setae, including on the posterior and lateral margins. Middorsal protuberances and lateral spines are lacking. Although S10 is short, it extends beyond S9 to form a smooth outline with the anal pyramid. The sterna are buff and paler than the dorsal surface. The female gonapophyses are vestigial, appearing as a pair of minute lobes barely visible between S8 and S9. The caudal appendages are short, thick, and setose and form a triangular anal pyramid (Fig. 10.70). The epiproct is triangular, sharply pointed, and longer than its basal width, and the dorsal surface is mostly smooth but with scattered, minute, thin setae. The paraprocts are longer than the epiproct, with the tips divergent and margins not serrated, and the lateral surface is covered with long setae and bristle-like setae (Fig. 10.70). The cerci are the shortest of the caudal appendages, are weakly curved outward at the apex, and are half as long as the epiproct.

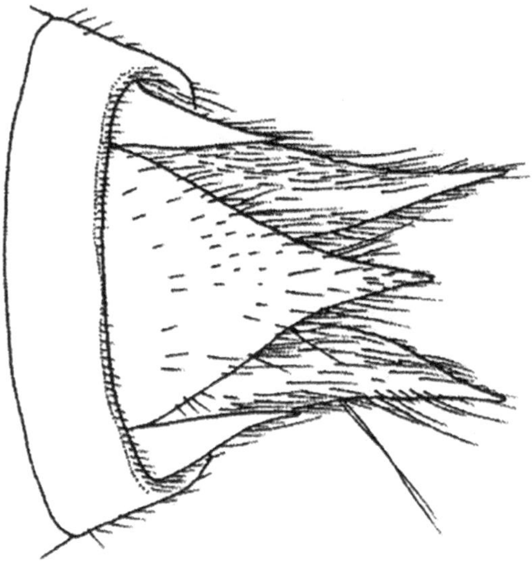

Fig. 10.70 The caudal appendages in dorsal view (after Sasamoto and Kawashima [2009])

Measurements (in mm). The total length including caudal appendages is 15.5; maximum width of the abdomen 3.8; and maximum width of the head 3.6.

10.13.3 Remarks

Hylaeothemis is a small genus with only four species, "three of these occur in India and Sri Lanka, whereas *H. clementia*, a rare and local species, was sporadically recorded from Borneo, Malaysia, Vietnam, Thailand and Laos" (Sasamoto and Kawashima 2009). *Hylaeothemis clementia* is found in "lowland alluvial swamps" (Orr 2005) and in "submontane humid forests … in muddy and grassy places … the larva was located in muddy and almost stagnant water" (Sasamoto and Kawashima 2009). In Thailand, only *H. clementia* has been found (Hämäläinen and Pinratana 1999).

10.13.4 Distribution

Hylaeothemis has been recorded only from the northern provinces of Chiang Mai (Hämäläinen and Pinratana 1999) and Chiang Rai (Katatani et al. 2004) (Fig. 10.71).

Fig. 10.71 Distribution of *Hylaeothemis*

10.14 ?*Indothemis* Ris, 1909

Fig. 10.72 Dorsal habitus of the fully grown nymph of ?*Indothemis*

Two fully grown nymphs could not be identified to any described or keyed genus. We are assigning them by supposition to *Indothemis* based upon details of wing venation, which is visible through the translucent cuticle.

10.14.1 General Appearance

The probable nymphs of *Indothemis* are small-sized (15–16 mm), with the body a complex color pattern of pale and dark bands and spots, head wide, legs banded, abdomen lacking dorsal protuberances but with small posterolateral spines on S8–S9, S10 annular and sunken into S9, and caudal appendages long and sharply pointed (Fig. 10.72).

10.14.2 Description

Head. The head is subtrapezoidal, almost twice wider than long (Fig. 10.73), and wider than the thorax (Fig. 10.72). The labrum is dark with a large, pale, central spot, and the anterior margin is biconcave, is slightly flattened ventrally, and has abundant stiff, long, golden setae. The clypeus is mostly bare, brown, and with a longitudinal pale stripe in the middle (Fig. 10.73). The frons anterior surface is slanting; its anterior margin is rounded, straight, and beset with short setae; and its dorsal surface is shallowly convex. The antennae are seven-segmented, the scape is ring-shaped, the pedicel is barrel-shaped, and both are brown; the flagellomeres are cylindrical and light-yellowish-brown, and antennomeres 3, 6, and 7 are the longest. The vertex is creamy-pale with brown spots, bare, and shallowly convex, and three ocelli are evident (Fig. 10.73).

Fig. 10.73 The head and pronotum in dorsal view

The compound eyes are rounded and prominent laterally (Fig. 10.73). The occiput is short, light-yellowish-brown, and covered with large, spiniform setae mainly on the posterior margin; it has nine oval, bare areas, the lateral margins are straight and converge posteriorly, and the posterior corners are rounded and have sparse, short, spiniform setae. The posterior margin of the head is straight. The mandibles have the formula L 1234 y- ab/R 1234 y abcd, with a > b in the left mandible and b > a > d > c in the right mandible. The maxilla's galeolacinia has seven teeth; the three dorsal teeth are slightly incurved, the two apicalmost teeth are of similar lengths and robustness, and the basal tooth is shorter and slender; the four ventral teeth are of different sizes and robustness, the basal tooth is the shortest, and the apical tooth is the largest. The ventral pad of the hypopharynx is wide and more or less kidney-shaped and has abundant long setae on most of its surface and the anterior and lateral margins; the posterior margin is strongly concave.

Fig. 10.74 The prementum in ventral view

Labium. The prementum-postmentum articulation reaches the middle of the mesosternum. The prementum is subrhomboidal and creamy-pale, and the lateral margins are mostly smooth, sinuate, and widely divergent to the apex (Fig. 10.74); 7 + 5 to 8 + 4 premental setae are on each side of the midline. The ligula is well-developed, translucent, triangular, and finely serrulated on the margin, it has a row of 8–10 submarginal spiniform setae on each side of the midline and 2 setae at the convex tip, and its dorsal surface is covered with small, regularly spaced setae. The labial palp is creamy-pale and has a large, light-brown, central spot (Fig. 10.74); the apical margin is finely serrulated and has small crenations, each bearing a single, large, spiniform seta; the dorsolateral margin is smooth; the ventral margin is finely serrulated and bears a submarginal row of elongate, spiniform setae; 8–9 palpal setae plus a group of 6–8 min, spiniform setae are at the basomesal surface of the palp. The movable hook is long, sharply pointed, and smooth. A row of four short, stout, spiniform setae is at the palp's articulation.

Thorax. The prothorax is narrower than the head (Fig. 10.73). The pronotal disc is mostly pale and has a longitudinal, inverted, "V-shaped," brown stripe on the midline; the sides are straight with rounded corners and beset with short, stout spiniform setae that become long, stiff setae on the posterolateral margins, and its posterior margin is widely convex. The proepisternum is short, and the ventral margin is straight and setose. The pterothorax is brown laterally and pale dorsally and has minute spiniform setae. The legs are moderately setose and pale; the fore and middle legs are relatively short, and the hind legs are relatively long (Figs. 10.72 and 10.75) (i.e., when fully extended, they far surpass S10); the femora have two dark bands and dorsal rows of short, spiniform setae and long, delicate setae; the tibiae have two diffuse dark bands and fringes of long setae on the dorsal (external) and ventral (internal) margins; most of the mesal surfaces of the fore and middle tibiae have tridentate setae, and the remainder of the fore and middle tibiae have simple, short and long setae; the ventral surface of the hind tibia has stout spiniform setae; the ventromesal rows of all tarsi bear tridentate setae; the pretarsal claws are simple and each has a pulvilliform empodium. The anterior and posterior wing sheaths are parallel, have alternating pale and brown transverse bands, and reach the basal half and posterior margin of S6, respectively (Fig. 10.72). The anterior margin of the mesolaterosternite is convex and beset with stiff, stout setae. The anterior margin of the metasternellum is straight and 0.60× as wide as the metasternellum midline length.

Fig. 10.75 Ventral habitus of the fully grown nymph of *?Indothemis*

Abdomen. The abdomen is fusiform, abruptly truncate posteriorly, and widest at S6 and has a complex color pattern (Fig. 10.72); S6–S10 each has a large, grayish, longitudinal, middorsal spot. Terga 1–10 posterior margins are fringed with spiniform setae. Middorsal protuberances are lacking. A row of small spiniform setae is on the lateral margins of S6–S9, with those on S8–S9 ending in pale, short, slightly incurved, sharply pointed spines, which are 0.33× and 0.50× as long as the dorsal length of their respective segments. S10 is very short and sunken into S9. Sterna 1–9 are creamy-pale, sterna 7–9 also have large brown spots, and sternum 10 is brown (Fig. 10.75). Sterna 1–10 are mostly smooth, and sterna 7–10 have small spiniform setae on the posterior margins, which increase in size and robustness posteriorly. Sterna 3–5 are each divided into five sternites and sterna 6–9 into three sternites. The caudal appendages are long, sharply pointed, and as long as the dorsal length of S8 + S9, and the tips are straight (Fig. 10.72). The epiproct and paraprocts are pyramidal, and their mesal surfaces are covered abundantly by short, stout, spiniform setae; the lateral margins of the epiproct and the dorsal and ventral margins of paraprocts have a row of spiniform setae intermingled with long, stiff setae. The paraprocts have a row of five, robust, spiniform setae on the lateral surfaces. The epiproct is 0.40× longer than its maximum basal width. The paraprocts are 0.20× longer than their basal width. The cerci are digitiform and 0.60× as long as the epiproct.

Measurements (in mm). The total length including caudal appendages is 15–16; abdomen (ventral, including appendages) 9.5–9.8; maximum width of the abdomen 5.3–5.6; maximum width of the head 5.2–5.3; and length of hind femur 5.0.

10.14.3 Remarks

Indothemis adults are found in open habitats, primarily open marshes and ponds, but have also been seen at reservoirs (Tang et al. 2010). In Thailand, only two species have been recorded (Hämäläinen and Pinratana 1999). We collected these nymphs in a pond in Loei Province and a vegetated lake in Phu Kradueng National Park, which is also in Loei Province.

10.14.4 Distribution

Indothemis is uncommon in Thailand (Hämäläinen and Pinratana 1999) and has been found in the north, east, and south. More specifically, it has been recorded from the provinces of Bangkok, Chaiyaphum, Chanthaburi, Chiang Mai, Chon Buri, Loei, Nakhon Nayok, Nakhon Si Thammarat, Phangnga, Sakon Nakhon, Trat (Hämäläinen and Pinratana 1999), Trang (Kitagawa and Katatani 2003), and Kanchanaburi (Ferro et al. 2009) and more recently from Nakhon Ratchasima, Phetchabun, and Ubon Ratchathani (Day et al. 2012) (Fig. 10.76).

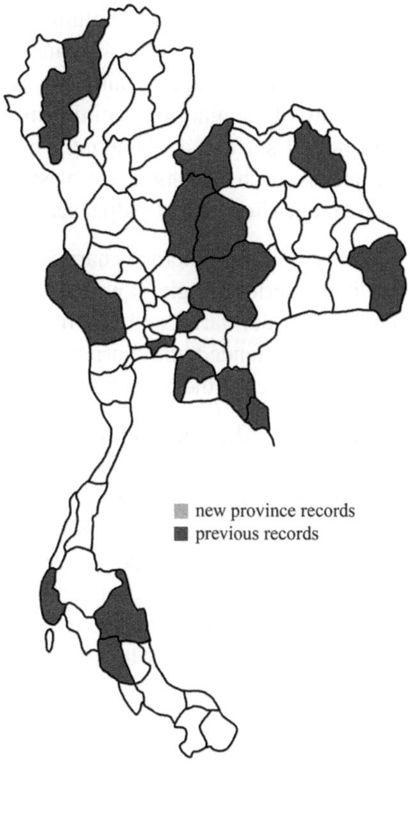

Fig. 10.76 Distribution of *Indothemis*

10.15 *Lathrecista* Kirby, 1889

In the absence of specimens to examine, the description presented here has been adapted from Adambukulam and Kakkassery (2013) and Theischinger and Hawking (2006).

10.15.1 General Appearance

The nymphs of *Lathrecista* are small-sized (16–19 mm), with the body light-brown, legs setose, femora pale with two dark bands, abdomen lacking dorsal protuberances or with only minute spines on S4–S6 and short posterolateral spines on S8–S9, and S10 short and sunken into S9.

10.15.2 Description

Head. The head is twice wider than long. The antennae are long, as long as the head, and seven-segmented, and the scape and pedicel are barrel-shaped. The compound eyes are small, rounded, and laterally prominent. The occiput is large, the lateral margins converge posteriorly, and the hind corners are widely rounded. The posterior margin of the head is shallowly concave.

Labium. The prementum-postmentum articulation reaches the posterior margin of the fore coxae. The prementum is subpentagonal, the lateral margins are widely divergent to the apex, and 14 premental setae are on each side of the midline. The ligula is well-developed, triangular, and finely serrulated on the apical margin and has a row of submarginal, spiniform setae on each side of the midline. The labial palp lateral surface is abundantly mottled; the apical margin has well-developed crenations, each bearing a pair of spiniform setae; the ventral margin is finely serrulated; and there are 11 palpal setae. The movable hook is sharply pointed.

Thorax. The prothorax is narrower than the head. The pronotal disc is transverse, and its posterior margin is widely convex. The legs are setose, spiny, and pale and have two wide, dark bands on the femora; the hind legs are relatively long (i.e., when fully extended, they surpass S10). The anterior and posterior wing sheaths are parallel and reach the basal half and posterior margin of S5, respectively.

Abdomen. The abdomen is widest at S6 and abruptly truncate posteriorly, and the lateral margins of S4–S9 are covered with small, hair-like setae. Middorsal protuberances are lacking or present only as minute spines on S4–S6. The posterolateral spines on S8–S9 are small and sharply pointed. S10 is very short and sunken into S9. The sterna are reddish-brown and have tufts of very long, golden setae on the distal half of S9. The caudal appendages are pyramidal, sharply pointed, and setose. The epiproct is apparently slightly longer than its basal width. The paraprocts are slightly longer than the epiproct. The cerci are triangular and about half the length of the epiproct.

Measurements (in mm). The total length including caudal appendages is 16.3–19.0 and the mean width of the abdomen 5.54.

10.15.3 Remarks

Lathrecista is a monotypic genus of lowland dragonflies found “in coastal marshes and swamps … also in landward margin of mangroves” (Orr 2005) and “in temporary ponds near paddy field” (Adambukulam and Kakkassery 2013).

10.15.4 Distribution

In Thailand, *Lathrecista* has been recorded from the provinces of Bangkok, Chachoengsao, Chaiyaphum, Chanthaburi, Chiang Mai, Chon Buri, Kanchanaburi, Krabi, Lampang, Nakhon Nayok, Nakhon Pathom, Nakhon Ratchasima, Nakhon Si Ayutthaya, Nakhon Si Thammarat, Nonthaburi, Pathum Thani, Prachuap Khiri Khan, Ranong, Ratchaburi, Rayong, Sakon Nakhon, Saraburi, Satun, Songkhla, and Trang (Hämäläinen and Pinratana 1999) and more recently from Nan and Surat Thani (Day et al. 2012) (Fig. 10.77).

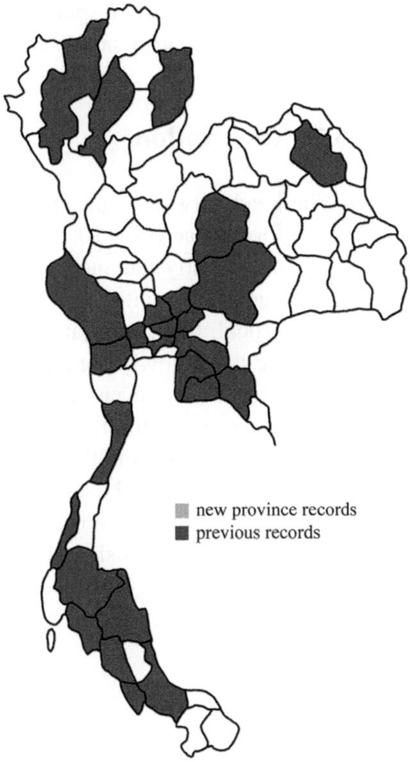

Fig. 10.77 Distribution of *Lathrecista*

10.16 *Lyriothemis* Brauer, 1868

Fig. 10.78 Dorsal habitus of the fully grown nymph of *Lyriothemis pachygastra*

10.16.1 General Appearance

The nymphs of *Lyriothemis* are small-sized (13–15 mm), with the body light-yellowish-brown; hind legs long; abdomen with dorsal protuberances on S4–S9 and large, incurved posterolateral spines on S8–S9; S10 annular and sunken into S9; and caudal appendages sharply pointed (Fig. 10.78).

10.16.2 Description

Head. The head is twice wider than long and covered with bristle-like setae and short, spiniform setae (Fig. 10.79). The labrum is yellow, the anterior margin is biconcave and slightly flattened ventrally, and it bears short setae on the surface and long setae on the anterior margin. The clypeus is bare. The frons anterior surface is vertical, high, and flat, its anterior margin is convex in frontal view, its dorsal surface is shallowly convex, and it is beset with long, bristle-like, yellow setae. The antennae are seven-segmented, the scape and pedicel are barrel-shaped, the flagellomeres are cylindrical, and antennomeres 6 and 7 are the longest. The vertex is yellow and has small, stout, reddish-brown spiniform setae; three ocelli are evident. The compound eyes are relatively small and prominent laterally, and the posterolateral corners are rounded (Fig. 10.79). The occiput is large, yellow, and beset with short, reddish-brown spiniform setae and long bristle-like setae, except on the nine oval, posterodorsal, glabrous areas. The cephalic lobes are large, but not bulging (Fig. 10.79). The posterior margin of the head is shallowly concave.

Fig. 10.79 The head in dorsal view

Mandibles have the formula L 1234 0 ab/R 1234 y abdx, with b > a in the left mandible and d > b > a > x in the right mandible. The maxilla's galeolacinia has seven teeth; the three dorsal teeth are slightly incurved, the two apicalmost teeth are of similar lengths and robustness, and the basal tooth is shorter and slender; the four ventral teeth are of different sizes and robustness, the basal tooth is minute, and the apical tooth is the largest. The ventral pad of the hypopharynx is wide and more or less kidney-shaped and has large setae on the surface and margins; the posterior margin is widely concave.

Labium. The prementum-postmentum articulation reaches the anterior margin of the middle coxae. The prementum is yellow and subrhomboidal, the lateral margins have a row of short setae and diverge to the apex (Fig. 10.80), and 7 + 3 to 6 or 8 + 4 premental setae are on each side of the midline. The ligula is well-developed and triangular (Fig. 10.80), the tip is rounded and beset with two enlarged spiniform setae, the apical margin is finely serrulated, and a row of 14–15 submarginal, spiniform setae is on each side of the midline. The labial palp apical margin is very finely serrulated and with 11–12 weakly developed crenations, each bearing a graded series of 4–5 spiniform setae; the dorsolateral margin is fringed with short, delicate setae; the ventral margin is smooth with a row of elongate, stout, regularly spaced, spiniform setae intermingled with small, spiniform setae; 7–10 palpal setae plus 1 setella and a group of 5–6 min spiniform setae are on the basomesal surface. The movable hook is relatively short, slender, sharply pointed, and smooth. A row of 3–4 small spiniform setae is on the palp's articulation.

Fig. 10.80 The prementum in ventral view

Thorax. The prothorax is narrower than the head (Fig. 10.78). The pronotal disc is light-yellowish-brown and has a longitudinal, posteriorly widening, creamy-pale band, and in the middle of which is a group of short, reddish-brown, stout spiniform setae; the sides are rounded and set with a tuft of long, reflexed, stiff, yellow setae; the posterolateral margins are straight and converge posteriorly, and the posterior margin is straight in the middle. The proepisternum is concave anteriorly, and its ventral margin is fringed with long setae. The pterothorax is yellow and has abundant short, spiniform setae and long bristle-like setae mainly on the dorsalmost area. The legs are yellow, spiny, and setose, have two dark rings on the femora, and are relatively long (Figs. 10.78 and 10.81) (i.e., when fully extended, the hind legs far surpass the posterior margin of S10); the femora and tibiae are fringed with long, stout, stiff, yellow setae on the ventral (internal) and dorsal (external) margins, respectively; the apex of the fore tibia bears tridentate setae, and the remainder of the legs have stout, short and long, spiniform setae; the pretarsal claws are simple and each has a pulvilliform empodium. The anterior and posterior wing sheaths are parallel, have a row of small setae on the costal and anal margins, and reach the posterior margin of S6 and basal half of S7, respectively. The anterior margin of the mesolaterosternite is shallowly convex and fringed with long, stiff setae. The anterior margin of the metasternellum is straight and as wide as the metasternellum midline length.

Fig. 10.81 Ventral habitus of the fully grown nymph of *Lyriothemis pachygastra*

Abdomen. The abdomen is semicircular and widest at S5 (Figs. 10.78 and 10.81). The terga are light-yellowish-brown on the lateral third, creamy-pale on the middle third of S1–S5, and reddish-brown on the middle third of S6–S10; the dorsal surfaces and posterior margins of S2–S10 and lateral margins of S2–S9 are covered with short and long stiff setae, which increase in number and robustness posteriorly. The middorsal protuberances on S4–S9 are well-developed, semierect, and spine-like (Fig. 10.82), those of S6–S9 are more or less of the same size and robustness, those on S4–S5 are smaller, that on S9 reaches the basal 0.40 of the epiproct, and all are covered with abundant short and long, stiff setae. The posterolateral spines on S8–S9 are sharply pointed and incurved and measure 0.5 and 0.8 mm in length, respectively (Fig. 10.80); those on S8 are as long as the middorsal length of S8, and those on S9 are one third longer than the middorsal length of S9 and reach the basal half of the paraprocts. S10 is short and sunken into S9.

Fig. 10.82 The abdomen and caudal appendages in dorsolateral view

The sterna are yellow and setose (Fig. 10.81), including their posterior margins. Sterna 3 and 6–9 are each divided into three sternites and sterna 4–5 into five sternites. The caudal appendages are longer than the middorsal length of S8 + S9 + S10 and sharply pointed and bear long, reddish-brown and bristle-like setae. The epiproct is triangular and longer than its basal width; its dorsal margin appears concave in profile view. The paraprocts are pyramidal and longer than the epiproct (Fig. 10.82), their dorsal and ventral margins have rows of stout spiniform setae and bristle-like setae, and their tips are long, slender, and divergent. The cerci are triangular with short tips; they are the shortest of the caudal appendages and half the length of the epiproct.

Measurements (in mm). The total length including caudal appendages is 13.5–14.5; abdomen (ventral, including appendages) 8.8–9.0; maximum width of the abdomen 5.8–5.9; maximum width of the head 4.7–4.9; and length of hind femur 4.0–4.1.

10.16.3 Remarks

Lyriothemis is found in "lowland and montane forest" (Hämäläinen and Pinratana 1999). The nymphs mainly inhabit phytotelmata (Kitching 1986, Orr 2005), especially in tree holes and bamboo stumps, but are also found "in marshes and the muddy bottom of ditches" (Lieftinck et al. 1984 cited in van der Poorten 2009). In Thailand, eight species have been found, although most are rare (Hämäläinen & Pinratana 1999, Kompier et al. 2021).

10.16.4 Distribution

Lyriothemis has been recorded from the provinces of Chanthaburi, Chumphon, Kanchanaburi, Loei, Nakhon Nayok, Nan, Phangnga, Ranong, Satun, Songkhla, Trang (Hämäläinen and Pinratana 1999), and Nakhon Si Thammarat (Day et al. 2012) and more recently from Trat (Kompier et al. 2021) (Fig. 10.83).

Fig. 10.83 Distribution of *Lyriothemis*

10.17 *Macrodiplax* Brauer, 1868

Fig. 10.84 Dorsal habitus of the nymph of *Macrodiplax cora*

10.17.1 General Appearance

The nymphs of *Macrodiplax* are medium-sized (18.2–20.3 mm), with the body robust and light-yellowish-brown; hind legs long; abdomen with dorsal protuberances on S3–S8 and large, straight posterolateral spines on S8–S9; S10 annular and sunken into S9; and caudal appendages relatively short and sharply pointed (Fig. 10.84).

10.17.2 Description

Head. The head is wider than long and mostly smooth (Fig. 10.85). The labrum is yellow and mostly bare, and the anterior margin is biconcave and slightly flattened ventrally and bears long setae. The clypeus is bare. The frons anterior surface is rounded, and the dorsal surface is shallowly convex and smooth. The antennae are seven-segmented, the scape and pedicel are barrel-shaped, the flagellomeres are cylindrical, and antennomeres 6 and 7 are the longest. The vertex is bare and has three ocelli. The compound eyes are relatively large and prominent laterally, and the posterolateral corners are rounded and sometimes slightly pointed (Fig. 10.85). The occiput is large and mostly bare; the cephalic lobes are large and rounded, but not bulging, and are beset with small, stout, spiniform setae. The posterior margin of the head is shallowly concave.

Fig. 10.85 The head and prothorax in dorsal view

Mandibles have the formula L 1234 0 abb'/R 1 + 234 y abcd, with b > a > b' in the left mandible and d > c > b > a in the right mandible. The maxilla's galeolacinia has seven teeth; the three dorsal teeth are slightly incurved, the two apicalmost teeth are of similar lengths and robustness, and the basal tooth is shorter and slender; the four ventral teeth are of different sizes and robustness, the basal tooth is small, and the apical tooth is the largest. The ventral pad of the hypopharynx is wide, more or less kidney-shaped, and setose and has large, golden setae on the surface and margins; the posterior margin is widely concave.

Labium. The prementum-postmentum articulation reaches the posterior margin of the mesosternum. The prementum is yellow, subrhomboidal, and slightly wider than long and has a ventrolateral longitudinal carina on each side; the lateral margins are strongly sinuate and divergent to the apex (Fig. 10.86) and bear a row of small, delicate setae; 8 + 4 or 8 + 5 premental setae are on each side of the midline. The ligula is well-developed, widely convex (Fig. 10.86), and finely serrulated on the apical margin, and the tip is rounded and beset with two large spiniform setae; a row of eight submarginal, large, spiniform setae is on each side of the midline. The labial palp is large, the apical margin is very finely serrulated, and it has 11–12 weakly developed crenations, each bearing a graded series of 3–4 spiniform setae; the basal half of the dorsolateral margin has a row of 9–10 short, stout, spiniform setae; the ventral margin is finely serrulated and has a row of large, stout, regularly spaced, spiniform setae; 8 palpal setae plus 1 setella and a group of 11–12 min spiniform setae are on the basomesal surface. The movable hook is relatively long, robust, sharply pointed, and smooth. A row of 4–5 small, spiniform setae is on the palp's articulation.

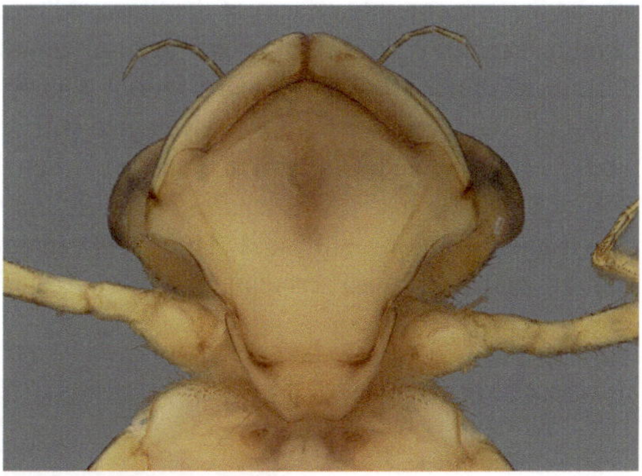

Fig. 10.86 The prementum in ventral view

Thorax. The prothorax is narrower than the head (Fig. 10.84). The pronotal disc is light-yellowish-brown and has a longitudinal, posteriorly widening, creamy-pale band (Fig. 10.85), and in the middle of which is a group of minute setae, and immediately posterior is an erect, short, spiniform seta on each side of the midline; the sides are slightly angled and beset with a row of stout, spiniform setae; the posterolateral margins are straight and convergent posteriorly, and the posterior margin is straight in the middle. The proepisternum is small, its anterior margin has a row of spiniform setae, and the ventral margin is fringed with long setae.

Fig. 10.87 Ventral habitus of the nymph of *Macrodiplax cora*

The proepimeron is produced ventrally as a triangular tubercle. The mesothorax is granulose anterodorsally and the metathorax is mostly bare. The legs are yellow, spiny, setose, and relatively long (Figs. 10.84 and 10.87) (i.e., when fully extended, the hind legs far surpass the posterior margin of S10); the femora have two dark rings; the femora and tibiae are fringed with long, stiff, yellow setae on the ventral (internal) and dorsal (external) margins, and these fringes are much more dense on the tibiae; tridentate setae are on the apicomesal surfaces of the tibiae, whereas the remainder of legs have stout, short and long, spiniform setae; the pretarsal claws are simple and each has a pulvilliform empodium. The anterior and posterior wing sheaths are parallel (Fig. 10.84); have a row of small, spiniform setae intermingled with long stiff setae on the costal margins and spiniform setae on the anal margins; and reach the posterior margin of S5 and basal half of S6, respectively. The anterior margin of the mesolaterosternite is widely convex and fringed with long, stiff setae. The anterior margin of metasternellum is straight and half as wide as the metasternellum midline length (Fig. 10.87).

Fig. 10.88 The abdomen and caudal appendages in lateral view. Notice the large dorsal and lateral spines

Abdomen. The abdomen is more or less fusiform and widest at S6 (Fig. 10.87). The terga are creamy-pale with brown, irregular spots forming a complex color pattern (Fig. 10.84); the dorsal surface and posterior margins of S2–S10 and lateral margins of S6–S9 are covered with minute and short, spiniform setae, which increase in number and robustness posteriorly. The middorsal protuberances on S3–S8 are well-developed and spine-like and diminish in height posteriorly (Fig. 10.88); the protuberances on S3–S5 are smaller, that on S8 reaches the posterior margin of S9, and all have a row of stout, spiniform setae. The posterolateral spines on S8–S9 are large, straight, sharply pointed, and 0.6 and 1.1 mm in length, respectively; those on S8 are half as long as the middorsal length of S8, and those on S9 are slightly longer than the middorsal length of S9 and surpass the tips of the cerci. S10 is short and sunken into S9 (Fig. 10.88). The sterna are creamy-pale and mostly smooth (Fig. 10.87); sterna 6–9 have minute, spiniform setae including on their posterior margins, and the distal third of sternum 9 bears long, brown, bristle-like setae. Sterna 3–5 are each divided into five sternites and sterna 6–9 into three sternites. The caudal appendages are pyramidal, shorter than the middorsal length of S8 + S9 + S10, sharply pointed (Fig. 10.88), and mostly smooth and have long, reddish-brown, bristle-like setae on the margins. The epiproct and paraprocts are triangular. The epiproct is 0.25× longer than its basal width, and its dorsal margin appears straight in lateral view. The paraprocts are slightly longer than the epiproct, with the tips long, slender, and parallel. The cerci are digitiform, the shortest of the caudal appendages, and approximately half as long as the epiproct, and their tips are short (Fig. 10.88).

Measurements (in mm). The total length including caudal appendages is 18.2–20.3; abdomen (ventral, including appendages) 10.9–12.4; maximum width of the abdomen 7.2–7.7; maximum width of the head 6.5–6.8; and length of hind femur 5.9–6.5.

10.17.3 Remarks

Macrodiplax is found in "open coastal areas, especially behind mangroves" (Orr 2005), and the nymphs inhabit "riverine pools and lagoons, ponds, swamps; it is salt-tolerant" (Theischinger and Hawking 2006). *Macrodiplax* is essentially a tropical genus with only two known species: *M. balteata* (Hagen, 1861), which occurs in America, and *M. cora* (Brauer, 1867), which occurs mainly from southern Asia to the western Pacific, Australia, and some Indian Ocean islands (Theischinger and Hawking 2006). Only *M. cora* is known to occur in Thailand (Hämäläinen and Pinratana 1999).

10.17.4 Distribution

Macrodiplax has been recorded from the provinces of Bangkok, Chon Buri, Nakhon Sawan, Songkhla (Hämäläinen and Pinratana 1999), Samut Sakhon, and Surat Thani (Day et al. 2012) and more recently from Trat (Kosterin 2015). ***New province record:*** Ranong (Fig. 10.89).

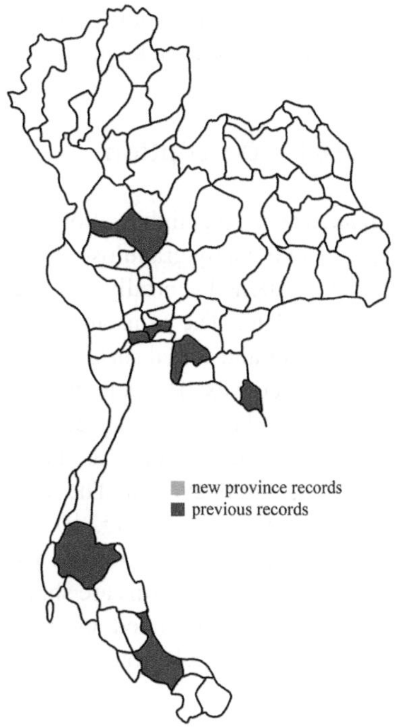

Fig. 10.89 Distribution of *Macrodiplax*

10.18 *Nannophya* Rambur, 1842

Fig. 10.90 Dorsal habitus of the nymph of *Nannophya pygmaea*

10.18.1 General Appearance

The nymphs of *Nannophya* are very small-sized (7.5–11.0 mm), with the body creamy-pale, robust, and moderately setose, compound eyes set well forward on the head, occiput large, hind legs long, abdomen lacking dorsal protuberances but with short posterolateral spines on S8–S9, S10 short and sunken into S9, and caudal appendages short, thick, and sharply pointed (Fig. 10.90).

10.18.2 Description

Head. The head is light-yellowish-brown, large, roughly subtrapezoidal, wider than long, and dorsally granulose (Fig. 10.91). The labrum anterior margin is biconcave and not flattened ventrally and bears abundant stiff, long, golden setae. The clypeus is mostly bare. The frons anterior surface is vertical; the anterior margin is convex; the anterior and dorsal surfaces are beset with long, stiff, yellow setae; and its dorsal surface is convex. The antennae are short and seven-segmented; the scape and pedicel are light-brown and barrel-shaped; the flagellomeres are yellow, cylindrical, and short; and antennomere 7 is the longest. The vertex is shallowly convex and three ocelli are barely evident. The compound eyes are rounded, prominent and set well forward on the head (Fig. 10.91), and the posterior margin is emarginated. The occiput is large and is covered with short, stiff setae except on the nine oval, bare areas; the lateral margins are shallowly convex and tend to converge posteriorly; the hind corners are rounded and bear short, spiniform setae. The posterior margin of the head is shallowly concave (Fig. 10.91).

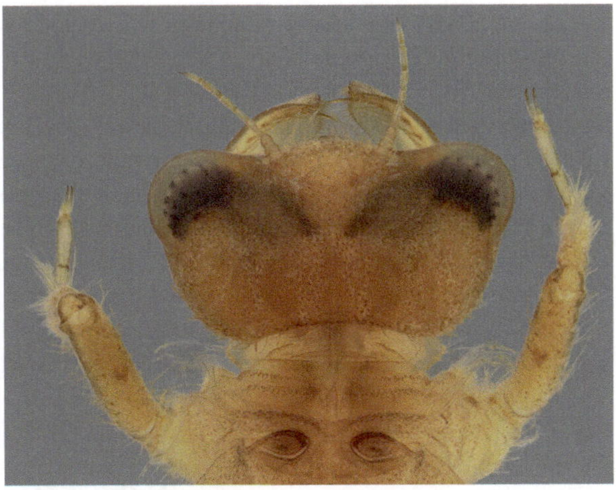

Fig. 10.91 The head and prothorax in dorsal view. Notice the large mesospiracles

Mandibles have the formula L 1234 0 ab'bd/R 123 + 4 y abd, with d > a = b > b' in the left mandible and d > b > a in the right mandible. The maxilla's galeolacinia has six teeth; two of the dorsal teeth are slightly incurved and of similar lengths and robustness, and the basal tooth is minute; the three ventral teeth are of different sizes and robustness, and the apical tooth is the largest. The ventral pad of the hypopharynx is wide and more or less kidney-shaped, and it has abundant long setae on the apical half of its surface and on the lateral margins; the posterior margin is strongly concave.

Fig. 10.92 The prementum in ventral view

Labium. The prementum-postmentum articulation reaches the posterior margin of the prosternum. The prementum is subpentagonal with the lateral margins setose, straight, and widely divergent to the apex (Fig. 10.92); 6 + 3 to 6 + 4 premental setae are on each side of the midline. The ligula is well-developed, triangular, and finely serrulated on the apical margin; it bears four regularly spaced submarginal setae and one seta at the tip. The labial palp apical margin is finely serrulated with weakly developed crenations (Fig. 10.93), and the middle crenations each bear a group of spiniform setae arranged in a graded series of three; the dorsolateral margin has a row of stout setae; the ventral margin is very finely serrulated and has a submarginal row of spiniform setae; there are six palpal setae. The movable hook is relatively short, slender, sharply pointed, and smooth. Two small, spiniform setae are at the palp's articulation.

Fig. 10.93 The head and prementum in frontal view

Thorax. The prothorax is narrower than the head (Fig. 10.91). The pronotal disc is transverse and subrectangular; its posterior margin is thick, widely convex, but straight in the middle, and the sides are straight and bear stiff setae; a group of small, stiff setae is in the middle of the pronotum. The proepisternum anterior margin is strongly convex and setose with long setae combined with spiniform setae. The dorsal part of the mesepisternum is flat, subtriangular, and beset with small, spiniform setae, whereas the remainder of the thorax is mostly bare. The mesospiracles are large (Fig. 10.91). The legs are moderately hairy and pale and lack dark bands (Fig. 10.90); the fore and middle legs are relatively short, and the hind legs are relatively long (Fig. 10.90) (i.e., when fully extended, they far surpass S10); the femora and tibiae have a row of long, stout, stiff setae mainly on the dorsal margins; the distal fourth of the ventral (internal) surface of the fore tibia and the apex of the middle tibia have tetradentate setae, and the remainder of the legs have simple, short, spiniform setae and long, delicate setae; the pretarsal claws are simple and each has a pulvilliform empodium. The anterior and posterior wing sheaths are parallel and reach the posterior margins of S6 and S7, respectively (Fig. 10.90); the posterior wing sheaths are half as wide as long. The anterior margin of the mesolaterosternite is widely convex and fringed with abundant long, stiff setae. The anterior margin of the metasternellum is straight and 0.75× as wide as the metasternellum midline length.

Fig. 10.94 Ventral habitus of the nymph of *Nannophya pygmaea*

Abdomen. The abdomen is semicircular, widest at S6, and abruptly truncate posteriorly (Figs. 10.90 and 10.94). The terga are covered with small spiniform setae and long, sparse, bristle-like setae. Middorsal protuberances are lacking (Fig. 10.95). The posterolateral spines on S8–S9 are short, slightly incurved, sharply pointed, 0.2 mm in length, and 0.50× and 0.75× as long as the middorsal length of their respective segments. S10 is very short and sunken into S9 (Fig. 10.95). The sterna are creamy-pale (Fig. 10.94); sterna 1–6 are mostly smooth, sterna 7–10 have minute spiniform setae, and sternum 9 bears a row of long, stiff, bristle-like setae on the posterior margin. Sterna 3 and 6–9 are each divided into three sternites and sterna 4–5 into five sternites. The caudal appendages are short, thick, and setose and form a compact anal pyramid (Fig. 10.95). The epiproct and paraprocts are triangular. The epiproct is as long as its basal width. The paraprocts are slightly longer than the epiproct and their tips are divergent. The cerci are digitiform, their mesal and lateral margins strongly converge in the distal halves, and they are 0.75× as long as the epiproct.

Fig. 10.95 The abdomen and caudal appendages in lateral view

Measurements (in mm). The total length including caudal appendages is 7.5; abdomen (ventral, including appendages) 4.0; maximum width of the abdomen 3.2; maximum width of the head 2.8; and length of hind femur 2.3.

10.18.3 Remarks

Nannophya is a small genus with only five species of small to tiny dragonflies known commonly as "pygmyflies." They occur in Asia and Australia and are found in "boggy seepages and swamps" (Theischinger and Hawking 2006) and also in "grassy borders of lakes, drains and marshes in open country or at forest margins" (Orr 2005). In Thailand, only *N. pygmaea* is known (Hämäläinen and Pinratana 1999).

10.18.4 Distribution

Nannophya has been recorded from only Loei Province (Hämäläinen and Pinratana 1999). ***New province records:*** Krabi, Phangnga, Phuket, and Trang (Fig. 10.96).

Fig. 10.96 Distribution of *Nannophya*

10.19 *Nannophyopsis* Lieftinck, 1935

In the absence of specimens to examine, the description presented here has been adapted from Yeh and Lien (1995).

10.19.1 General Appearance

The nymphs of *Nannophyopsis* are small-sized (9 mm), with the body pale-green to light-brown with a complex pattern of dark spots and dots, hind legs long, abdomen broadly oval with dorsal protuberances on S4–S9 and posterolateral spines on S7–S9, and S10 annular and sunken into S9.

10.19.2 Description

Head. The head is twice wider than long. The anterior margin of the labrum is prominent and angular. The clypeus has slender, light-brown setae. The anterior surface of the frons is vertical and has two small tubercles slightly above the antennae. The antennae are seven-segmented, the scape and pedicel are thick and barrel-shaped, the flagellomeres are cylindrical and subequal in length, and antennomere 6 is the longest. The vertex is apparently small. The compound eyes are rounded and strongly prominent laterally and have several dark streaks laterally. The occiput is relatively large and covered with slender, pale setae, and a tuft of blackish, bristle-like setae is close to the mesal end of the non-functional part of the compound eyes; the lateral margins are more or less sinuate and strongly convergent posteriorly; the hind corners are rounded. The posterior margin of the head is widely concave.

Labium. The prementum-postmentum articulation reaches the posterior margin of the mesosternum. The prementum is subrhomboidal, the ventral surface is sparsely mottled, and the lateral margins are widely concave and divergent to the apex; 7 + 5 to 8 + 4 premental setae are on each side of the midline. The ligula is moderately developed and triangular, its apical margin is weakly sinuate, and a submarginal row of spiniform setae is on each side of the midline, although the tip lacks setae. The labial palp apical margin is finely serrulated and has six strong crenations, each bearing one long, spiniform seta; the ventral margin has a row of elongate, regularly spaced, spiniform setae; there are nine palpal setae plus one setella. The movable hook is relatively long, sharply pointed, and smooth.

Thorax. The thorax is narrower than the head. The pronotal disc is narrow, the anterolateral corners slightly protrude laterally, and the posterolateral and posterior margins are widely convex. The proepimeron bears long setae on the ventral margin. The legs are long, robust, pale, setose, spiny, and relatively long (i.e., when fully extended, the hind legs surpass the caudal appendages); the femora and tibiae have dark rings. The anterior and posterior wing sheaths are parallel and reach the basal half of S6.

Abdomen. The abdomen is broadly oval, slightly depressed, widest at S7, and strongly truncate posteriorly. Terga 6–9 have a dark, longitudinal, middorsal stripe flanked by a pale spot; the posterior margins of S5–S9 and lateral margins of S4–S9 are covered with small, spiniform setae, which increase in number and robustness posteriorly. The middorsal protuberances are well-developed and spine-like on S4–S7 and low and reduced on S8–S9. The lateral margins of S7–S9 each end in a short, incurved, sharply pointed spine, much shorter than the dorsal length of the segment; the lateral spines of S9 reach the tips of the cerci. S10 is very short, annular, and sunken into S9. The caudal appendages are short, thick, and sharply pointed. The epiproct is triangular and as long as its basal width. The paraprocts are pyramidal and slightly longer than the epiproct; their dorsal and ventral margins have rows of stout, spiniform setae and bristle-like setae; and their tips are long, slender, and straight. The cerci are more or less digitiform, the shortest of the caudal appendages, and approximately 0.5× as long as the epiproct.

Measurements (in mm). The total length including caudal appendages is 9.0; abdomen (ventral, including appendages) 6.7; maximum width of the abdomen 5.1; maximum width of the head 3.2; and length of hind femur 3.7.

10.19.3 Remarks

Nannophyopsis is found in lakes, ponds, and sluggish, shaded streams (Yeh and Lien 1995, Orr 2001, Sribal et al. 2018); the nymphs live among aquatic grass roots (Yeh and Lien 1995). In Thailand, only *N. clara* (Needham, 1930) has been recorded (Sribal et al. 2018).

10.19.4 Distribution

In Thailand, *Nannophyopsis* was recently recorded from Bueng Kan Province (Sribal et al. 2018) (Fig. 10.97).

Fig. 10.97 Distribution of *Nannophyopsis*

10.20 *Neurothemis* Brauer, 1867

Fig. 10.98 Dorsal habitus of the nymph of *Neurothemis*

10.20.1 General Appearance

The nymphs of *Neurothemis* are small-sized (14.3–16.5 mm), with the body yellowish-brown to dark-brown; legs pale; abdomen with a middorsal, longitudinal, pale stripe, dorsal protuberances lacking, and posterolateral spines on S8–S9 small; S10 annular and sunken into S9 (Fig. 10.98); and caudal appendages sharply pointed and slightly decurved (Fig. 10.102).

10.20.2 Description

Head. The head is roughly subtrapezoidal, wider than long, and wider than the thorax and has a complex color pattern (Fig. 10.99). The labrum is light-brown, and the anterior margin is biconcave and rounded ventrally and bears short, golden setae. The clypeus is mostly bare and light-brown and has a large, central, pale spot. The frons is yellow, shallowly convex dorsally, straight on the anterior margin, and beset with delicate, golden setae. The antennae are short, thin, and seven-segmented (Fig. 10.99); the scape is light-brown and the pedicel brown, both are barrel-shaped, the flagellomeres are cylindrical and yellow, and antennomeres 6 and 7 are the longest. The vertex is yellow with a complex pattern of brown spots and dots, and three ocelli are evident. The compound eyes are rounded, prominent anterolaterally, and diffusely striped (Fig. 10.99).

Fig. 10.99 The head in dorsal view

The occiput is large, yellow with brown dots (Fig. 10.99), and covered with short and long, stout, spiniform setae and bristle-like setae on the posterior margin, and it has three oval, bare areas in the middle; the lateral margins are more or less straight, not bulging, and tend to converge posteriorly. The posterior margin of the head is shallowly convex in the middle. Mandibles have the formula L 1234 y- ab'b/R 1234 y abcdx, with a > b > b' in the left mandible and b > d > a > c > x in the right mandible; a rugose, dull, oval area is immediately below the basalmost molar tooth. The maxilla's galeolacinia has seven teeth; the three dorsal teeth are slightly incurved, the two apicalmost teeth are of similar lengths and robustness, and the basal tooth is shorter and slender; the four ventral teeth are of different sizes and robustness, the basal tooth is minute, and the apical tooth is the largest. The ventral pad of the hypopharynx is light-brown, wide, and more or less kidney-shaped, and abundant long, anteriorly directed setae are on most of its surface and margins; the posterior margin is strongly concave.

Fig. 10.100 Ventral habitus of the nymph of *Neurothemis*

Labium. The prementum-postmentum articulation reaches the anterior margin of the mesosternum (Fig. 10.100). The prementum is subpentagonal, light-yellowish-brown, and sparsely mottled on the ventral surface; the lateral margins are smooth, concave, and widely divergent to the apex (Fig. 10.100), and a ventrolateral, longitudinal carina is on each side; 8 + 4 premental setae are on each side of the midline. The ligula is well-developed, translucent, triangular, and finely serrulated on the apical margin and bears a row of 8–10 small, submarginal, spiniform setae on each side of the midline and 2 small, submarginal setae at the tip; its dorsal surface is covered with abundant regularly spaced, minute, spiniform setae.

Fig. 10.101 The head and labial palps in frontal view showing the mottled pattern

The labial palp is creamy-pale and abundantly mottled on the lateral surface; the apical margin is finely serrulated and has weakly developed crenations (Fig. 10.101), each bearing a single, large, spiniform seta; the dorsolateral margin is smooth; the ventral margin is finely serrulated and has a row of regularly spaced, elongate, spiniform setae; there are 9–11 palpal setae and a row of 3–5 min setae is at the basomesal surface. The movable hook is relatively short (i.e., shorter than the preceding palpal seta), sharply pointed, and smooth; a dorsal row of 2–3, small, spiniform setae is at the palp's articulation.

Thorax. The prothorax is narrower than the head. The pronotal disc is mostly pale, and a wide, longitudinal, brown stripe is in the middle (Fig. 10.98); its posterior margin is convex; the anterolateral margins are straight and end posteriorly in small, slightly angled tubercles; and the posterolateral margins are rounded and beset with a group of large and short spiniform setae intermingled with bristle-like setae. The proepisternum anterior margin is straight and furnished with a row of large, stout, spiniform setae, which are followed by elongate setae. The thoracic pleuron is mostly pale with a longitudinal, wide, brown stripe. The legs are setose and pale and the hind legs are relatively long (Figs. 10.98 and 10.100) (i.e., when fully extended, they surpass S10); the femora have two dark rings and a dorsal row of long, spiniform setae; the fore and middle tibiae bear rows of long, stout, stiff setae mainly on the dorsal and ventral margins, tridentate setae are on the mesal surfaces, and the remainder of the fore and middle tibiae have simple, short and long setae; the hind tibia has short, stout, spiniform setae on the ventral surface; all tarsi have tridentate setae in the ventromesal rows; the pretarsal claws are simple and each has a pulvilliform empodium. The anterior and posterior wing sheaths are parallel (Fig. 10.98) and light-brown and reach the posterior margins of S5 and S6, respectively. The anterior margin of the mesolaterosternite is convex and fringed with stout setae; the setal sockets give a granular appearance. The anterior margin of the metasternellum is straight and 0.75× as wide as the metasternellum midline length.

Abdomen. The abdomen is ovate, pyriform, truncate posteriorly, widest at S6 (Fig. 10.98), elevated along the middorsal line, and more or less flattened ventrally, so it appears triangular in cross section. S1–S5 are mostly creamy-pale; S6–S10 are reddish-brown to dark-brown; S6–S10 have a longitudinal, pale stripe that extends to the basal half of the epiproct or slightly farther (Fig. 10.98); and S1–S9 tergal surfaces and posterior margins are covered with small, spiniform setae and long, delicate, whitish setae. Middorsal protuberances are lacking (Fig. 10.102). A row of spiniform setae on the lateral margins of S5–S9 end in small, slightly incurved, and sharply pointed posterolateral spines on S8–S9; these spines are 0.50× and 0.75× the dorsal lengths of S8 and S9, respectively. S10 is very short and sunken into S9. The sterna (Fig. 10.100) are variable in coloration, from pale on 1–5, brown on 6–10, to uniformly brown with white spots on laterosternites 2–9, or pale with three transverse, reddish-brown stripes.

Fig. 10.102 The abdomen and caudal appendages in lateral view

The posterior margins of sterna 1–5 and 10 are smooth, and those of 6–9 have a row of small, spiniform setae. Sterna 3–5 are each divided into five sternites and sterna 6–9 into three sternites. The caudal appendages are long, sharply pointed (Fig. 10.102), and as long as the dorsal length of S8 + S9, and the tips are slightly decurved. The epiproct and paraprocts are pyramidal, and the mesal surfaces are covered abundantly by short, stout, spiniform setae. The paraprocts have a row of 5–6 robust, spiniform setae along the midline of the lateral surface. The epiproct is one third longer than its maximum basal width. The cerci are digitiform and half as long as the epiproct.

Measurements (in mm). The total length including caudal appendages is 14.3–16.5; abdomen (ventral, including appendages) 8.5–10.0; maximum width of the abdomen 4.5–5.7; maximum width of the head 4.3–5.0; and length of hind femur 3.5–4.4.

10.20.3 Remarks

Neurothemis is found in "lakes, marshes, drains, swampy grasslands, padi, sluggish streams, and well vegetated ponds in open and disturbed habitats" (Orr 2005). In Thailand, four species have been recorded: *N. fluctuans* (Fabricius, 1793), *N. fulvia* (Drury, 1773), *N. intermedia* (Rambur, 1842), and *N. tullia* (Drury, 1773) (Hämäläinen and Pinratana 1999).

10.20.4 Distribution

Neurothemis is widespread and very common, especially in lowlands, and has been recorded from the provinces of Ang Thong, Bangkok, Chachoengsao, Chai Nat, Chaiyaphum, Chanthaburi, Chiang Mai, Chiang Rai, Chon Buri, Chumphon, Kalasin, Kamphaeng Phet, Kanchanaburi, Krabi, Lampang, Loei, Lop Buri, Mae Hong Son, Mukdahan, Nakhon Nayok, Nakhon Pathom, Nakhon Ratchasima, Nakhon Sawan, Nakhon Si Ayutthaya, Nakhon Si Thammarat, Narathiwat, Nong Khai, Nonthaburi, Pathum Thani, Pattani, Phangnga, Phetchaburi, Phrae, Phuket, Prachuap Khiri Khan, Ranong, Rayong, Sa Kaew, Sakon Nakhon, Samut Prakan, Samut Sakhon, Saraburi, Satun, Sing Buri, Songkhla, Tak, Trang, Trat, Ubon Ratchathani, Uttaradit, Yala (Hämäläinen and Pinratana 1999), Buri Ram, Khon Kaen, Phatthalung, Phayao, Prachin Buri, Roi Et, Suphan Buri, Surat Thani, Yasothon (Thipaksorn et al. 2003), Nan, Phitsanulok, and Ratchaburi (Ferro et al. 2009) and more recently from Phetchabun (Day et al. 2012) (Fig. 10.103).

Fig. 10.103 Distribution of *Neurothemis*

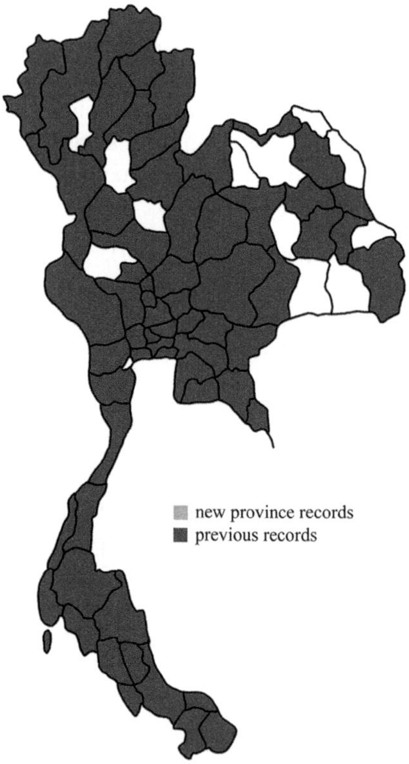

10.21 *Onychothemis* Brauer, 1868

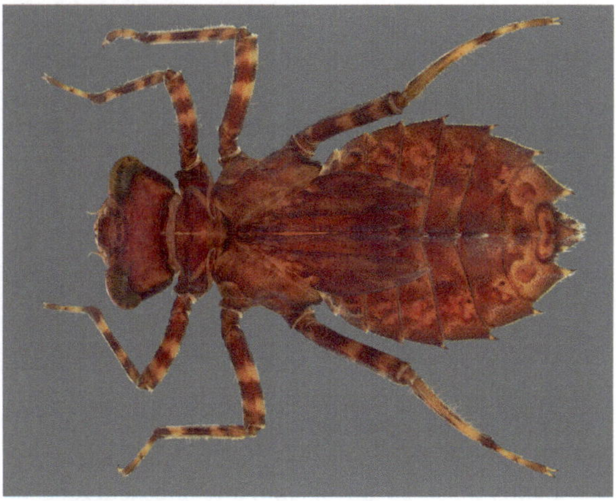

Fig. 10.104 Dorsal habitus of the nymph of *Onychothemis*

10.21.1 General Appearance

The nymphs of *Onychothemis* are large-sized (20–26 mm), with the body robust and stocky, legs long, abdomen strongly convex dorsally and highly arched, dorsal protuberances on S2–S8 as strong hooks, posterolateral spines on S5–S9 large, S10 annular and sunken into S9, and caudal appendages sharply pointed and straight (Fig. 10.104).

10.21.2 Description

Head. The head is subtrapezoidal and twice wider than long (Fig. 10.105). The labrum is light-brown, and the anterior margin is straight and beset with a brush of long, golden setae on each side of the midline. The clypeus is glabrous. The frons anterior surface is flat and high, its anterior margin is straight, and the dorsal surface is shallowly concave and granulose. The antennae are seven-segmented and very short and almost reach the basal half of the labrum; the scape and pedicel are brown and subspherical, the flagellomeres are light-yellow (Fig. 10.105), antennomere 3 is the longest, and antennomeres 1 and 6 are the shortest. The vertex is small and the ocelli are almost undistinguishable. The compound eyes are rounded, prominent anterolaterally, and diffusely striped (Fig. 10.105). The occiput is large; the posterolateral margins are straight and covered with minute spiniform setae, tend to converge posteriorly, and end in a posterolateral, small, swollen, blunt, posteriorly directed tubercles. The posterior margin of the head is widely concave, but straight in the middle.

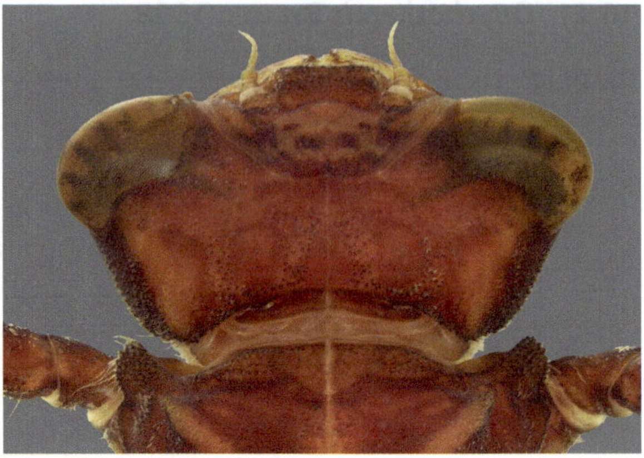

Fig. 10.105 The head in dorsal view

Mandibles have the formula L 1234 0 b/R 1234 y b, and the tooth “a” in both mandibles is lacking. The maxilla’s galeolacinia has six teeth; the two dorsal teeth are slightly incurved and of similar lengths and robustness; the four ventral teeth are of different sizes and robustness, the basal tooth is the shortest, and the apical tooth is the largest. The ventral pad of the hypopharynx is light-brown, wide, and more or less kidney-shaped and has abundant long, anteriorly directed setae on most of its surface and margins; the posterior margin is strongly concave.

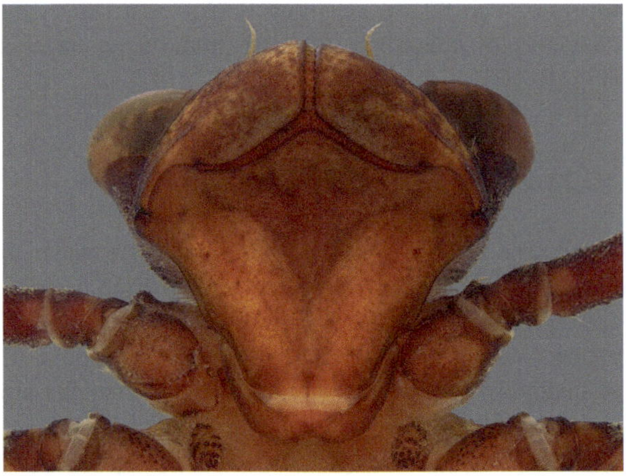

Fig. 10.106 The prementum in ventral view

Labium. The prementum-postmentum articulation reaches the basal half of the mesosternum. The prementum is subrhomboidal; the lateral margins are setose, sinuate, and divergent to the apex (Fig. 10.106); and it bears a ventrolateral, longitudinal carina on each side; 8 + 3 or 8 + 4 premental setae are arranged in a more or less straight line on each side of the midline. The ligula is well-developed, triangular, widely rounded at the tip (Fig. 10.106), and finely serrulated on the apical margin and has a submarginal row of large setae intermingled with short setae on each side of the midline, as well as 2–3 large setae at the tip. The labial palp is mottled on the lateral surface; the apical margin is finely serrulated and has weakly developed crenations (Fig. 10.107), each bearing stout, spiniform setae arranged in a graded series of 4–5 setae; the dorsolateral margin has a row of short and long, spiniform setae; the ventral margin is finely serrulated and has a submarginal row of elongate, spiniform setae; 9 palpal setae plus 2 setellae and a group of 5–6 min, spiniform setae are present at the basomesal surface of the palp. The movable hook is short, thick, sharply pointed, and smooth; a row of 2–3 small, robust, spiniform setae is at the palp's articulation.

Fig. 10.107 The head and labial palps in frontal view

Thorax. The prothorax is narrower than the head (Fig. 10.104). The pronotal disc has the lateral and posterior margins thick, raised, and roughly granulose (Fig. 10.105). The proepisternum is strongly convex anteriorly and has abundant small, stout, spiniform setae and long setae on the anteroventral corner. The pterothorax is granulose. The legs are long (i.e., when fully extended, the hind leg surpasses the tip of the abdomen) (Fig. 10.108), uniformly colored, granulose, spiny, and scaly and have diffuse dark bands on the femora and tibiae and long setae on the femora; the fore tibia has tridentate setae on the apical fifth of the mesal surface, and the middle and hind tibiae bear stout, spiniform setae; only the fore tarsus has tridentate setae on the ventromesal row, and the middle and hind tarsi have a double row of short, robust, spiniform setae on the ventral surface; the pretarsal claws are simple and each has a pulvilliform empodium. The anterior and posterior wing sheaths are parallel and reach the basal half and posterior margin of S6, respectively (Fig. 10.104). The mesolaterosternite has a large, bulky tubercle, which is densely covered with short, stout, scale-like setae (Figs. 10.106 and 10.108) and fringed anteriorly with stiff setae; the metalaterosternite has a low tubercle on the anterior margin, which is mostly covered with short, scale-like setae. The anterior margin of the metasternellum is straight and 0.60× as wide as the metasternellum mid-line length.

Fig. 10.108 Ventral habitus of the nymph of *Onychothemis*

Abdomen. The abdomen is almost circular and highly arched resembling a turtle shell (Fig. 10.109), the highest point is at S5, the sides are strongly convex, and it is widest at S6 (Fig. 10.108). Terga 3–10 have small spiniform setae, including on the posterior margins of S5–S10. The middorsal protuberances are covered with small, spiniform setae and thick setae on the tips on S2–S8, and the protuberances are expressed as vertical spines on S2–S3 and strong hooks on S4–S8 (Fig. 10.109); the hooks on S4–S7 surpass the anterior margin of the next segment, and the hook on S8 only reaches the basal half of its own segment (Fig. 10.109); the posterior half of S8 and S9–S10 form a slanting, middorsal ridge. The lateral margins of S3–S9 bear small, closed-packed, spiniform setae; those of S5–S9 are keeled and end in strong spines increasing in size and robustness posteriorly (Fig. 10.108); the spines on S5–S6 are divergent, on S7–S9 are moderately incurved, and on S9 are 0.60× as long as the middorsal length of S9 and reach the tips of the cerci. S10 is very short and sunken into S9. S7–S10 are tightly fused and without intersegmental membranes. The sterna are smooth; sterna 3–4 are each divided into five sternites and sterna 5–9 into three sternites. The female gonapophyses are vestigial and gomphid-like. The caudal appendages are pyramidal, short, sharply pointed (Fig. 10.109), and almost as long as half the dorsal length of S9, and the tips are straight. The epiproct is 0.10× longer than its basal width, and strong spiniform setae are on the lateral margins and ventral surface. The paraprocts' mesal surface and dorsal and ventrolateral margins are furnished with strong spiniform setae, and the ventral surface is smooth. The cerci are digitiform, smooth, and half as long as the epiproct.

Fig. 10.109 The abdomen and caudal appendages in lateral view

Measurements (in mm). The total length (including caudal appendages) is 20.5–25.7; abdomen (ventral, including appendages) 12.5–14.2; maximum width of the abdomen 10.0–10.2; maximum width of the head 6.3–6.6; and length of hind femur 6.3–6.5.

10.21.3 Remarks

Onychothemis is found in lowlands and hill country in "clear forest streams … with nymphs inhabiting in riffles or slower water with a sandy bottom" (Orr 2005). In Thailand, only two species have been recorded: *O. culminicola* Förster, 1904, and *O. testacea* Laidlaw, 1902 (Hämäläinen and Pinratana 1999).

10.21.4 Distribution

Onychothemis has been recorded from the provinces of Chaiyaphum, Chanthaburi, Chiang Mai, Chumphon, Kanchanaburi, Krabi, Nakhon Nayok, Phetchaburi, Prachuap Khiri Khan, Ranong, Saraburi, Songkhla, Tak, Trang (Hämäläinen and Pinratana 1999), Nakhon Si Thammarat, Phangnga, Surat Thani (Ferro et al. 2009), Nan, and Phrae (Katatani et al. 2010) and more recently from Nakhon Ratchasima, Phitsanulok, and Ratchaburi (Day et al. 2012). ***New province records:*** Mae Hong Son, Phatthalung, and Yala (Fig. 10.110).

Fig. 10.110 Distribution of *Onychothemis*

10.22 *Orchithemis* Brauer, 1878

In the absence of specimens to examine, the description presented here has been adapted from Butler (2013).

10.22.1 General Appearance

The nymphs of *Orchithemis* are small-sized (15 mm), with the body robust and light-brown, head and abdomen with dark markings, head rounded and with long setae, compound eyes small, abdomen with dorsal protuberances on S4–S9 and posterolateral spines on S8–S9, S10 set off from S9, and caudal appendages long and sharply pointed.

10.22.2 Description

Head. The head is small, rounded, wider than long and wider than the prothorax, and light-brown with darker patches around the ocelli and on the occiput, and it has patches of long setae. The anterior margin of the frons is slightly protuberant. The antennae are seven-segmented with scattered, long setae on antennomeres 1–6, and antennomere 3 is the longest. The compound eyes are small, rounded, very slightly prominent laterally, and slightly curved posteriorly. The posterolateral margins of the occiput tend to converge posteriorly. The posterior margin of the head is shallowly concave.

Labium. The prementum-postmentum articulation reaches the anterior margin of the mesosternum. The prementum is subpentagonal and slightly wider than long, and the distal half of the lateral margin is straight and widely divergent to the apex; long premental setae on each side of the midline are lacking, and approximately 15 small, stout setae are on each side of a central fossa arranged as an inverted "U." The ligula is poorly developed and sinuate on the apical margin and has a series of seven shallow dentations on each side of the midline with a small seta between each dentation. The labial palp distal margin is straight and bears eight indistinct crenations, with two spiniform setae between crenations; the ventral margin is weakly serrulated and has small setae; there are six short palpal setae. The movable hook is longer and stouter than the palpal setae and sharply pointed.

Thorax. The prothorax is narrower than the head. The pronotal disc is saddle-shaped, raised on the anterior margin, and fringed with long setae on the anterolateral corners, and its posterior margin is widely convex and thick. The legs have no visible banding; the hind legs are relatively long; the femora and tibiae have abundant long and short setae, except on the hind legs where they become sparser and more spiny. The anterior and posterior wing sheaths are parallel and reach the basal half of S5.

Abdomen. The abdomen is ovate and widest at S6–S7, and the pale coloration on the basal segments gradually becomes darker on the posterior segments, and S10 is dark-brown; longitudinal dark stripes, which darken posteriorly, are on the dorsum of S6–S9; the surface appears finely granulose and is covered with minute setae and long, delicate setae, mainly laterally. The posterior margins of S2–S9 are fringed with spiniform setae, which become longer and flattened posteriorly. Well-developed middorsal protuberances are on S4–S9, which are densely beset with setae that gradually increase in length posteriorly; the lateral margins of S8–S9 end in stout, sharp, and slightly incurved posterolateral spines, with those on S9 the longest. S10 is very short and set off from S9. The caudal appendages are long, sharply pointed, and as long as the middorsal length of S9 + S10. The epiproct is longer than its basal width and slightly shorter than the paraprocts. The paraproct tips are incurved. The cerci are straight and less than half the length of the paraprocts.

Measurements (in mm). The total length excluding caudal appendages is 15; maximum width of the abdomen 5.25; and maximum width of the head 4.25.

10.22.3 Remarks

Orchithemis is widespread in Sundaland (Lieftinck 1954) and is known from Cambodia (Kosterin 2012), Thailand, Vietnam, and the Philippines (Orr 2005); it is found in "forest marshes of the plains and hills … and in tiny rivulets" (Lieftinck 1954). In Thailand, only *Orchithemis pruinans* (Selys, 1878) and *O. pulcherrima* Brauer, 1878, have been recorded (Hämäläinen and Pinratana 1999, Pinratana 2003).

10.22.4 Distribution

Orchithemis has been recorded from the provinces of Chanthaburi (Hämäläinen and Pinratana 1999), Narathiwat (Pinratana 2003), and Phatthalung (Ferro et al. 2009) and more recently from Phangnga and Sa Kaew (Day et al. 2012) (Fig. 10.111).

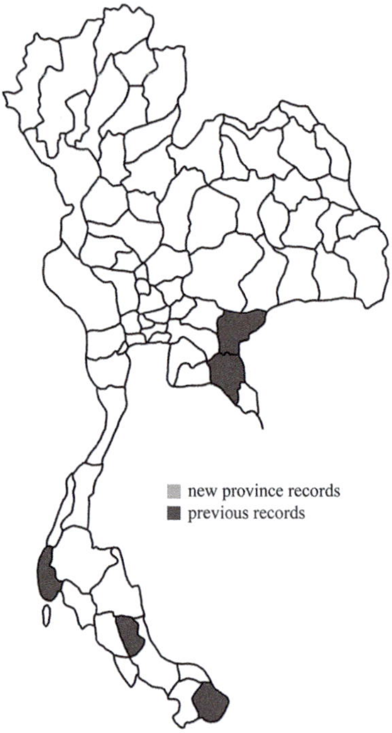

Fig. 10.111 Distribution of *Orchithemis*

10.23 *Orthetrum* Newman, 1833

Fig. 10.112 Dorsal habitus of the nymph of *Orthetrum sabina*

10.23.1 General Appearance

The nymphs of *Orthetrum* are medium-sized (18–21 mm) and highly variable, with the body light-brown anteriorly, dark-brown posteriorly, robust, and setose, head subrectangular, compound eyes small and produced dorsally, abdomen with dorsal protuberances large or reduced and posterolateral spines on S8–S9 short, S10 annular and set off from S9, and caudal appendages long and sharply pointed (Fig. 10.112).

10.23.2 Description

Head. The head is small, subrectangular (Fig. 10.113), almost twice wider than long, and narrower than the thorax. The labrum is pale with three brown spots; the anterior margin is biconcave and slightly flattened ventrally and bears long, golden setae. The clypeus is mostly bare. The frons anterior surface is vertical and shallowly convex; the anterior margin is beset with very long, golden setae; and its dorsal surface is shallowly convex and setose. The antennae are seven-segmented; the scape is wide and cylindrical, the pedicel is barrel-shaped, and both are longitudinally bicolored with brown and yellow; the flagellomeres are cylindrical, yellow to light-brown, and subequal in length, and antennomeres 3, 6, and 7 are the longest (Fig. 10.113).

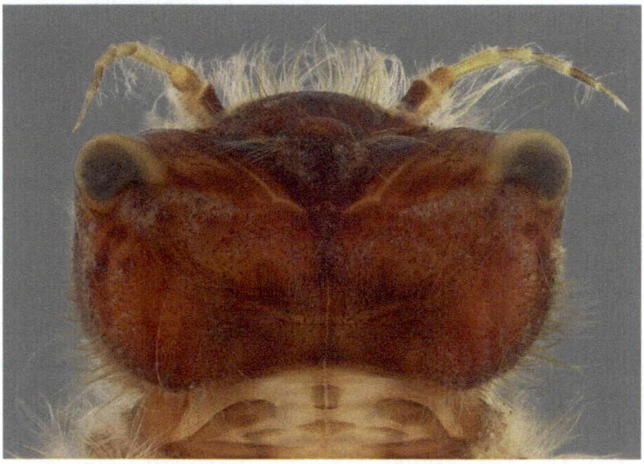

Fig. 10.113 The head in dorsal view

The vertex is yellow with brown spots and has long bristle-like setae; three ocelli are visible. The compound eyes are small, rounded, and prominent dorsally (Figs. 10.113 and 10.115). The occiput is large, with long, reddish-brown, bristle-like setae and 7–9 large, oval, bare areas; the lateral margins are shallowly convex; the posterior corners are rounded and fringed with very long, stiff, golden setae (Fig. 10.113). The posterior margin of the head is slightly concave.

Mandibles have the formula L 1234 0 or y- ab/R 123 + 4 or 1 + 2, 3 + 4 y- abd, with b > a in the left mandible and b > d = a or d > b > a in the right mandible. The maxilla's galeolacinia is mostly smooth and has seven teeth; the three dorsal teeth are slightly incurved, the two apicalmost teeth are of similar lengths and robustness, and the basal tooth is shorter and slender; the four ventral teeth are of different sizes and robustness, the basal tooth is the shortest, and the apical tooth is the largest. The ventral pad of the hypopharynx is more or less kidney-shaped and has abundant long setae on the basal half of its surface and on the anterior and lateral margins, and the anterior half of its surface is glabrous and shining; the posterior margin is concave.

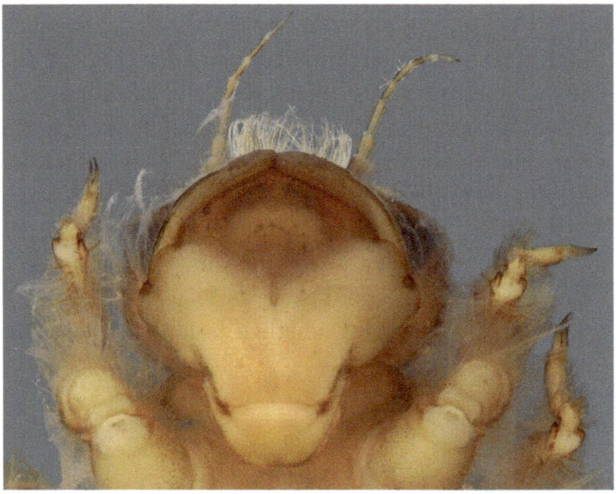

Fig. 10.114 The prementum in ventral view

Labium. The prementum-postmentum articulation reaches the posterior margin of the prosternum. The prementum is subrhomboidal and slightly longer than its maximum width, and the distal half of the lateral margin is straight, fringed with delicate setae, and widely divergent to the apex (Fig. 10.114); some species have three long premental setae on each side of the midline, with many small, spiniform setae in between, arranged as an inverted "U," whereas others have 4 + 5 long setae arranged in "U" plus a group of 8–10 small setae without any particular arrangement, and others more have 7 + 5 long setae arranged in "U." The ligula is well-developed and strongly crenate on the apical margin (Fig. 10.115) with the central crenation the largest and extending onto the dorsal surface as a swollen tubercle; a minute, spiniform seta is between each crenation; the ventral surface of the ligula is covered with long, stiff, regularly spaced setae.

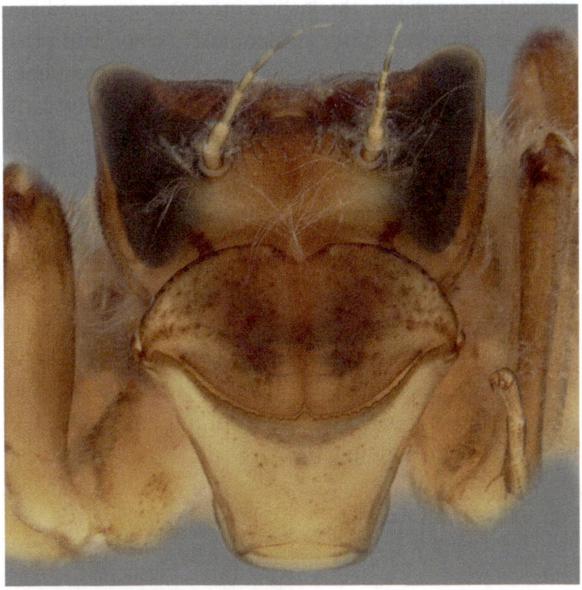

Fig. 10.115 The head and prementum in frontal view. Notice the dorsally prominent compound eyes

The labial palp is yellow and densely mottled (Fig. 10.115) and bears long, white, delicate setae on the lateral surface; the apical margin is smooth and has nine indistinct crenations, each bearing two spiniform setae of different sizes; the dorsolateral margin is smooth; the ventral margin is smooth and has a row of minute, regularly spaced, spiniform setae; there are 5–9 palpal setae. The movable hook is short and smooth, its basal half is brown and thick, and the apical half is slender and sharply pointed. A row of 5–8 short, stout, spiniform setae is at the palp's articulation.

Fig. 10.116 Ventral habitus of the nymph of *Orthetrum sabina*

Thorax. The prothorax is shorter than the head and about as wide (Fig. 10.112). The pronotal disc is mostly light-brown and has a middorsal, longitudinal, pale stripe bearing or lacking a group of short, stiff setae; the anterolateral corners have a pale spot, the posterior margin is widely convex and thick, and the sides are straight or rounded, thick, bulging anteriorly, and sparsely set with short, brown setae or abundant long, golden, stiff setae. The proepisternum is very short, its anterior margin is straight, and its ventral margin has both minute and long setae. The pterothorax is yellow to light-brown and setose, but smooth in the creamy-pale areas. The legs are yellow and relatively short (Fig. 10.116); the femora and tibiae have mesal and lateral brushes of long, golden setae (Figs. 10.112 and 10.114); the tibiae have a row of long, stout, stiff, brown setae on the ventral (internal) surface; most of the mesal surface of the fore tibia and the ventromesal row of the fore tarsus have small, tridentate setae, and the remainder of the legs have simple, short and long setae; the pretarsal claws are simple and each has a pulvilliform empodium. The anterior and posterior wing sheaths are parallel and reach the basal half and the posterior margin of S5, respectively (Fig. 10.112). The anterior margin of the mesolaterosternite is widely convex and beset with stiff setae. The anterior margin of the metasternellum is straight and 0.20–0.40× as wide as the metasternellum midline length.

Fig. 10.117 The abdomen and caudal appendages in lateral view

Abdomen. The abdomen is elongate, fusiform, and widest at S6; S1–S5 are mostly creamy-pale and S6–S10 dark-brown. Terga 2–9 bear short and long setae, including on the posterior margins of 2–5; the posterior margins of 6–9 and the lateral margins of 2–10 are fringed with spiniform setae. Middorsal protuberances are well-developed on S4–S7 (Fig. 10.117) or S4–S8 or strongly reduced on S4–S7 or only S4–S5, and all are beset with stiff setae. The posterolateral spines on S8–S9 are pale, short, sharply pointed, and 0.25–0.50× and 0.35–0.67× as long as the dorsal lengths of their respective segments. S10 is very short and set off from S9 (Fig. 10.117). Sterna 1–6 are yellow and sterna 7–10 light-brown (Fig. 10.116). The sterna have minute and long setae and are fringed with spiniform setae only on 7–9. Sterna 3–5 are each divided into five sternites and sterna 6–9 into three sternites. The caudal appendages are long, sharply pointed (Fig. 10.117), and as long as or longer than the dorsal length of S8 + S9 + S10, and the tips are long and straight. The epiproct and paraprocts are pyramidal, and their mesal surfaces are covered by stout, spiniform setae; the lateral margins of the epiproct and the dorsal and ventral margins of the paraprocts bear a row of spiniform setae intermingled with long, stiff setae. The paraprocts have long, stout setae on the lateral surfaces. The epiproct and paraprocts are slightly more than twice longer than their respective basal widths. The cerci are digitiform and 0.40× as long as the epiproct.

Measurements (in mm). The total length including caudal appendages is 18–21; abdomen (ventral, including appendages) 12.8–13.0; maximum width of the abdomen 5.8–6.2; maximum width of the head 4.1–4.3; and length of hind femur 4.5–4.8.

10.23.3 Remarks

Orthetrum is widespread in the temperate and tropical areas of the Old World (Theischinger and Hawking 2006). It is found in forested and cultivated areas (Hämäläinen and Pinratana 1999), with the nymphs occurring in "a wide range of situations including clear forest streams to disturbed open habitats as ponds and marshes … over a range of current speeds and substrates" (Orr 2005). In Thailand, eight species have been recorded (Hämäläinen and Pinratana 1999).

10.23.4 Distribution

Orthetrum is widespread and very common in Thailand, especially in lowlands, and has been recorded from the provinces of Bangkok, Chachoengsao, Chai Nat, Chaiyaphum, Chanthaburi, Chiang Mai, Chiang Rai, Chon Buri, Chumphon, Kalasin, Kanchanaburi, Khon Kaen, Krabi, Lampang, Lamphun, Loei, Lop Buri, Mae Hong Son, Maha Sarakham, Nakhon Nayok, Nakhon Pathom, Nakhon Ratchasima, Nakhon Sawan, Nakhon Si Ayutthaya, Nakhon Si Thammarat, Narathiwat, Nonthaburi, Pathum Thani, Pattani, Phangnga, Phetchabun, Phetchaburi, Phrae, Phuket, Prachuap Khiri Khan, Ranong, Rayong, Roi Et, Sa Kaew, Sakon Nakhon, Samut Prakan, Saraburi, Satun, Songkhla, Tak, Trang, Ubon Ratchathani, Uttaradit, Yala, Yasothon (Hämäläinen and Pinratana 1999), Phatthalung (Thipaksorn et al. 2003), Kamphaeng Phet, Phitsanulok, Ratchaburi, Surat Thani, Uthai Thani (Ferro et al. 2009), and Nan (Katatani et al. 2010) and more recently from Prachin Buri, Samut Sakhon, Sukhothai, and Trat (Day et al. 2012). ***New province records:*** Si Sa Ket, Suphan Buri, and Udon Thani (Fig. 10.118).

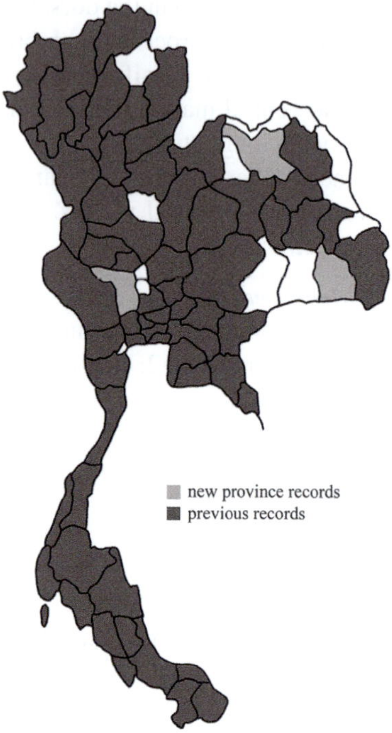

Fig. 10.118 Distribution of *Orthetrum*

10.24 *Palpopleura* Rambur, 1842

Fig. 10.119 Dorsal habitus of the nymph of *Palpopleura* (left fore and hind legs broken)

10.24.1 General Appearance

The nymphs of *Palpopleura* are small-sized (13–16 mm), with the body pale-brownish-yellow and with distinct whitish stripes, head subpentagonal, legs banded, abdomen lacking dorsal protuberances but with small posterolateral spines on S8–S9, S10 annular and sunken into S9, and caudal appendages sharply pointed (Fig. 10.119).

10.24.2 Description

Head. The head is subpentagonal, almost twice wider than long, and wider than the thorax (Fig. 10.120). The labrum is mostly pale with sides dark brown. The clypeus is mostly brown and bare. The anterior surface of the frons is vertical and bare; its anterior margin is straight in dorsal view and beset with long, stiff setae; and its dorsal surface is slightly convex. The antennae are seven-segmented, the scape and pedicel are barrel-shaped, the flagellomeres are cylindrical, and the 6th and 7th antennomeres are the longest. The vertex is slightly convex with some sparse bristle-like setae; three ocelli are distinguishable. The compound eyes are rounded and prominent laterally (Fig. 10.120). The occiput is relatively short, pale, mottled, and covered with long spiniform setae and some bristle-like setae on posterior margin, and the lateral margins are brown and straight and tend to converge posteriorly; the posterior corners are rounded. The posterior margin of the head is concave. Mandibles have the formula L 1234 0 abb'/R 1234 -y abd, with b > a = b' in the left mandible and d > b > a in the right mandible. The maxilla's galeolacinia has seven teeth; the three dorsal teeth are slightly incurved, the two apicalmost teeth are of similar lengths and robustness, and the basal tooth is shorter and slender; the four ventral teeth are of different sizes and robustness, the basal tooth is minute, and the apical tooth is the largest. The ventral pad of the hypopharynx is wide and more or less kidney-shaped and has large setae on the surface and margins; the posterior margin is widely concave.

Fig. 10.120 The head in dorsal view

Labium. The prementum-postmentum articulation reaches the anterior margin of the middle coxae. The prementum is subrhomboidal and yellow (Fig. 10.121); the lateral margins have a row of minute, white setae; and the lateral margins are widely divergent to the apex; 9 + 3 long premental setae are on each side of the midline. The ligula is well-developed, triangular, and finely serrate, and a submarginal row of spiniform setae is along its apical margin, including the tip.

Fig. 10.121 The prementum in ventral view

The labial palp apical margin is straight with obsolete crenations, but it has alternating long and short setae; the dorsolateral margin has a row of 3–7 small, spiniform setae on basal third; the ventral margin bears a submarginal row of 12 elongate, spiniform setae; there are 9–10 palpal setae. The movable hook is long and sharply pointed. A row of 2–4 small spiniform setae is on the palp's articulation.

Thorax. The prothorax is narrower than the head. The pronotal disc is mostly pale, and its sides are rounded, thick (Fig. 10.120), and set with three short, stout, spiniform setae. The proepisternum is shallowly concave anteriorly, and its ventral margin is fringed with long setae. The pterothorax is brown laterally and paler dorsally. The legs are yellowish-brown and moderately setose, and the hind legs are relatively long (i.e., when fully extended, they far surpass abdominal segment 10); the femora have two dark bands and bear 3–5 long, robust, spiniform setae; the tibiae have two dark bands and are furnished with a row of fine spines and long bristle-like setae, which gradually increase in length from the distal to the proximal end. The anterior and posterior wing sheaths are parallel and reach the posterior margin of S5 and basal third of S6, respectively. The anterior margin of the mesolaterosternite is convex and fringed with long, stiff setae. The anterior margin of the metasternellum is straight and 0.75× the metasternellum midline length.

Fig. 10.122 Ventral habitus of the nymph of *Palpopleura*

Abdomen. The abdomen is semicircular and widest at S6 and has a light-brown coloration (Figs. 10.119 and 10.122). Terga 2–10 have a whitish, wide, longitudinal, middorsal stripe bordered by a diffuse, dark stripe; the dorsal surfaces and posterior margins of S2–S9 and lateral margins of S8–S9 are covered with minute spiniform setae. Middorsal protuberances are lacking (Fig. 10.123). Posterolateral spines are short and sharply pointed on S8–S9 and measure 0.1 and 0.2 mm in length, respectively. S10 is very short and is sunken into S9. Sterna 1–9 are yellow and sparsely mottled, and sterna 7–9 are light-brown with some transverse, darker spots (Fig. 10.122). Sterna 3 and 6–9 are each divided into three sternites and sterna 4–5 into five sternites. The caudal appendages are as long as the middorsal length of 8 + 9 and sharply pointed, and their tips are straight. The epiproct and paraprocts are pyramidal. The epiproct is longer than its basal width. The paraprocts are longer than the epiproct. The cerci are digitiform and 0.60× the length of epiproct (Fig. 10.123).

Fig. 10.123 The abdomen and caudal appendages in ventrolateral view

Measurements (in mm). The total length including caudal appendages is 13–16; abdomen 7.5–10.0; maximum width of the abdomen 5.0–5.1; maximum width of the head 4.3–5.5; and length of hind femur 5.5–6.0.

10.24.3 Remarks

Palpopleura is widespread in Asia and is found in disturbed open habitats, breeding in marshy areas and small pools (Dow 2009); the nymphs cling to submerged plants (Tze-wai et al. 2011). In Thailand, only *Palpopleura sexmaculata* (Fabricius, 1787) has been recorded (Hämäläinen and Pinratana 1999).

10.24.4 Distribution

Palpopleura sexmaculata is a common local species in mountains and upland areas (Hämäläinen and Pinratana 1999). It is found in Northern and Central Thailand, but not in Southern Thailand (Dow 2009). More specifically, it has been recorded from the provinces of Bangkok, Chaiyaphum, Chanthaburi, Chiang Mai, Chiang Rai, Kanchanaburi, Lampang, Loei, Mae Hong Son, Nakhon Nayok, Rayong, Sakon Nakhon, Saraburi, Sa Kaew, and Uttaradit (Hämäläinen and Pinratana 1999) and more recently from Phetchabun (Day et al. 2012) (Fig. 10.124).

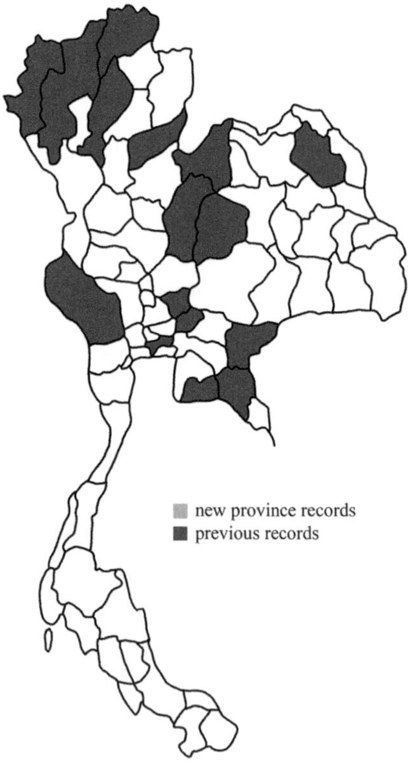

Fig. 10.124 Distribution of *Palpopleura*

10.25 *Pantala* Hagen, 1861

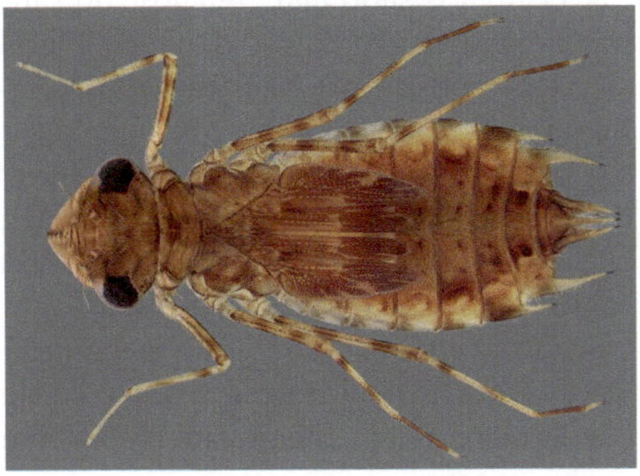

Fig. 10.125 Dorsal habitus of the nymph of *Pantala flavescens*

10.25.1 General Appearance

The nymphs of *Pantala* are medium-sized (22–24 mm), with the body yellowish-brown, hind legs short, abdomen with small dorsal protuberances on S2–S4 and very large posterolateral spines on S8–S9, S10 annular, and caudal appendages long and sharply pointed (Fig. 10.125).

10.25.2 Description

Head. The head is wider than long and yellowish-brown (Fig. 10.126). The labrum is yellow and bordered with brown, and its anterior margin is biconcave and slightly flattened ventrally in the middle and has long setae. The clypeus is light-brown, bare, and creamy-pale in the middle. The frons anterior surface is vertical and flat, its anterior margin is rounded and beset with short setae, and its dorsal surface is convex. The antennae are seven-segmented and light-brown, the scape is annular, the pedicel is barrel-shaped, the flagellomeres are cylindrical, and antennomere 3 is the longest. The vertex is bare and shallowly convex and has three ocelli.

Fig. 10.126 The head in dorsal view

The compound eyes are rounded and strongly prominent laterally (Fig. 10.126). The occiput is short and granulose dorsally and has nine oval, smooth areas; the posterior lobes are poorly developed, not bulging, and beset with strong, short, spiniform setae. The posterior margin of the head is straight. Mandibles have the formula L 1234 0 ab/R 1234 y abcd, with b > a in the left mandible and b > a > d > c in the right mandible. The maxilla's galeolacinia has seven teeth; the three dorsal teeth are slightly incurved, the two apicalmost teeth are of similar lengths and robustness, and the basal tooth is shorter and slender; the four ventral teeth are of different sizes and robustness, the basal tooth is minute, and the apical tooth is the largest. The ventral pad of the hypopharynx is wide and more or less kidney-shaped, and large setae are on the surface and margins; the lateral margins have long, stout, spiniform setae; the posterior margin is shallowly concave.

Fig. 10.127 The prementum in ventral view

Labium. The prementum-postmentum articulation reaches the anterior margin of the metasternum. The prementum is subrhomboidal and yellow (Fig. 10.127); its lateral margins are sinuate, divergent to the apex, and fringed with short setae; 10 + 6 to 12 + 5 premental setae are on each side of the midline. The ligula is well-developed and triangular, and the apical margin is finely serrulated and lacks marginal setae. The labial palp is yellow, and its apical margin is finely serrulated and has 9 strongly developed crenations (Fig. 10.128), each bearing a graded series of 3–4 spiniform setae; the dorsolateral margin has a row of small, blunt, spiniform setae; the ventral margin is finely serrulated and bears a submarginal row of elongate, stout, regularly spaced, spiniform setae; 13–14 palpal setae and some minute spiniform setae are at the basomesal surface. The movable hook is relatively short, stout, sharply pointed, and smooth. A row of four short, stout, regularly spaced, spiniform setae is at the palp's articulation.

Fig. 10.128 The head and labial palps in frontal view

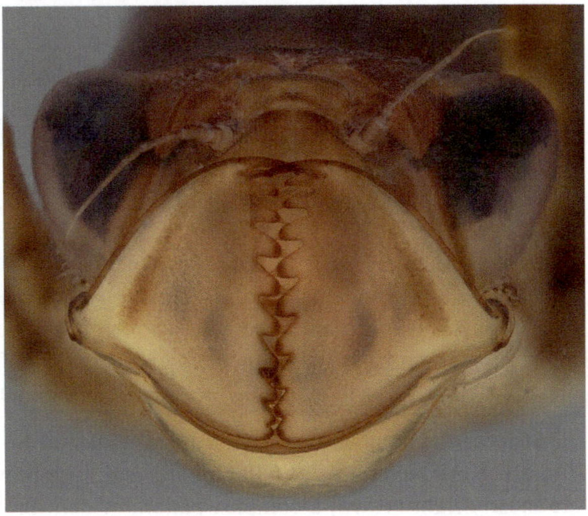

Thorax. The prothorax is narrower than the head. The pronotal disc is mostly bare and pale, and a pair of round, brown spots is on each side of the midline; the lateral margins are black, rounded, and set with a row of small, spiniform setae; its posterior margin is widely convex and mostly smooth. The proepisternum anterolateral corner is strongly convex and bears a few spiniform setae and a tuft of long setae. The pterothorax is pale dorsally, dark-brown laterally, and granulose. The legs are pale, spiny, and with four dark bands on the femora and two dark bands on the tibiae (Fig. 10.125); the hind legs are relatively short (i.e., when fully extended, they reach the tips of paraprocts) (Figs. 10.125 and 10.129); the tibiae are fringed with long, delicate, whitish setae on the dorsal (external) margins; the distal ends of the fore and middle tibiae and ventrolateral row of the tarsi bear tridentate setae, and the remainder of legs have stout, short, spiniform setae; the fore tarsus is yellow, and the middle and hind tarsi are brown; the pretarsal claws are simple and each has a pulvilliform empodium.

Fig. 10.129 Ventral habitus of the nymph of *Pantala flavescens*

The wing sheaths are parallel and light-brown and have a transverse, yellow stripe at the basal 0.70 and 0.50 of the length of the anterior and posterior wing sheaths, respectively (Fig. 10.125); they are mostly bare but have a row of spiniform setae on the costal and anal margins, and they reach the basal half and the posterior margin of S6, respectively. The anterior margin of the mesolaterosternite is shallowly convex and has short, stout, spiniform setae. The anterior margin of the metasternellum is straight and 0.70× as wide as the metasternellum midline length.

Abdomen. The abdomen is ovoid and widest at S6 (Figs. 10.125 and 10.129). Terga 1–5 are creamy-pale in the middle and light-brown laterally, and terga 6–10 are mostly light-brown; a complex pattern of dark-brown spots and dots is on the sides of 4–5 and over the entire surface of S6–S9 (Fig. 10.125). Most of the dorsal surface of S5–S9 and posterior margins of S5–S10 are covered with small, spiniform setae; the lateral margins of S3–S9 are fringed with small, spiniform setae, which increase in number and robustness posteriorly.

Fig. 10.130 The abdomen and caudal appendages in lateral view

The middorsal protuberances on S2–S4 are small and spine-like; on S5–S9, they are lacking; and on S10, they are reduced to a minute to small, low, triangular to convex projection on the posterior margin (Fig. 10.130). The posterolateral spines on

S8–S9 are very large (1.5 and 2.5 mm in length, respectively), sharply pointed (Figs. 10.125 and 10.129), straight on S8, incurved and reaching the tips of the cerci on S9, and 1.3× and >2.0× longer than the middorsal lengths of their respective segments (Fig. 10.130). S10 is annular, short, and set off from S9. The sterna are creamy-pale and have diffuse, light-brown spots on S9–S10 (Fig. 10.129); sterna 2–9 are glabrous, but have minute to small spiniform setae on their posterior margins. Sterna 3–4 are each divided into five sternites and sterna 5–9 into three sternites. The caudal appendages are pyramidal, long, slender distally, sharply pointed (Fig. 10.130), and strongly spiny on the margins and have long, white, bristle-like setae; they are mostly creamy-pale, but basally light-brown with the tips dark-brown. The epiproct is almost twice longer than its basal width. The paraprocts are 0.98× as long as the epiproct (Fig. 10.130); their dorsal and ventral margins have rows of stout, spiniform setae and bristle-like setae; and their tips are long, slender, slightly recurved, and almost twice longer than their basal width. The cerci are narrowly elongate, the shortest of the caudal appendages, and 0.80× as long as the epiproct and have long tips (Fig. 10.130).

Measurements (in mm). The total length including caudal appendages is 22.0–23.5; abdomen (ventral, including appendages) 14.5–16.4; maximum width of the abdomen 7.8–8.3; maximum width of the head 5.7–6.1; and length of hind femur 6.3–7.0.

10.25.3 Remarks

Pantala contains only two species: *P. flavescens* (Fabricius, 1798) and *P. hymenaea* (Say, 1839). *Pantala hymenaea* occurs in only the New World; however, *P. flavescens* is a migrant species widely distributed in the tropical and subtropical areas of both the New and Old Worlds; it breeds in practically any lentic habitat in open country, "from ephemeral desert pools to waters on small isolated tropical islands" (Suhling and Martens 2007). The nymphs of *P. flavescens* exhibit fast development time and can tolerate high water temperature (Suhling and Martens 2007).

10.25.4 Distribution

Pantala flavescens is very common and widely distributed in Thailand; it has been recorded from the provinces of Bangkok, Chachoengsao, Chai Nat, Chaiyaphum, Chanthaburi, Chiang Mai, Chon Buri, Kalasin, Kamphaeng Phet, Kanchanaburi, Krabi, Loei, Lop Buri, Nakhon Nayok, Nakhon Pathom, Nakhon Ratchasima, Nakhon Si Ayutthaya, Nakhon Si Thammarat, Pathum Thani, Phangnga, Phetchaburi, Phuket, Prachuap Khiri Khan, Ranong, Ratchaburi, Samut Sakhon, Satun, Songkhla, Surat Thani, Trang, Trat, Ubon Ratchathani (Hämäläinen and Pinratana 1999), Chiang Rai (Katatani et al. 2004), Khon Kaen, Lop Buri, Phatthalung (Thipaksorn et al. 2003), Mukdahan, Phitsanulok, Uthai Thani (Ferro et al. 2009), and Phrae (Katatani et al. 2010). ***New province records:*** Mae Hong Son, Si Sa Ket, and Udon Thani (Fig. 10.131).

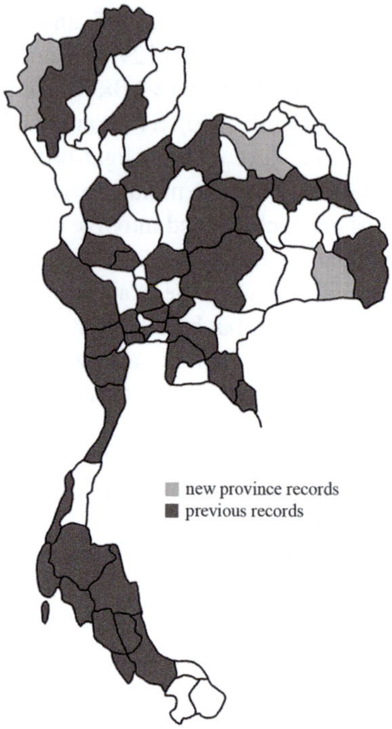

Fig. 10.131 Distribution of *Pantala*

10.26 *Potamarcha* Karsch, 1890

Fig. 10.132 Dorsal habitus of the nymph of *Potamarcha congener*

10.26.1 General Appearance

The nymphs of *Potamarcha* are small-sized (16.5–18.0 mm), with the body light-brown anteriorly, dark-brown posteriorly, and setose, compound eyes small, hind legs long, abdomen with small dorsal protuberances on S4–S5 and small posterolateral spines on S8–S9, and S10 short and sunken into S9 (Fig. 10.132).

10.26.2 Description

Head. The head is large, twice wider than long, and mostly brown with a complex pattern of dark and pale markings (Fig. 10.133). The labrum is light-brown laterally and creamy-pale in the middle, and the anterior margin is biconcave and beset with long, golden setae. The clypeus is mostly yellowish-brown and bare. The frons anterior surface is flat and vertical; the anterior margin is straight in dorsal view and fringed with very long, golden setae and scattered, red, bristle-like setae. The antennae are long, almost as long as the head, and seven-segmented; the scape and pedicel are barrel-shaped, the flagellomeres are cylindrical, the scape is dark-brown, the remainder of the antennomeres are yellow, and antennomeres 3, 6, and 7 are the longest. The vertex is shallowly convex and has tufts of dark, long, bristle-like setae and three ocelli. The compound eyes are small and dorsally prominent (Figs. 10.133 and 10.135). The occiput is relatively large and covered with small, brown, spiniform setae; the lateral margins tend to converge posteriorly and are furnished with stout, spiniform setae; the cephalic lobes are widely rounded. The posterior margin of the head is straight and beset with short, spiniform setae.

Fig. 10.133 The head in dorsal view. Notice the long antennae

Mandibles have the formula L 12’234 0 ab/R 123 + 4 y- abd, with the tooth “b” huge and >a in the left mandible and b > a > d in the right mandible. The maxilla’s galeolacinia has seven teeth almost straight; the three dorsal teeth are very slightly incurved, the two apicalmost teeth are of similar lengths and robustness, and the basal tooth is shorter and slender; the four ventral teeth are of different sizes and robustness, the basal tooth is small, and the subapical tooth is largely fused with the largest apical tooth in a manner that it appears as a small cuspid at the midlength of the apical tooth. The ventral pad of the hypopharynx is wide and more or less kidney-shaped and bears large setae on the surface and margins; the posterior margin is shallowly concave.

Fig. 10.134 The prementum in ventral view

Labium. The prementum-postmentum articulation reaches the anterior margin of the mesosternum (Fig. 10.136). The prementum is subpentagonal and yellow, and its lateral margins are smooth, sinuate, and widely divergent to the apex (Fig. 10.134); 8 + 3 or 8 + 4 premental setae are on each side of the midline. The ligula is translucent, more or less triangular, and moderately crenate on the apical margin, and it has a spiniform seta between each crenation, with the largest crenation at the tip. The labial palp is yellow, its apical margin is strongly crenate (Fig. 10.135), and it has 10 well-developed crenations, each bearing a graded series of 3–4 spiniform setae; the dorsolateral margin of the palp is smooth; the ventral margin has a row of spiniform setae; 8 palpal setae plus a row of 3–4 min spiniform setae are at the basomesal surface. The movable hook is slender, slightly thicker basally, sharply pointed, and smooth. A tight row of 4–5 small, spiniform setae is at the palp's articulation.

Fig. 10.135 The head and prementum in frontal view. Notice the strong crenations on the labial palps and the dorsally protuberant compound eyes

Thorax. The prothorax is narrower than the head (Fig. 10.132). The pronotal disc sides are straight, the posterior corners are rounded, the posterolateral margins are shallowly convex, the middle part of the posterior margin is short and straight, and all the margins are beset with small, reddish-brown setae and long, bristle-like setae. The ventral margin of the pleuron is strongly setose with long, golden, delicate setae; the proepisternum is very short and strongly convex ventrally. The pterothorax is light-brown to light-reddish-brown and has scattered, long and short, reddish-brown, bristle-like setae. The legs are setose; the coxae, the trochanters, the distal halves of tibiae, and the tarsi are creamy-pale; the femora are light-brown with a subbasal dark band; the hind legs are relatively long (Fig. 10.132) (i.e., when fully extended, the tip of the tibia reaches the posterior margin of S10); the ventral (internal) surfaces of the tibiae have numerous scale-like setae and tridentate setae; the tarsi have tridentate setae in the ventrolateral row; the pretarsal claws are simple and each has a pulvilliform empodium. The anterior and posterior wing sheaths are parallel (Fig. 10.132) and reach the basal half and posterior margin of S5, respectively. The anterior margin of the mesolaterosternite is convex and beset with abundant stiff, stout setae. The anterior margin of the metasternellum is narrowly convex and 0.17× as wide as the metasternellum midline length.

Fig. 10.136 Ventral habitus of the nymph of *Potamarcha congener*

Abdomen. The abdomen is elongate, fusiform, and widest at S6. Terga 1–4 are creamy-pale, 5–10 are brown to dark-brown, most of the surface of 2–10 is smooth and shining, and scattered setae are on the dorsolateral areas (Fig. 10.137); the posterior margin of 5 has a row of short, reddish-brown setae, and those of 6–9 have a row of pale, small, spiniform setae; the lateral margins are setose and bear minute, spiniform setae only on 6–9. Small, low, middorsal protuberances are on S4–S5; a tuft of short, reddish-brown, bristle-like setae diminishing in length posteriorly is on the midline of S4–S9. The posterolateral spines on S8–S9 are small, sharply pointed, 0.2 mm in length, and 0.28× and 0.33× as long as the middorsal length of their respective segments. S10 is very short and sunken into S9.

Fig. 10.137 The abdomen and caudal appendages in lateral view

Sterna 1–6 are creamy-pale, glabrous, and light-reddish-brown on 7–10 (Fig. 10.136); a row of small spiniform setae is on the posterior margins of 8–9, and tufts of very long, golden setae are on the distal half of 9. Sterna 3–5 are each divided into five sternites and sterna 6–9 into three sternites. The caudal appendages (Fig. 10.137) are dark-brown basally and creamy-pale distally with the margins dark-brown, pyramidal, sharply pointed, setose, and spiny on the lateral margins and mesal surfaces. The epiproct is 0.30× longer than its basal width, and its dorsal margin is straight in lateral view. The paraprocts are slightly longer than the epiproct (Fig. 10.137), the dorsal and ventral margins are straight in lateral view, and the tips are parallel. The cerci are digitiform and 0.45× as long as the epiproct.

Measurements (in mm). The total length including caudal appendages is 16.5–18.0; abdomen (ventral, including appendages) 10.1–12.2; maximum width of the abdomen 5.5–6.5; maximum width of the head 4.8–5.0; and length of hind femur 4.0–4.4.

10.26.3 Remarks

Potamarcha is found in "open disturbed habitats … in standing and slowly flowing water" (Orr 2005). The nymphs are sluggish and bottom-dwellers (De Fonseka 2000). In Thailand, only *P. congener* (Rambur, 1842) has been recorded (Hämäläinen and Pinratana 1999).

10.26.4 Distribution

Potamarcha congener is widespread and very common, especially in lowlands, and has been recorded from the provinces of Ang Thong, Bangkok, Chachoengsao, Chai Nat, Chaiyaphum, Chanthaburi, Chiang Mai, Chiang Rai, Chon Buri, Kanchanaburi, Krabi, Lampang, Lamphun, Mae Hong Son, Nakhon Nayok, Nakhon Pathom, Nakhon Sawan, Nakhon Si Ayutthaya, Nakhon Si Thammarat, Narathiwat, Pathum Thani, Phetchaburi, Prachuap Khiri Khan, Ranong, Ratchaburi, Sakon Nakhon, Saraburi, Satun, Songkhla, Tak, Trang, Ubon Ratchathani, Uttaradit, Yala (Hämäläinen and Pinratana 1999), and Phatthalung (Ferro et al. 2009) and more recently from Khon Kaen, Loei, Nan, Phangnga, Phitsanulok, Samut Sakhon, and Surat Thani (Day et al. 2012). ***New province record:*** Phetchabun (Fig. 10.138).

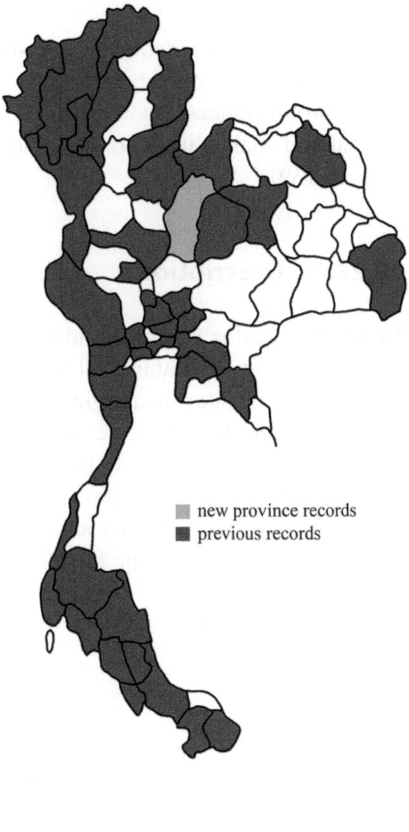

Fig. 10.138 Distribution of *Potamarcha*

10.27 *Rhodothemis* Ris, 1909

Fig. 10.139 Dorsal habitus of the nymph of *Rhodothemis rufa*

10.27.1 General Appearance

The nymphs of *Rhodothemis* are small-sized (15.5–19.3 mm), with the body yellowish-brown to dark-brown; head wide; legs long and banded; abdomen lacking dorsal protuberances but with short, decurved posterolateral spines on S9; S10 annular and sunken into S9; and caudal appendages decurved and sharply pointed (Fig. 10.139).

10.27.2 Description

Head. The head is subtrapezoidal, twice wider than long, wider than the thorax, and light-brown with abundant fine, dark dots (Fig. 10.140). The labrum is dark-brown, and the anterior margin is biconcave and slightly flattened ventrally and bears short and golden setae. The clypeus is mostly bare and dark-brown. The frons anterior surface is flat and vertical, and the anterior margin is straight and fringed with stiff, yellow setae. The antennae are short, thin, and seven-segmented; the scape is brown and ring-shaped, the pedicel is barrel-shaped and brown with a pale spot, the flagellomeres are cylindrical and yellow, and antennomere 3 is the longest. The vertex is yellow with brown dots and three ocelli are evident (Fig. 10.140).

Fig. 10.140 The head in dorsal view. Notice the large occiput

The compound eyes are rounded, strongly prominent laterally (Fig. 10.140), and with seven black, longitudinal stripes. The occiput is large and covered with tufts of bristle-like setae, and seven oval, bare areas are on the posterior margin (Fig. 10.140); the lateral margins are slightly rounded and tend to converge posteriorly. The posterior margin of the head is straight. Mandibles have the formula L 1234 y- ab'b/R 123 + 4 y abcd, with b > a in both mandibles and b > d > c in the right mandible. The maxilla's galeolacinia has seven teeth; the three dorsal teeth are slightly incurved, the two apicalmost teeth are of similar lengths and robustness, and the basal tooth is short and slender; the four ventral teeth are of different sizes and robustness, the basal tooth is minute, and the apical tooth is the largest. The ventral pad of the hypopharynx is wide, pale-brown laterally, and more or less kidney-shaped and has abundant long, anteriorly directed setae on most of its surface and margins; the posterior margin is strongly concave.

Fig. 10.141 The prementum in ventral view

Labium. The prementum-postmentum articulation reaches the anterior margin of the mesosternum. The prementum is subrhomboidal, mostly dark-reddish-brown (Fig. 10.141), and variegated on the ventral surface; the lateral margins are sinuate and widely divergent to the apex and have a row of spiniform setae; 8 + 4, 8 + 5, or 9 + 4 premental setae are on each side of the midline, and sometimes the outermost long seta is separate from the others by a wide interval. The ligula is well-developed (Fig. 10.141) and mottled, its tip is widely convex, and the apical margin is finely serrulated and bears a submarginal row of large, spiniform setae.

Fig. 10.142 The head and prementum in frontal view

The labial palp is creamy-pale and abundantly mottled on both the mesal and lateral surfaces (Fig. 10.142); the apical margin is brown and finely serrulated and has small crenations, each bearing a graded series of 4–5 large, spiniform setae; the dorsolateral margin is yellow and has a row of stout, large, spiniform setae; the ventral margin is finely serrulated and has a row of elongate, spiniform setae; 8–9 palpal setae plus 1 setella and a group of small, spiniform setae are present at the basomesal surface of the palp. The movable hook is relatively short (i.e., shorter than the preceding palpal seta), sharply pointed, and smooth; two short, spiniform setae and a tuft of long, delicate setae are at the palp's articulation.

Fig. 10.143 Ventral habitus of the nymph of *Rhodothemis rufa*

Thorax. The prothorax is narrower than the head (Fig. 10.139). The pronotal disc is uniformly light-brown, the posterolateral margins are rounded and beset with a group of large and short spiniform setae intermingled with bristle-like setae, and its posterior margin is widely convex and elevated; a pair of long, stout, erect, spiniform setae and a group of minute, spiniform setae are on each side of the midline. The proepisternum anterior margin has a row of 4–5 large, stout, spiniform setae followed by long setae. The pterothorax is mostly brown; the mesospiracles are large and elevated, representing the body's highest point in lateral view. The legs are yellow and setose (Figs. 10.139 and 10.143); the hind legs are relatively long (i.e., when fully extended, they far surpass S10); the femora have a subapical dark band; the tibiae are fringed with long, stout, stiff setae mainly on the dorsal (external) and ventral (internal) margins; the mesal surface of the fore tibia and the distal fourth of the middle tibia have tridentate setae, and the remainder of the fore and middle tibiae bear simple, short and long setae; the ventral surface of the hind tibia has short, stout, spiniform setae; the tarsi are creamy-pale except on the tips; only the fore tarsus has tridentate setae in the ventromesal row; the pretarsal claws are simple and each has a pulvilliform empodium. The anterior and posterior wing sheaths are parallel (Fig. 10.139), light-brown, and with long, stout, stiff setae on the margins, and they reach the basal halves of S6 and S7, respectively. The mesolaterosternite is strongly produced anteroventrally, and its anterior margin is more or less straight and fringed with long, stiff setae. The anterior margin of the metasternellum is straight and as wide as the metasternellum midline length.

Fig. 10.144 The abdomen and caudal appendages in lateral view

Abdomen. The abdomen is short, semicircular, widest at S6, truncate posteriorly, and mostly light-brown with a wide, longitudinal, middorsal pale band that is not always discernible, and it bears abundant minute, spiniform setae and brown setal sockets that give it a granulose appearance; the posterior margins of terga 1–9 are fringed with small, spiniform setae. Middorsal protuberances are lacking, but rows of short, stout, stiff setae are along the midline of S4–S9 (Fig. 10.144); a row of small, spiniform setae intermingled with long, brown, stiff, bristle-like setae is on the lateral margins of S5–S10; those on S9 end in a pale, short, decurved, hook-like, sharply pointed, posterolateral spine. S10 is very short and sunken into S9. The sterna are reddish-brown to blackish-brown (Fig. 10.143), the lateral margins are creamy-pale, and a pale spot is on the lateral side of the submedian sutures on sterna 3–7. The posterior margins of sterna 1–4 are smooth and those of 5–9 have a row of spiniform setae. Sterna 3–5 are each divided into five sternites and sterna 6–9 into three sternites. The caudal appendages are sharply pointed (Fig. 10.144), strongly setose, spiny, and as long as the dorsal length of S8 + S9. The epiproct and paraprocts are pyramidal. The epiproct is 0.25× longer than its maximum basal width, and the tip is slightly decurved. The paraprocts have long, hook-like, strongly decurved tips (Fig. 10.144). The cerci are digitiform and divergent, the tips are decurved, and they are 0.65× as long as the epiproct.

Measurements (in mm). The total length including caudal appendages is 15.5–19.3; abdomen (ventral, including appendages) 9.3–10.0; maximum width of the abdomen 6.8–7.2; maximum width of the head 6.0–6.5; and length of hind femur 5.7–6.6.

10.27.3 Remarks

Rhodothemis is found in lakes, marshes, and ponds with abundant floating plants in open and disturbed habitats. In Thailand, only *R. rufa* (Rambur, 1842) has been recorded (Hämäläinen and Pinratana 1999).

10.27.4 Distribution

Rhodothemis rufa is widespread and very common in tropical Asia, with scattered records throughout Thailand (Hämäläinen and Pinratana 1999). It has been recorded from the provinces of Bangkok, Chachoengsao, Chaiyaphum, Chiang Mai, Chon Buri, Kalasin, Nakhon Nayok, Nakhon Si Ayutthaya, Nakhon Si Thammarat, Nonthaburi, Phangnga, Phuket, Rayong, Sa Kaew, Saraburi, Satun, Songkhla (Hämäläinen and Pinratana 1999), Phatthalung, Phetchaburi, and Surat Thani (Ferro et al. 2009) and more recently from Khon Kaen, Nakhon Ratchasima, and Trat (Day et al. 2012). ***New province records:*** Chumphon, Krabi, Narathiwat, Prachuap Khiri Khan, Trang, Ubon Ratchathani, and Yala (Fig. 10.145).

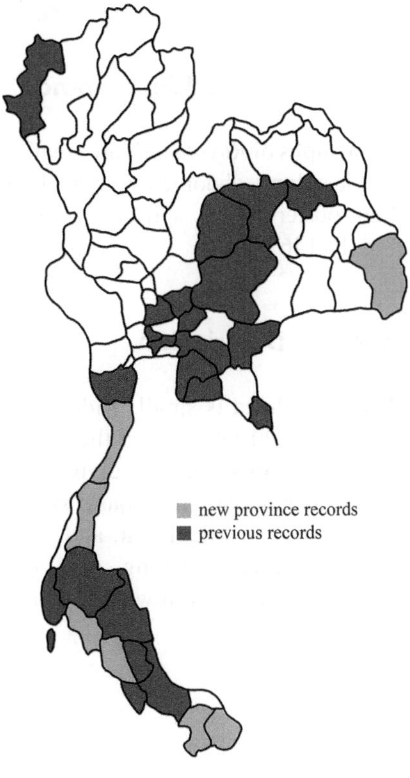

Fig. 10.145 Distribution of *Rhodothemis*

10.28 *Rhyothemis* Hagen, 1867

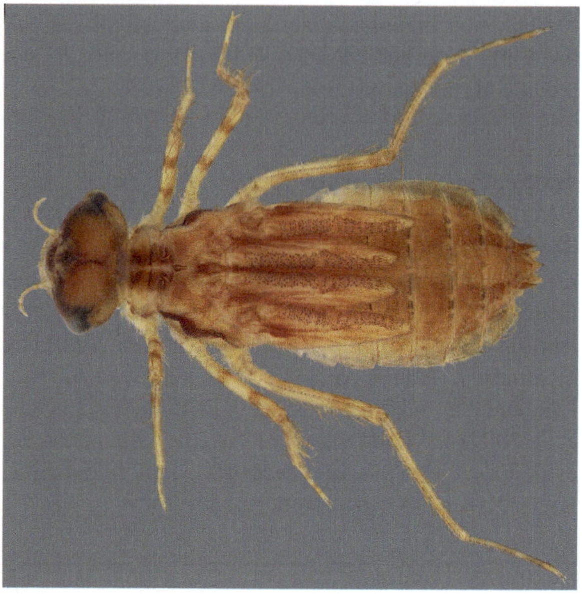

Fig. 10.146 Dorsal habitus of the nymph of *Rhyothemis*

10.28.1 General Appearance

The nymphs of *Rhyothemis* are small-sized (13–15 mm), with the body light-brown, hind legs long, abdomen with well-developed dorsal protuberances on S4–S9 and short posterolateral spines on S8–S9, and S10 annular and sunken into S9 (Fig. 10.146).

10.28.2 Description

Head. The head is small, wider than long, and light-brown (Fig. 10.147). The labrum is light-brown, and the anterior margin is biconcave ventrally and bears abundant short setae. The clypeus is bare and light-brown and has a large, diffuse, brown, rectangular, basal spot on each side of the midline. The frons anterior surface is vertical, flat, and yellow; its anterior margin is convex in frontal view and beset with long, stiff, yellow, posteriorly directed setae; its dorsal surface is convex, yellowish-brown in the middle, and brown laterally.

Fig. 10.147 The head in dorsal view

The antennae are relatively long, thick, and seven-segmented (Fig. 10.147), the scape and pedicel are barrel-shaped, and the flagellomeres are cylindrical but widening distally; antennomere 6 is the longest; antennomeres 1, 2, and 6 are light-brown; and the remainder are yellow. The vertex is convex and light-brown and has three ocelli. The compound eyes are relatively small, rounded, and prominent laterally (Fig. 10.147). The occiput is moderately sized and light-brown and has small spiniform setae and delicate setae on the dorsum; the posterior and lateral margins have long, stiff setae; an anterior, central, shallow concavity is at the union with the vertex; the lateral margins are straight and tend to converge posteriorly. The posterior margin of the head is shallowly concave. Mandibles have the formula L 1234 0 ab/R 1234 y- abd, with a = b> in the left mandible and a = b > d in the right mandible. The maxilla's galeolacinia has seven teeth; the three dorsal teeth are slightly incurved, the two apicalmost teeth are of similar lengths and robustness, and the basal tooth is shorter and slender; the four ventral teeth are of different sizes and robustness, the basal tooth is minute, and the apical tooth is the largest. The ventral pad of the hypopharynx is wide and more or less kidney-shaped, and abundant long setae are on the surface and margins; the posterior margin is strongly concave.

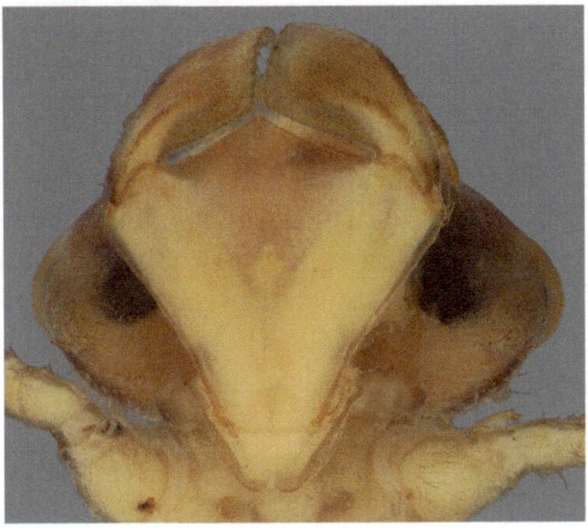

Fig. 10.148 The prementum in ventral view

Labium. The prementum-postmentum articulation reaches the posterior margin of the prosternum. The prementum is short, subrhomboidal, and creamy-pale, its ventral surface is bare, and the lateral margins are smooth, sinuate, and divergent to the apex (Fig. 10.148); 1–2 long premental setae plus 3–4 small setae are on each side of the midline. The ligula is triangular and well-developed (Fig. 10.148), and the apical margin is finely serrulated and has a submarginal row of spiniform setae. The labial palp is light-brown but translucent in the ventral area; the apical margin is finely serrulated and has weak crenations (Fig. 10.149), and between each crenation is one large, spiniform seta; the dorsolateral margin of the palp has 5–6 spiniform setae followed by a row of delicate setae; the ventral margin is finely serrulated and has a row of submarginal, regularly spaced, spiniform setae on the distal half; 5–6 palpal setae plus 3–4 min spiniform setae are at the basomesal surface of the palp's articulation. The movable hook is relatively long, sharply pointed, and smooth. A row of 2–3 short, stout, spiniform setae and a tuft of hair-like setae are at the palp's articulation.

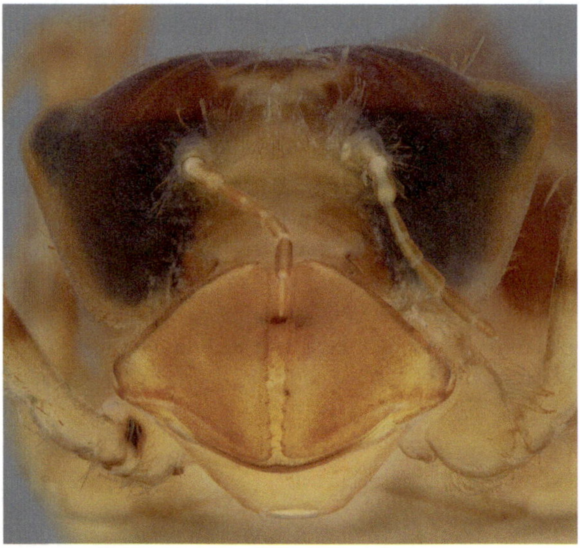

Fig. 10.149 The head and prementum in frontal view

Thorax. The prothorax is narrower than the head (Fig. 10.146). The pronotal disc is pale and has a wide, longitudinal, brown stripe on each side of a middorsal creamy-pale stripe; the sides are widely rounded and beset with short, spiniform setae and long, stout, stiff setae; the posterior margin is thick, widely convex, and elevated at the middle. The proepisternum is strongly protruding ventrally and has stout, hook-like setae and long setae.

Fig. 10.150 Ventral habitus of the nymph of *Rhyothemis*

The pterothorax is largely brown, but is pale dorsally, and appears granular. The legs are setose and pale; the hind legs are relatively long (i.e., when fully extended, the tip of the hind tibia far surpasses the caudal appendages) (Figs. 10.146 and 10.150); two dark rings are on the fore and middle femora and on the ventral (internal) surface of the fore tibia; the distal fourth of the middle and hind tibiae and the ventrolateral row of the fore tarsus bear tridentate setae, whereas the remainder of legs have simple, short and long, spiniform setae and long, hair-like setae; the pretarsal claws are simple and each has a pulvilliform empodium. The anterior and posterior wing sheaths are parallel (Fig. 10.146) and brown and almost reach the basal half of S7 and anterior margin of S8, respectively. The anterior margin of the mesolaterosternite is shallowly convex and beset with long, stiff setae. The anterior margin of the metasternellum is straight and as wide as the metasternellum midline length.

Fig. 10.151 The abdomen and caudal appendages in lateral view

Abdomen. The abdomen is more or less pyriform, widest at S6, and truncate posteriorly (Figs. 10.146 and 10.150). The terga are creamy-pale on S1–S5, and the remainder are light-brown with a pale, middorsal, longitudinal stripe that widens posteriorly; the dorsal surface and posterior margins of S4–S10 and the lateral margins of S6–S10 are covered with small, spiniform setae. The middorsal protuberances are well-developed and spine-like on S4–S9, with that on S9 reaching beyond the posterior margin of S10 (Fig. 10.151). The posterolateral spines on S8–S9 are short and sharply pointed and 0.3 mm in length, their tips are slightly convergent on S8 and strongly convergent on S9, and they are 0.50× and 0.60× as long as the middorsal lengths of their respective segments. S10 is very short and sunken into S9. Sterna 1–7 are light-yellow, sterna 8–9 are light-yellowish-brown (Fig. 10.150), sternum 10 is light-brown, and all have minute, delicate setae; sterna 8–9 and the posterior margins of sterna 7–9 have a row of small, spiniform setae. Sterna 3–5 and 9 are each divided into five sternites and sterna 6–8 into three sternites. The caudal appendages are thick, sharply pointed (Fig. 10.151), and as long as S8 + S9. The epiproct and paraprocts are pyramidal and yellow. The epiproct is as long as its maximum basal width and has a relatively long and straight tip; its dorsal surface and the lateral margins are beset with spiniform setae and have a row of long, stiff setae. The paraprocts are slightly longer than the epiproct (Fig. 10.151); their dorsal and ventral margins have a row of long, stiff setae and spiniform setae on the lateral surfaces; and their tips are widely divergent. The cerci are digitiform, the shortest of the caudal appendages, and 0.60–0.75× as long as the epiproct, and the lateral surfaces are brown and the tips are parallel.

Measurements (in mm). The total length including caudal appendages is 13.5–14.2; abdomen (ventral, including appendages) 7.5–8.3; maximum width of the abdomen 5.6–6.1; maximum width of the head 4.3–4.5; and length of hind femur 4.9–5.1.

10.28.3 Remarks

Rhyothemis is found "in marshy ponds, lakes, drains and sluggish streams … including slightly brackish waters" (Orr 2005); some nymphs cling to aquatic plants, and others are sprawlers among bottom leaf debris (Orr 2005, Tse-wai et al. 2011). In Thailand, five species of *Rhyothemis* have been recorded (Hämäläinen and Pinratana 1999).

10.28.4 Distribution

Rhyothemis is widely distributed throughout Thailand. It has been recorded from the provinces of Bangkok, Chachoengsao, Chaiyaphum, Chanthaburi, Chiang Mai, Chon Buri, Chumphon, Kalasin, Kanchanaburi, Krabi, Lampang, Loei, Lop Buri, Maha Sarakham, Nakhon Nayok, Nakhon Pathom, Nakhon Ratchasima, Nakhon Sawan, Nakhon Si Ayutthaya, Nakhon Si Thammarat, Narathiwat, Nonthaburi, Pathum Thani, Pattani, Phangnga, Phetchaburi, Phuket, Prachuap Khiri Khan, Ranong, Ratchaburi, Rayong, Roi Et, Sakon Nakhon, Saraburi, Sing Buri, Si Sa Ket, Songkhla, Tak, Trang, Trat, Ubon Ratchathani, Udon Thani (Hämäläinen and Pinratana 1999), Chiang Rai

(Katatani et al. 2004), Lop Buri (Thipaksorn et al. 2003), Phatthalung, Satun (Ferro et al. 2009), Buri Ram, Khon Kaen, Nan, Phetchabun, Prachin Buri, Samut Sakhon, and Surat Thani (Day et al. 2012) and more recently from Trat (Kosterin 2015). ***New province record:*** Uttaradit (Fig. 10.152).

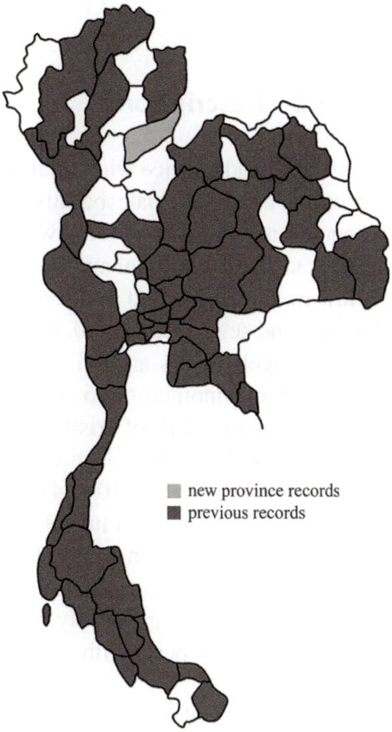

Fig. 10.152 Distribution of *Rhyothemis*

10.29 *Tetrathemis* Brauer, 1868

Fig. 10.153 Dorsal habitus of the nymph of *Tetrathemis*

10.29.1 General Appearance

The nymphs of *Tetrathemis* are small-sized (12 mm), with the body yellowish-brown; head mostly pale; hind legs long; abdomen with short, low, spine-like dorsal protuberances on S4–S9 and short posterolateral spines on S8–S9; and S10 short and sunken into S9 (Fig. 10.153).

10.29.2 Description

Head. The head is twice wider than long and mostly pale dorsally (Fig. 10.154). The labrum is bicolored, pale basally, and brown distally, and its anterior margin is biconcave and has sparse, long, white setae. The clypeus is pale and bare. The frons anterior surface is vertical and flat, the anterior margin is straight, and dorsally it is yellow in the middle and dark-brown laterally. The antennae are almost as long as the head and seven-segmented, the scape and pedicel are yellow and barrel-shaped, and the flagellomeres are cylindrical, brown in the basal half, and yellow in the distal half; antennomeres 3, 6, and 7 are subequal, although 3 is slightly longer. The vertex is convex, a tuft of bristle-like setae is on each side of the midline, and three ocelli are barely evident. The compound eyes are relatively small, rounded, and laterally prominent (Fig. 10.154). The occiput is relatively short and covered with minute, spiniform setae; a tuft of short bristle-like setae intermingled with a group of dorsally directed, elongate scales is on each side of the midline; the lateral margins are straight and tend to converge posteriorly and are beset with elongate scales; the posterior corners are widely rounded. The posterior margin of the head is shallowly concave and beset with stout, spiniform setae.

Fig. 10.154 The head in dorsal view

Mandibles have the formula L 1234 y- ab/R 123 + 4 y- abdx, with b > a in the left mandible and b > d > a > x in the right mandible. The maxilla's galeolacinia has seven teeth; the three dorsal teeth are slightly incurved, the two apicalmost teeth are of similar lengths and robustness, and the basal tooth is shorter and slender; the four ventral teeth are of different sizes and robustness, the basal tooth is small, and the apical tooth is the largest. The ventral pad of the hypopharynx is wide and more or less kidney-shaped, and large setae are on the surface and margins; the posterior margin is widely concave.

Fig. 10.155 The prementum in ventral view

Labium. The prementum-postmentum articulation reaches the anterior margin of the mesosternum. The prementum is subrhomboidal; the lateral margins are smooth, sinuate, and widely divergent to the apex (Fig. 10.155); and its ventral surface is indistinctly mottled; 8 + 4 or 9 + 3 premental setae are on each side of the midline. The ligula is well-developed, triangular (Fig. 10.155), and with abundant minute, spiniform setae on the dorsal surface; its apical margin is finely serrulated and bears a submarginal row of regularly spaced, spiniform setae, with one seta at the tip. The labial palp is abundantly mottled on the lateral surface; its apical margin is finely serrulated and has 7–8 well-developed crenations (Fig. 10.156), each bearing a large, spiniform seta and a graded series of three spiniform setae; the dorsolateral margin of the palp has a row of small, spiniform setae becoming minute setae on the distal half; the ventral margin is finely serrulated and has a submarginal row of large, regularly spaced, spiniform setae; eight palpal setae plus one basal setella and a group of 4–6 min spiniform setae are at the basomesal surface. The movable hook is long, slender, sharply pointed, and smooth. A row of three small, spiniform setae is at the palp's articulation.

Fig. 10.156 The head and prementum in frontal view

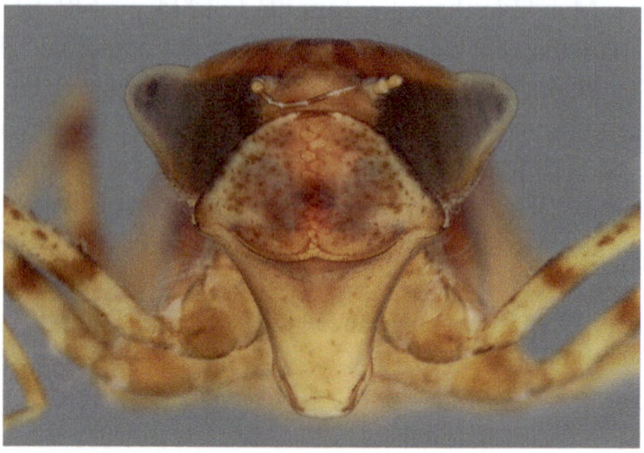

Thorax**.** The notum is mostly pale and the pleuron is brown. The prothorax is much narrower than the head (Fig. 10.153). The pronotal disc is bare and pale, and a longitudinal, jagged stripe is on each side of the midline; the sides are widely rounded, moderately elevated, and beset with small, spiniform setae; the posterolateral margins are thick, sinuate, and slightly elevated; the posterior margin is thick, short, and shallowly concave. The proepisternum has long setae. The pterothorax is mostly brown and spiny. The legs are relatively long (i.e., when fully extended, the hind legs far surpass S10), pale, and covered with long, stiff setae; the femora have two wide, dark bands; the fore and middle tibiae have three dark bands, and the hind tibia is entirely pale (Figs. 10.153 and 10.157); the femora and tibiae have rows of long, stout, stiff, bristle-like setae; the ventral (internal) surfaces of the tibiae have numerous tridentate setae; the tarsi are light-yellowish-brown and have tridentate setae in the ventrolateral row; the pretarsal claws are simple and each has a pulvilliform empodium.

Fig. 10.157 Ventral habitus of the nymph of *Tetrathemis*

The anterior and posterior wing sheaths are parallel (Fig. 10.153), pale dorsally, and dark-brown ventrally and reach the posterior margin and basal half of S5 and S6, respectively. The anterior margin of the mesolaterosternite is shallowly convex and beset with stiff, stout setae. The anterior margin of the metasternellum is straight, and its width is half the metasternellum midline length.

Fig. 10.158 The abdomen and caudal appendages in lateral view

Abdomen. The abdomen is more or less fusiform, widest at S6–S7, and truncate posteriorly (Fig. 10.157). The terga of S1–S5 are creamy-pale in the middle third and light-brown laterally; the terga of S6–S10 are brown in the middle third and light-brown laterally; a complex pattern of dark-reddish-brown spots and abundant spiniform setae are present on the surfaces and posterior margins of S5–S10; the lateral margins of S2–S9 are fringed with small, spiniform setae, which increase in size and robustness posteriorly. The middorsal protuberances are short, low, and spine-like on S4–S9 (Fig. 10.158), with that on S6 the largest and then decreasing in size posteriorly. The posterolateral spines on S8–S9 are sharply pointed and short, measuring 0.2 and 0.3 mm, respectively, and 0.5× and 0.7× as long as the middorsal lengths of their respective segments. S10 is very short and sunken into S9. The sterna are mostly yellowish-brown (Fig. 10.157); sterna 1–5 are mostly bare; sterna 6–9 have minute, spiniform setae on their surfaces; and sterna 7–9 posterior margins have small spiniform setae. Sterna 3–5 are each divided into five sternites and sterna 6–9 into three sternites. The caudal appendages are yellow, short, bulky, sharply pointed (Fig. 10.158), setose, spiny on the margins and mesal surfaces, and shorter than S8 + S9. The epiproct is slightly longer than its basal width, and its dorsal margin is straight in lateral view. The paraprocts are slightly longer than the epiproct (Fig. 10.158), the dorsal margin is convex in lateral view, and their tips are minute and slightly decurved. The cerci are digitiform and 0.75× as long as the epiproct.

Measurements (in mm). The total length including caudal appendages is 12; abdomen (ventral, including appendages) 8.5; maximum width of the abdomen 4; maximum width of the head 3.7; and length of hind femur 4.2.

10.29.3 Remarks

Tetrathemis is found "in alluvial swamp and slow streams at forest margins … Nymphs are amongst bottom detritus" (Orr 2005). In Thailand, only two species have been recorded: *T. irregularis* Brauer, 1868, and *T. platyptera* Selys, 1878 (Hämäläinen and Pinratana 1999).

10.29.4 Distribution

Tetrathemis has been recorded from the provinces of Chaiyaphum, Chanthaburi, Chiang Mai, Chiang Rai, Chon Buri, Kanchanaburi, Krabi, Lampang, Loei, Nakhon Nayok, Nakhon Ratchasima, Nakhon Si Thammarat, Phangnga, Ranong, Rayong, Sa Kaew, Sakon Nakhon, Songkhla, Tak, Trang, Yala (Hämäläinen and Pinratana 1999), Chumphon, Phetchaburi, Prachuap Khiri Khan (Ferro et al. 2009), and Nan (Katatani et al. 2010) and more recently from Khon Kaen, Phetchabun, Ratchaburi, Surat Thani, and Trat (Day et al. 2012) (Fig. 10.159).

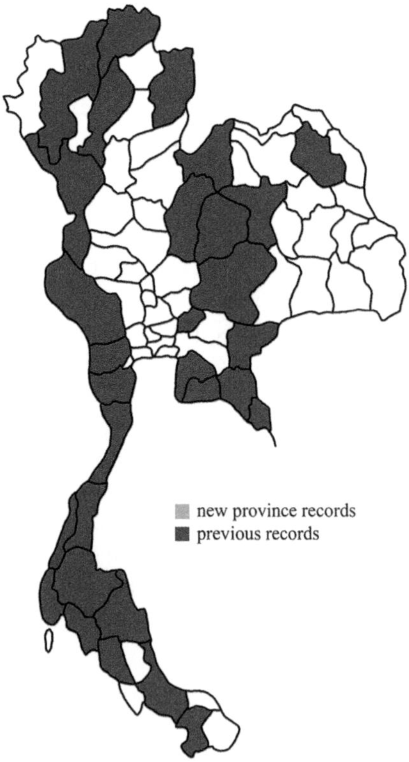

Fig. 10.159 Distribution of *Tetrathemis*

10.30 *Tholymis* Hagen, 1867

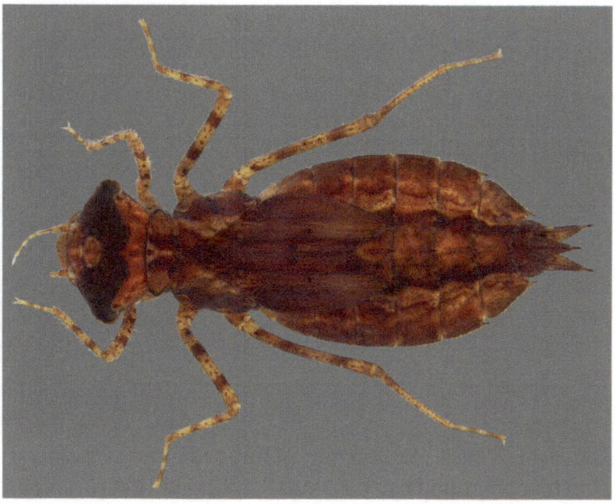

Fig. 10.160 Dorsal habitus of the nymph of *Tholymis tillarga*

10.30.1 General Appearance

The nymphs of *Tholymis* are small- to medium-sized (17.5–20.5 mm), with the body brown to dark-reddish-brown, hind legs short, abdomen with low dorsal protuberances on S3–S10 and short posterolateral spines on S8–S9, S10 set off from S9, and caudal appendages long (Fig. 10.160).

10.30.2 Description

Head. The head is subpentagonal, wider than long (Fig. 10.161), and finely granulose with the infraocular area roughly granulose. The labrum is brown, and the anterior margin is biconcave with abundant short setae. The clypeus is bare and moderately rugose, light-brown anteriorly, and dark-brown posteriorly. The frons anterior surface is vertical and flat; the anterior margin is straight and beset with long, stiff, white setae; and its dorsal surface is brown and shallowly convex. The antennae are long and seven-segmented (Fig. 10.161); the scape and pedicel are brown, setose, and barrel-shaped; the flagellomeres are yellow and cylindrical; and antennomere 3 is the longest. The vertex is convex and has three ocelli.

Fig. 10.161 The head in dorsal view

The compound eyes are small, rounded, and prominent dorsolaterally (Figs. 10.161 and 10.162). The occiput is large, moderately granulose, and covered with minute spiniform setae and elongate, scale-like setae mainly on the posterior and lateral margins, and seven oval, bare areas are on the posterior surface; the lateral margins are straight and strongly converge posteriorly (Fig. 10.161); the posterior corners are rounded. The posterior margin of the head is shallowly concave (Fig. 10.161). Mandibles have the formula L 1234 0 ab/R 1234 y- abcdx, with b > a in the left mandible and d > a > b > c > x in the right mandible. The maxilla's galeolacinia has seven teeth; the three dorsal teeth are slightly incurved, the two apicalmost teeth are of similar lengths and robustness, and the basal tooth is shorter and slender; the four ventral teeth are of different sizes and robustness, the basal tooth is minute, and the apical tooth is the largest. The ventral pad of the hypopharynx is wide and more or less kidney-shaped; short, regularly spaced setae are on its surface; and abundant long, reddish setae are on the apical and lateral margins; the posterior margin is strongly concave at the middle.

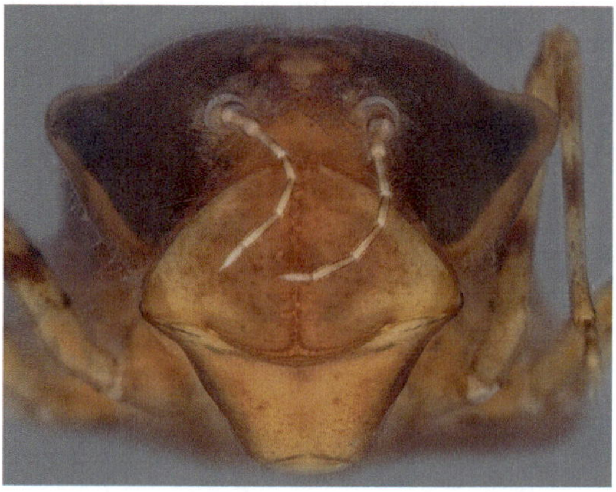

Fig. 10.162 The head and prementum in frontal view

Labium. The prementum-postmentum articulation surpasses the anterior margin of the mesosternum. The prementum is subrhomboidal and yellowish-brown on the ventral surface, and the lateral margins are smooth, sinuate, and divergent to the apex (Fig. 10.163); 5 + 2, 5 + 3, or 7 + 2 premental setae are on each side of the midline; a group of lateral, basidorsal, stout, spiniform setae is behind the premental setae. The ligula is well-developed, triangular (Fig. 10.163), and finely serrulated on the apical margin, a row of submarginal spiniform setae is on each side of the midline, and the tip is studded with two setae. The labial palp lateral surface is creamy-pale and mottled; the apical margin has 12 weakly developed crenations (Fig. 10.162), each bearing a graded series of 4 spiniform setae; the dorsolateral margin has a row of minute spiniform setae combined with small, delicate setae; the ventral margin is finely serrulated and has a submarginal row of elongate, spiniform setae, with those on the distal corner the largest; 5 palpal setae and a group of minute spiniform setae are present on the basomesal surface. The movable hook is relatively short, slender, sharply pointed, and smooth. A row of 3–4 short, stout, spiniform setae is on the palp's articulation.

Fig. 10.163 The prementum in ventral view

Thorax. The thorax is narrower than the head (Fig. 10.160). The pronotal disc is pale, and a longitudinal, wide, brown stripe is on each side of the midline (Fig. 10.160); the sides are rounded and beset with short, stout, spiniform setae and a row of long, stiff setae; the posterior margin is widely convex and thick. The proepisternum protrudes strongly as a conical tubercle beset with long setae. The pterothorax is mostly brown, pale dorsally, and finely granulose. The legs are moderately setose, spiny, pale, and relatively short (i.e., when fully extended, the tips of hind legs slightly surpass the caudal appendages) (Figs. 10.160 and 10.164); two dark bands are on the femora and three dark bands on the fore and middle tibiae; the distal fourth of the ventral (internal) surfaces of the fore and middle tibiae and the ventrolateral row of the fore tarsus bear tridentate setae, whereas the remainder of legs have simple, short and long, setae; the pretarsal claws are simple and each has a pulvilliform empodium.

Fig. 10.164 Ventral habitus of the nymph of *Tholymis tillarga*

The anterior and posterior wing sheaths are parallel (Fig. 10.160) and reddish-brown, the costal and anal margins are smooth, and they reach the posterior margin of S5 and basal half of S6, respectively. The mesolaterosternite is inflated, its anterior margin is convex, and it is beset with stiff, long setae. The anterior margin of the metasternellum is straight, and its width is 0.20× longer than the metasternellum midline length.

Abdomen. The abdomen is fusiform and widest at S6–S7 and then gradually tapers posteriorly (Figs. 10.160 and 10.164). The terga are reddish-brown and have a longitudinal, middorsal, dark, wide stripe with sinuate lateral margins; this stripe is flanked on each side by a pale, mottled, longitudinal stripe, and the remainder of the abdomen is uniformly brown (Fig. 10.160). The dorsal surface, including the posterior margins of S2–S10 and lateral margins of S2–S4, is bare; the lateral margins of S5–S9 are fringed with small, spiniform setae, which increase in size and robustness posteriorly.

Fig. 10.165 The abdomen and caudal appendages in lateral view

The middorsal protuberances on S3–S10 are low and spine-like (Fig. 10.165), and that on S10 is smaller; in dorsal view, the protuberances on S6–S10 form a continuous ridge onto the epiproct. The posterolateral spines on S8–S9 are short, sharply pointed, slightly convergent, and 0.4 and 0.5 mm in length, respectively, and are 0.25× and 0.40× as long as the dorsal lengths of their respective segments. S10 is short and set off from S9. The sterna are yellow-brown in the middle and reddish-brown laterally (Fig. 10.164); laterosternites 4–8 have a pale spot close to the longitudinal suture; sterna 2–10 are glabrous, including on the posterior margins. Sterna 3–5 and 8–9 are each divided into five sternites and sterna 6–7 into three sternites. The caudal appendages are long (Fig. 10.165), pyramidal, sharply pointed, and reddish-brown with a subapical pale ring; their margins and mesal surfaces are beset with long, stiff, bristle-like, dark setae. The epiproct is twice longer than its basal width, it has a full-length middorsal carina, and the tip is straight. The paraprocts are as long as the epiproct, they have a full-length lateral carina (Fig. 10.165), and their tips are straight and parallel. The cerci are digitiform, glabrous, brown basally, and creamy-pale distally; they are the shortest of the caudal appendages and 0.42× as long as the epiproct, and their tips are straight.

Measurements (in mm). The total length including caudal appendages is 17.5–20.5; abdomen (ventral, including appendages) 11.5–13.5; maximum width of the abdomen 6.3–6.5; maximum width of the head 5.0–5.2; and length of hind femur 5.2–5.5.

10.30.3 Remarks

Tholymis is found in open habitats, and the nymph is a bottom-dweller "in standing and slow flowing water" (Orr 2005), including brackish waters. Only two species are in this genus: *T. citrina* Hagen, 1867, inhabiting the tropical and subtropical areas of the New World, and *T. tillarga* (Fabricius, 1798), which is widespread in Asia, Africa, and Australasia. In Thailand, *T. tillarga* is a "common species found all the year throughout the country" (Hämäläinen and Pinratana 1999).

10.30.4 Distribution

Tholymis tillarga has been recorded from the provinces of Bangkok, Chachoengsao, Chaiyaphum, Chanthaburi, Chiang Mai, Chon Buri, Chumphon, Kanchanaburi, Krabi, Loei, Lop Buri, Mae Hong Son, Nakhon Nayok, Nakhon Pathom, Nakhon Ratchasima, Nakhon Sawan, Nonthaburi, Phayao, Phetchaburi, Phrae, Phuket, Prachuap Khiri Khan, Ranong, Ratchaburi, Sa Kaew, Sakon Nakhon, Songkhla, Tak, Trang, Ubon Ratchathani, Uttaradit (Hämäläinen and Pinratana 1999), Khon Kaen, Maha Sarakham, Phatthalung (Thipaksorn et al. 2003), Phayao (Ferro et al. 2009), Nan (Katatani et al. 2010), Phangnga, Sukhothai, and Surat Thani (Day et al. 2012) and more recently from Trat (Kosterin 2015). ***New province records:*** Kalasin, Phetchabun, and Phitsanulok (Fig. 10.166).

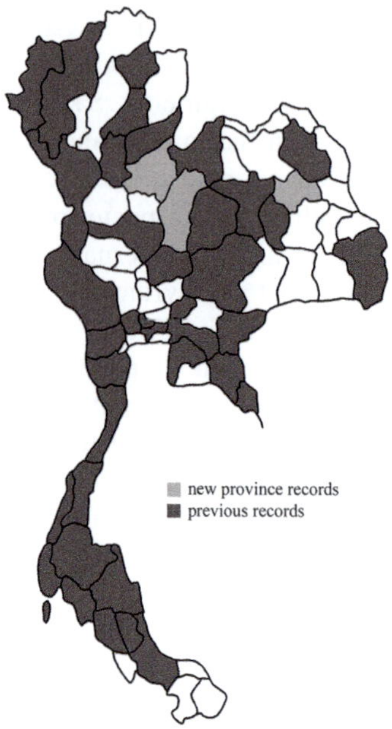

Fig. 10.166 Distribution of *Tholymis*

10.31 *Tramea* Hagen, 1861

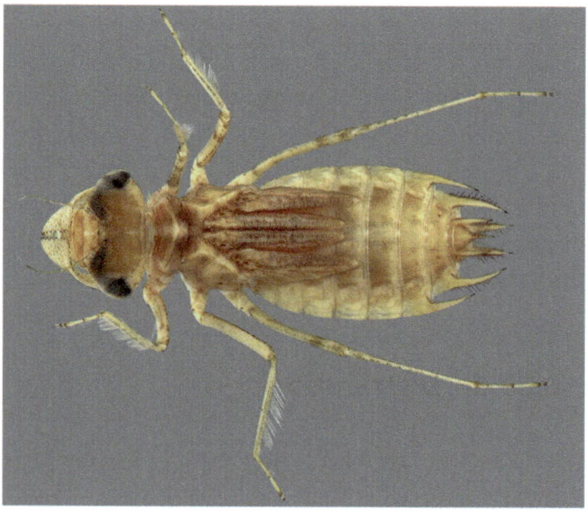

Fig. 10.167 Dorsal habitus of the nymph of *Tramea*

10.31.1 General Appearance

The nymphs of *Tramea* are characterized by their large size (26–28 mm), with the body robust and yellowish-brown, hind legs long, abdomen lacking dorsal protuberances but with very large posterolateral spines on S8–S9, S10 annular and sunken into S9, and caudal appendages long and sharply pointed (Fig. 10.167).

10.31.2 Description

Head. The head is much wider than long and yellowish-brown (Fig. 10.168). The labrum is creamy-pale bordered with brown, and the anterior margin is biconcave with long setae. The clypeus is bare, light-brown anteriorly, and creamy-pale posteriorly. The frons anterior surface is vertical and flat, its anterior margin is rounded and beset with short setae, and the dorsal surface is convex. The antennae are as long as the head, seven-segmented, and yellow, the scape and pedicel are barrel-shaped, the flagellomeres are cylindrical, and antennomeres 3, 5, and 7 are the same length and the longest. The vertex is bare and shallowly convex and has three ocelli. The compound eyes are large, rounded, and prominent laterally (Fig. 10.168). The occiput is rather short and glabrous dorsally; the cephalic lobes are poorly developed and bear large, strong, spiniform setae and sparse, bristle-like setae. The posterior margin of the head is straight.

Fig. 10.168 The head in dorsal view

Mandibles have the formula L 1234 y ab'bd/R 1234 y ab'bcd, with b > b' > d > a in the left mandible and b > b' > d > a > c in the right mandible. The maxilla's galeolacinia has seven teeth; the three dorsal teeth are slightly incurved, the two apicalmost teeth are of similar lengths and robustness, and the basal tooth is shorter and slender; the four ventral teeth are of different sizes and robustness, the basal tooth is smaller, and the apical tooth is the largest. The ventral pad of the hypopharynx is wide and more or less kidney-shaped, and large setae are on the surface and margins; the posterior margin is shallowly concave.

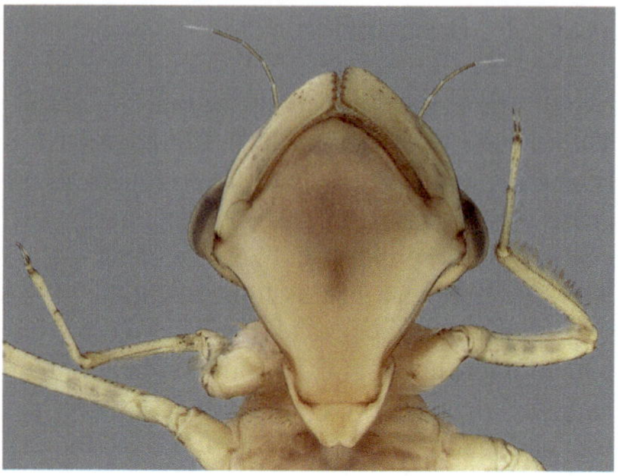

Fig. 10.169 The prementum in ventral view

Labium. The prementum-postmentum articulation reaches the anterior margin of the metasternum. The prementum is subrhomboidal and creamy-pale, and the lateral margins are sinuate, bare, and divergent to the apex (Fig. 10.169); 10 + 3 to 11 + 3 premental setae are on each side of the midline. The ligula is large and triangular; the apex is widely rounded (Fig. 10.169); the dorsal surface bears abundant small, spiniform setae; and the apical margin is finely serrulated and has a submarginal row of spiniform setae. The labial palp is yellow and weakly mottled on the lateral surface; the apical margin is serrulated and has 11–12 crenations (Fig. 10.170), each bearing a double graded series of 4–5 spiniform setae; the dorsolateral margin of the palp is glabrous; the ventral margin is finely serrulated and has a submarginal row of elongate, stout, regularly spaced, spiniform setae; 9–10 palpal setae plus 1 setella and minute spiniform setae are on the basomesal surface. The movable hook is relatively long, slender, sharply pointed, and smooth. A row of 4–5 short, stout, spiniform setae is at the palp's articulation.

Fig. 10.170 The head and labial palps in frontal view

Thorax. The thorax is narrower than the head (Fig. 10.167). The pronotal disc is mostly bare and pale, and a subrectangular, transverse, brown spot is on each side of the midline; the anterior margin is sinuate and bears a row of small, spiniform setae; the posterior margin is widely convex and has spiniform setae, except in the middle third; the lateral margins are rounded and set with a row of small, spiniform setae. The proepisternum is short, and its anterolateral corner has a few, long spiniform setae and a tuft of long setae. The pterothorax is yellowish-dark-brown and finely granulose.

Fig. 10.171 Ventral habitus of the nymph of *Tramea*

The legs are relatively long (Figs. 10.167 and 10.171) (i.e., when fully extended, the tips of the hind tibiae reach the tips of the paraprocts), pale, and spiny and have two poorly defined, dark bands on the femora; the tibiae are fringed with long, delicate, whitish setae on the dorsal (external) and ventral (internal) margins; the distal ends of the fore and middle tibiae and ventrolateral row of the fore tarsus have tridentate setae, whereas the remainder of legs have stout, short, spiniform setae; the tarsi are yellow; the pretarsal claws are simple and each has a pulvilliform empodium. The wing sheaths are parallel (Fig. 10.167), light-yellowish-brown, and mostly bare; they have a row of spiniform setae on the costal and anal margins, and the wing venation is well-marked; the anterior and posterior wing sheaths reach the posterior margin of S5 and basal half of S6, respectively. The anterior margin of the mesolaterosternite is shallowly convex and has short, stout setae. The anterior margin of the metasternellum is straight, and its width is half the metasternellum midline length.

Abdomen. The abdomen is elongate, fusiform, and widest at S6–S7 (Figs. 10.167 and 10.171). The terga of S1–S5 are creamy-pale in the middle and light-brown laterally; terga 6–10 are light-brown, with a poorly defined pattern of dark-brown spots and dots laterally on S4–S5 and on the entire surface of S6–S9; and occasionally, a diffuse, pale, longitudinal, wide stripe is on S6–S8; terga 5–9 have minute, spiniform setae; the posterior margins of S5–S10 and lateral margins of S3–S9 are fringed with small, spiniform setae, which increase in number and robustness posteriorly. Middorsal protuberances are lacking (Fig. 10.172).

Fig. 10.172 The abdomen and caudal appendages in lateral view

The posterolateral spines of S8–S9 are straight, sharply pointed, and very large (Fig. 10.167); they are 2.7 and 3.2 mm in length and twice and three times longer than the middorsal lengths of their respective segments; those of S9 reach the tips of the paraprocts. S10 is annular and sunken into S9. The sterna are creamy-pale with diffuse, light-brown spots on S9 (Fig. 10.171); sterna 2–4 and most of 5 are glabrous; most of the surfaces of S6–S9 and posterior margins of S5–S9 have minute, spiniform setae. Sterna 3–5 and 9 are each divided into five sternites and sterna 6–8 into three sternites. The caudal appendages are creamy-pale, long, slender distally, sharply pointed (Figs. 10.167 and 10.172), and strongly spiny on the margins and have long, reddish-brown, bristle-like setae. The epiproct is triangular and 0.32× longer than its basal width; it strongly narrows at midlength to form a long, slender tip; and a longitudinal row of 4–5 large, robust setae is along the dorsal margin in the distal half. The paraprocts are pyramidal, 0.30× longer than the epiproct (Fig. 10.172), and with rows of strong, long, spiniform setae combined with long, hair-like setae on the dorsal and ventral margins, and their tips are long, slender, straight, and parallel. The cerci are narrowly elongate; taper gradually to a long, sharp tip; and are the shortest of the caudal appendages and 0.95× as long as the epiproct (Fig. 10.172).

Measurements (in mm). The total length including caudal appendages is 26.3–27.5; abdomen (ventral, including appendages) 15.8–16.5; maximum width of the abdomen 8.7–9.3; maximum width of the head 8.3–8.5; and length of hind femur 8.0–8.3.

10.31.3 Remarks

Tramea can be found breeding in "well vegetated lakes, ponds and drains" (Orr 2005); the nymphs cling on submerged or floating vegetation. Three species have been recorded in Thailand: *T. basilaris* (Palisot de Beauvois, 1817); *T. transmarina* Brauer, 1867; and *T. virginia* (Rambur, 1842) (Hämäläinen and Pinratana 1999).

10.31.4 Distribution

Tramea has been recorded from the provinces of Chanthaburi, Chiang Mai, Chon Buri, Kalasin, Nakhon Pathom, Pattani, Phangnga, Phuket, Ranong, Songkhla, Ubon Ratchathani (Hämäläinen and Pinratana 1999), Chiang Rai (Katatani et al. 2004), Kamphaeng Phet, Kanchanaburi, Krabi, and Trang (Ferro et al. 2009) and more recently from Nakhon Ratchasima, Phetchaburi, and Surat Thani (Day et al. 2012) (Fig. 10.173).

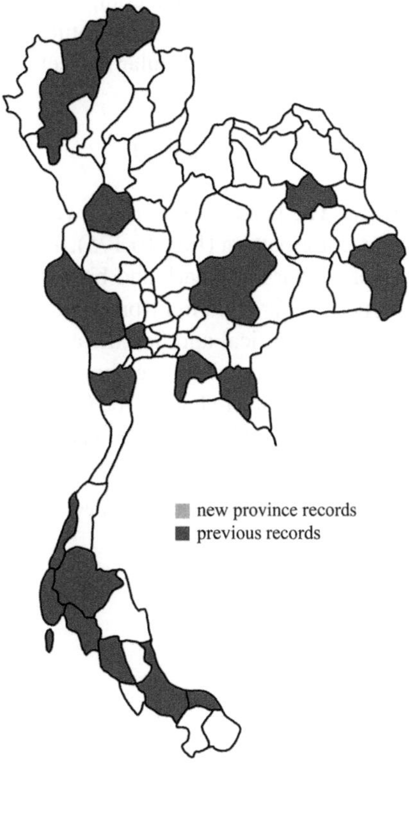

Fig. 10.173 Distribution of *Tramea*

10.32 *Trithemis* Brauer, 1868

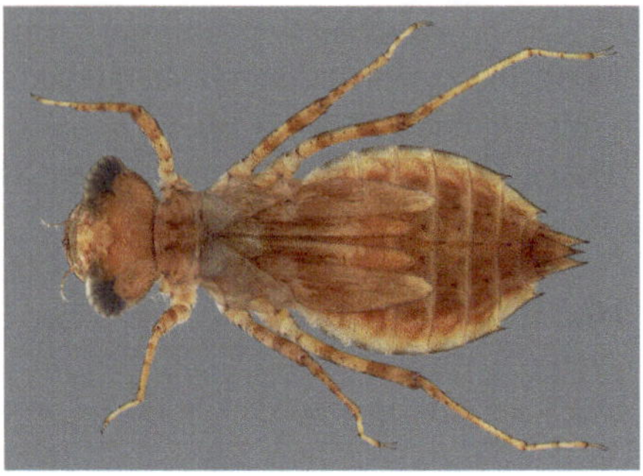

Fig. 10.174 Dorsal habitus of the nymph of *Trithemis*

10.32.1 General Appearance

The nymphs of *Trithemis* are characterized by their small size (15–17 mm), with the body light-brown with a complex pattern of dark and pale spots and dots, hind legs short, abdomen with dorsal protuberances on S3–S9 and short posterolateral spines on S8–S9, and S10 annular and set off from S9 (Fig. 10.174).

10.32.2 Description

Head. The head is more than twice wider than long, light-brown, and with a complex color pattern (Fig. 10.175). The labrum is pale with three brown spots, and the anterior margin is dark and biconcave ventrally and has abundant short setae. The clypeus is bare and light-brown and has a large, pale, distal spot. The frons anterior surface is vertical and flat, the anterior margin is rounded, and the dorsal surface is convex. The antennae are short and seven-segmented (Fig. 10.175), the scape and pedicel are barrel-shaped, the flagellomeres are cylindrical and subequal, and antennomere 3 is the longest. The vertex is convex and spotted and has three ocelli.

Fig. 10.175 The head in dorsal view

The compound eyes are large, rounded, and prominent laterally (Figs. 10.175 and 10.177). The occiput is rather large, light-brown dorsally, and dark-brown laterally and has small, spiniform setae and long, stiff setae on the posterior and lateral margins; the lateral margins are straight and tend to converge posteriorly; the posterior corners are rounded (Fig. 10.175). The posterior margin of the head is straight. Mandibles have the formula L 1234 0 ab/R 12 + 34 y- abcd, with b > a in the left mandible and d > b > a > c in the right mandible. The maxilla's galeolacinia has seven teeth; the three dorsal teeth are slightly incurved, the two apicalmost teeth are of similar lengths and robustness, and the basal tooth is shorter and slender; the four ventral teeth are of different sizes and robustness, the basal tooth is minute, and the apical tooth is the largest. The ventral pad of the hypopharynx is wide and more or less kidney-shaped and has abundant long setae on the margins and a transverse row of long setae in the middle; the posterior margin is shallowly concave.

Fig. 10.176 The prementum in ventral view

Labium. The prementum-postmentum articulation reaches the anterior margin of the mesosternum. The prementum is subrhomboidal (Fig. 10.176), sparsely and weakly mottled on the ventral surface, creamy-pale laterally, and translucent in the middle; the lateral margins are sinuate and divergent to the apex and bear delicate, small setae; 5 + 3 to 6 + 4 premental setae arranged in a semicircle are on each side of the midline. The ligula is well-developed, triangular, translucent, and finely serrulated on the apical margin and has a row of submarginal, spiniform setae and two setae at the tip. The labial palp lateral surface is moderately to densely mottled (Fig. 10.177), and the distal margin is finely serrulated and has 10 weakly developed crenations (Fig. 10.177), each bearing 2–3 large, spiniform setae; the dorsolateral margin has a row of small, spiniform setae; the ventral margin is rather smooth and has a submarginal row of elongate, spiniform setae; 6–7 palpal setae plus 5 min, spiniform setae are at the basomesal surface, close to the palp's articulation. The movable hook is relatively long, slender, sharply pointed, and smooth. A row of four short, spiniform setae is at the palp's articulation.

Fig. 10.177 The head and labial palps in frontal view

Thorax. The thorax is narrower than the head (Fig. 10.174). The pronotal disc is pale; a longitudinal brown stripe is on each side of the midline; the sides are rounded and set with rows of white, short, spiniform setae; and the posterior margin is widely convex. The proepisternum is short, strongly protruding ventrally, and beset with long setae. The pterothorax is light-brown and minutely setose.

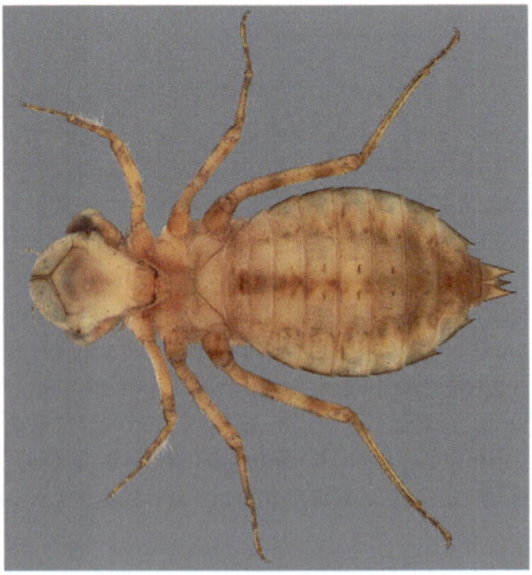

Fig. 10.178 Ventral habitus of the nymph of *Trithemis*

The legs are spiny, moderately setose, relatively short (i.e., when fully extended, the apex of the hind tibia does not reach the tip of the paraproct) (Fig. 10.174), pale, and with dark rings on the femora and tibiae; the distal half of the ventral (internal) surface of the fore tibia and the ventrolateral row of the fore tarsus have tridentate setae, whereas the remainder of the legs have simple, short, spiniform setae and long, stiff setae; the pretarsal claws are simple and each has a pulvilliform empodium. The anterior and posterior wing sheaths are parallel (Fig. 10.174), light-brown, and with a transverse pale band at the basal 0.60 of their length, and they reach the basal half of S5 and surpass the anterior margin of S6, respectively. The anterior margin of the mesolaterosternite is convex and beset with stiff, stout setae. The anterior margin of the metasternellum is straight, and its width is half the metasternellum midline length.

Abdomen. The abdomen is fusiform and widest at S6 and then tapers posteriorly (Figs. 10.174 and 10.178). The terga are creamy-pale on S2–S4 and the middle third of S5, whereas the remainder have a complex pattern of dark and pale spots and dots, which are more pronounced on S6–S9; most of the dorsal surface and posterior margins of S3–S9 and the lateral margins of S4–S9 are covered with small, spiniform setae.

Fig. 10.179 The abdomen and caudal appendages in lateral view

The middorsal protuberances on S3–S9 are spine-like and well-developed, and those on S3–S5 are vertical and gradually decrease in height on S6–S9 (Fig. 10.179); the dorsal margin is more or less straight and parallel to the long axis of the abdomen, and that on S9 reaches the basal 0.10 of the epiproct. The posterolateral spines on S8–S9 are short (Figs. 10.174 and 10.179), 0.3 mm in length, sharply pointed, with the tips subparallel on S8 and convergent on S9, and 0.35× and 0.50× the middorsal length of their respective segments. S10 is short and set off from S9. The sterna are light-yellow-brown, mostly with a complex color pattern (Fig. 10.178), and bare, and only the posterior margins of S7–S9 are fringed with spiniform setae. Sterna 3–5 are each divided into five sternites and sterna 6–9 into three sternites. The caudal appendages are thick, sharply pointed (Fig. 10.179), and almost as long as the dorsal length of S8 + S9 + S10. The epiproct and paraprocts are pyramidal and light-brown. The epiproct is 0.40× longer than its maximum basal width, a middorsal carina extends along the basal 0.70, and its relatively long tip is slightly decurved in lateral view (Fig. 10.179); its ventrolateral margins are set with a row of small, stout, spiniform setae. The paraprocts are as long as the epiproct; a row of long, spiniform setae is on both the dorsal and ventral margins; and a row of long bristle-like setae is on the ventral margins; the tips are slightly recurved in lateral view and incurved in ventral view. The cerci are conical, with the tips elongate, parallel, or slightly convergent, and they are the shortest of the caudal appendages and 0.45× as long as the epiproct (Fig. 10.179).

Measurements (in mm). The total length including caudal appendages is 15.3–17.2; abdomen (ventral, including appendages) 9.5–10.6; maximum width of the abdomen 5.0–5.4; maximum width of the head 4.3–4.5; and length of hind femur 4.0–4.3.

10.32.3 Remarks

Trithemis is widely distributed in tropical Asia, preferring open areas; it is common in "ponds, lakes, weedy drains, sluggish streams and larger streams" (Orr 2005) and also in rice fields (Thipaksorn et al. 2003). In Thailand, three species have been recorded: *T. aurora* (Burmeister, 1839), *T. festiva* (Rambur, 1842), and *T. pallidinervis* (Kirby, 1889) (Hämäläinen and Pinratana 1999).

10.32.4 Distribution

Trithemis is common and widespread in Thailand and has been recorded from the provinces of Bangkok, Chai Nat, Chaiyaphum, Chanthaburi, Chiang Mai, Chon Buri, Chumphon, Kanchanaburi, Khon Kaen, Krabi, Lampang, Loei, Lop Buri, Mae Hong Son, Nakhon Nayok, Nakhon Pathom, Nakhon Ratchasima, Nakhon Sawan, Nakhon Si Ayutthaya, Nakhon Si Thammarat, Narathiwat, Nong Khai, Nonthaburi, Phangnga, Pathum Thani, Phetchaburi, Phrae, Phuket, Prachuap Khiri Khan, Ranong, Ratchaburi, Sa Kaew, Sakon Nakhon, Saraburi, Satun, Si Sa Ket, Songkhla, Surat Thani, Tak, Trang, Trat, Ubon Ratchathani, Uttaradit, Yala, Yasothon (Hämäläinen and Pinratana 1999), Chiang Rai (Katatani et al. 2004), Rayong (Kosterin 2006), Kamphaeng Phet, Kalasin, Mukdahan, Phatthalung, Phayao, Phitsanulok (Ferro et al. 2009), and Nan (Katatani et al. 2010) and more recently from Prachin Buri and Sukhothai (Day et al. 2012). ***New province record:*** Phetchabun (Fig. 10.180).

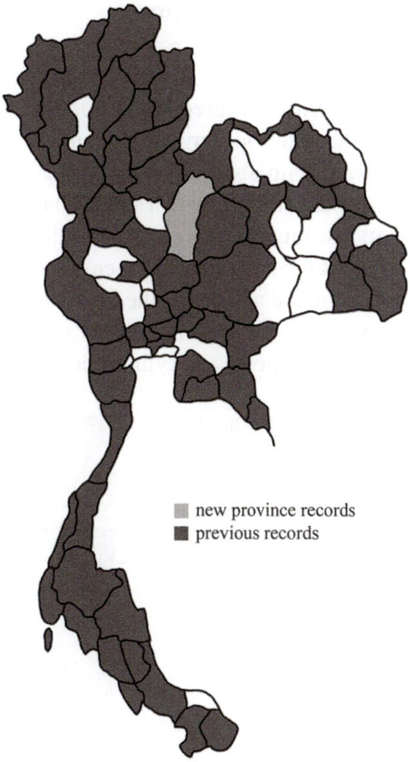

Fig. 10.180 Distribution of *Trithemis*

10.33 *Urothemis* Brauer, 1868

Fig. 10.181 Dorsal habitus of the nymph of *Urothemis abbotti*

10.33.1 General Appearance

The nymphs of *Urothemis* are characterized by their medium size (19–24 mm), with the body yellowish-brown with a complex filigree-like color pattern, hind legs long, abdomen with dorsal protuberances on S3–S8 and large posterolateral spines on S8–S9, S10 annular and sunken into S9, and caudal appendages short and sharply pointed (Fig. 10.181).

10.33.2 Description

Head. The head is almost twice wider than long and light-yellow to yellowish-brown with a filigree-like color pattern (Fig. 10.182). The labrum is creamy-pale, and the anterior margin is biconcave, slightly flattened ventrally in the middle, and covered with abundant white, long setae. The clypeus is creamy-pale and bare and has a pair of light-brown submedian spots. The frons anterior surface is vertical and flat; its anterior margin is rounded, shallowly concave, and set with long, stiff, brown setae; its dorsal surface is shallowly convex and yellow with brown dots. The antennae are as long as the head and seven-segmented (Fig. 10.182); antennomeres 1–6 are brown, antennomere 7 is yellow, the scape and pedicel are barrel-shaped, the flagellomeres are cylindrical and subequal, and antennomere 5 is the longest.

Fig. 10.182 The head in dorsal view

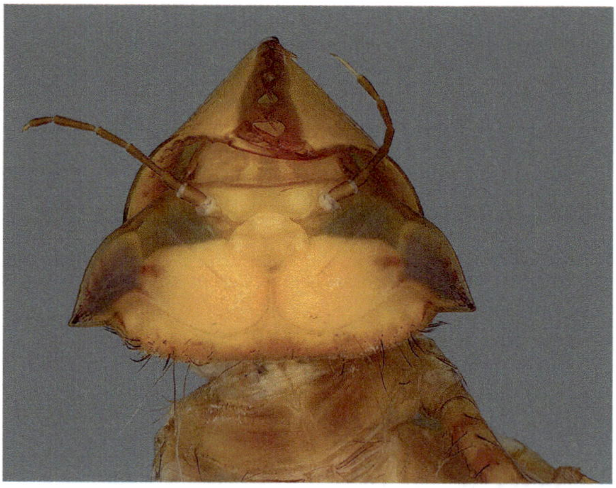

The vertex is bare, shallowly convex, yellow with brown dots, and with three ocelli. The compound eyes are strongly prominent laterally (Figs. 10.182 and 10.184), the posterolateral corners are acute, and the posterior margin is straight. The occiput is large and mostly bare dorsally and has a filigree-like color pattern; the posterior lobes are poorly developed, not bulging, and beset with stout, black and brown, bristle-like setae of different sizes (Fig. 10.182). The posterior margin of the head is straight. Mandibles have the formula L 123'34 0 abb'/R 1234 y abcd, with $b > a$ in the left mandible and $b > d > a = c$ in the right mandible. The maxilla's galeolacinia has seven teeth; the three dorsal teeth are slightly incurved, the two apicalmost teeth are of similar lengths and robustness, and the basal tooth is shorter and slender; the four ventral teeth are of different sizes and robustness, the basal tooth is small, and the apical tooth is the largest. The ventral pad of the hypopharynx is wide and more or less kidney-shaped, and large setae are on its surface and margins; the lateral margins have long, stout setae; the posterior margin is strongly concave.

Labium. The prementum-postmentum articulation reaches the anterior margin of the metasternum. The prementum is subrhomboidal and densely mottled on the ventral surface, and the lateral margins are smooth, sinuate, and divergent to the apex (Fig. 10.183); 7 + 4 premental setae are on each side of the midline. The ligula is well-developed, widely rounded at the tip (Fig. 10.183), and finely serrulated on the apical margin, and a submarginal row of alternating large and small, stout, spiniform setae extends along the margin with 2–4 large, spiniform setae at the tip.

Fig. 10.183 The prementum in dorsal view

The labial palp lateral surface is densely mottled (Fig. 10.184); the apical margin is finely serrulated and has 9–10 moderately developed crenations (Fig. 10.184), each bearing a graded series of 3–4 stout, enlarged, spiniform setae; the dorsolateral margin of the palp has a short row of minute, spiniform setae on the basal 0.30; the ventral margin is finely serrulated and bears a submarginal row of elongate, stout, regularly spaced, spiniform setae intermingled with small, spiniform setae; 7–8 long palpal setae and a row of 6–8 min, spiniform setae are present on the basomesal surface. The movable hook is relatively long, stout, sharply pointed, and smooth. A row of 4–6 short, stout, spiniform setae is on the palp's articulation.

Fig. 10.184 The head and palps in frontal view

Thorax. The prothorax is narrower than the head (Fig. 10.181). The pronotal disc is pale and has a wide, longitudinal, irregular, reddish-brown stripe on each side of the midline; the anterior margin is sinuate and bears a row of spiniform setae except in the middle; the posterior margin is strongly convex and bears a row of spiniform setae and regularly spaced, brown, long, stiff, bristle-like setae; the sides are straight and set with a row of minute, spiniform setae. The proepisternum is short and has a tuft of long setae on the ventral margin. The pterothorax is granulose and pale with thin, vertical, brown stripes.

Fig. 10.185 Ventral habitus of the nymph of *Urothemis abbotti*

The legs are relatively long (i.e., when fully extended, the hind legs far surpass the tips of the caudal appendages) (Figs. 10.181 and 10.185), pale, spiny, and setose; the femora have three dark rings (the distal ring is diffuse); the femora and tibiae are fringed with long, stout, stiff, brown setae on the ventral (internal) and dorsal (external) margins, which are intermingled with long, black, thick, bristle-like setae; the distal half of the ventral (internal) surface of the fore tibia, distal fourth of the middle tibia, and ventrolateral row of the fore and middle tarsi have tridentate setae, whereas the remainder of the legs have stout, short and long, spiniform setae; the pretarsal claws are simple and each has a pulvilliform empodium. The anterior and posterior wing sheaths are parallel (Fig. 10.181), light-brown, and with a row of spiniform setae on the costal and anal margins and reach the basal halves of S6 and S7, respectively. The anterior margin of the mesolaterosternite is shallowly convex and has long, stout setae. The anterior margin of the metasternellum is narrowly convex, and its width is 0.16× the metasternellum midline length.

Abdomen. The abdomen is elongate, more or less parallel-sided, widest at S6–S7, and truncate posteriorly. The terga are creamy-pale on S1–S4 and have a complex, filigree-like color pattern of brown dots and pale spots on S5–S9; the dorsolateral surfaces of S2–S4, the entire surface of S5–S10, and all their posterior margins bear minute, spiniform setae; the posterior margins of S7–S9 have a row of long, sparse, dark-reddish-brown, bristle-like setae, and the lateral margins of S2–S7 are smooth.

Fig. 10.186 The abdomen and caudal appendages in lateral view

The middorsal protuberances on S3–S8 are small on S3 and spine-like and increasing in size and robustness posteriorly on S4–S8, that on S8 reaches the posterior margin of S9 (Fig. 10.186) or beyond, and those on S4–S8 have a longitudinal, dorsal row of spiniform setae; a pair of long, dark, bristle-like setae is on each side of the dorsal protuberances on S8–S9. The lateral margins of S8–S9 bear a row of short, stout, spiniform setae, which end in large, sharply pointed posterolateral spines (Figs. 10.181, 10.185, and 10.186); the spines on S8 are parallel, 0.6 mm, and 0.70× as long as the middorsal length of S8; the spines on S9 are slightly convergent, 1.0 mm, and almost twice longer than the middorsal length of S9 and almost reach the tips of the paraprocts. S10 is short and sunken into S9. The sterna are mostly yellow or yellowish-brown (Fig. 10.185) and densely mottled; sterna 1–6 are mostly smooth but have sparse, long, stiff setae mainly on the laterosternites; sterna 7–9 have minute, spiniform setae on the surfaces, and their posterior margins are fringed with small, spiniform setae and long, reddish-brown, bristle-like setae. Sterna 3–5 are each divided into five sternites and sterna 6–9 into three sternites. The caudal appendages are short (Fig. 10.186), shorter than the middorsal length of S8 + S9, straight, sharply pointed, strongly spiny on the margins, and yellow to yellowish-brown and have long, stout, reddish-brown, bristle-like setae. The epiproct is triangular and as long as its basal width. The paraprocts are pyramidal, as long as their basal widths, and 0.30× longer than the epiproct, and a longitudinal row of stout, spiniform setae and bristle-like setae is on their lateral surfaces; the tips are long, slender, and straight. The cerci are yellow, digitiform, the shortest of the caudal appendages, and 0.85× as long as the epiproct, and the tips are straight and parallel.

Measurements (in mm). The total length including caudal appendages is 18.8–24.0; abdomen (ventral, including appendages) 10.3–15.0; maximum width of the abdomen 7.3–8.5; maximum width of the head 6.6–7.4; and length of hind femur 6.7–7.5.

10.33.3 Remarks

Urothemis is "fairly common in lowland lakes, ponds and slow streams in open country and swamp forest" (Orr 2005). Only two species inhabit Thailand: *U. abbotti* Laidlaw, 1927 and *U. signata* (Rambur, 1842) (Hämäläinen and Pinratana 1999).

10.33.4 Distribution

Urothemis is widespread and has been recorded from the provinces of Bangkok, Chaiyaphum, Chiang Mai, Chon Buri, Kalasin, Kanchanaburi, Nakhon Nayok, Nakhon Sawan, Nakhon Si Ayutthaya, Nakhon Si Thammarat, Nonthaburi, Pathum Thani, Phetchaburi, Prachuap Khiri Khan, Phuket, Rayong, Sakon Nakhon, Samut Prakan, Trang, Ubon Ratchathani, Udon Thani (Hämäläinen and Pinratana 1999), Chumphon, Kamphaeng Phet, Nakhon Pathom, Nakhon Ratchasima, Roi Et, Suphan Buri, and Surat Thani (Thipaksorn et al. 2003) and more recently from Khon Kaen, Phangnga, Prachin Buri, and Trat (Day et al. 2012). ***New province records:*** Krabi, Loei, Phetchabun, Ranong, and Songkhla (Fig. 10.187).

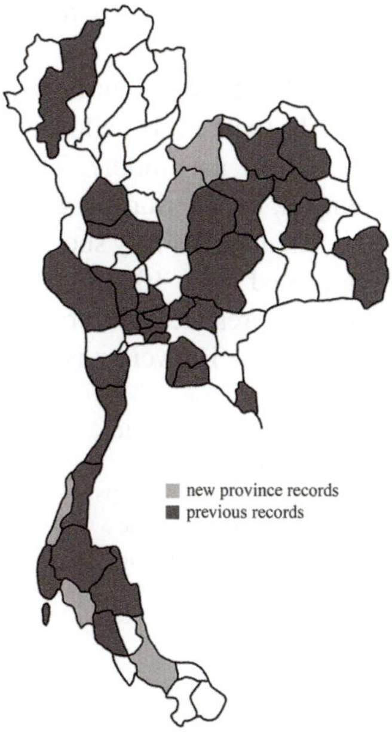

Fig. 10.187 Distribution of *Urothemis*

10.34 *Zygonyx* Hagen, 1867

Fig. 10.188 Dorsal habitus of the nymph of *Zygonyx iris*

10.34.1 General Appearance

The nymphs of *Zygonyx* are medium- to large-sized (21–30 mm), with the body robust and stocky, legs long, abdomen strongly convex dorsally with dorsal protuberances on S2–S10 and posterolateral spines on S7–S9 or S8–S9, S10 annular and not sunken into S9, and caudal appendages obtusely or sharply pointed and straight (Fig. 10.188).

10.34.2 Description

Head. The head is subtrapezoidal, twice wider than long (Fig. 10.188), and mostly bare. The labrum is yellow and setose on the distal half, and the anterior margin is concave in the middle. The clypeus is glabrous. The frons anterior surface is rounded, its anterior margin is sinuate, and the dorsal surface is shallowly convex. The antennae are very short, barely reaching the basal half of the labrum, and seven-segmented (Fig. 10.189); the scape and pedicel are brown and subspherical, the flagellomeres are light-brown, antennomere 3 is the longest, and antennomere 6 is the shortest. The vertex is small and convex and has three ocelli.

Fig. 10.189 The head in frontal view. Notice the short antennae, labial palps lacking crenations on the distal margins, and blunt posterolateral tubercles

The compound eyes are rounded, prominent anterolaterally (Fig. 10.189), and strongly concave on the posterior margin (Fig. 10.188). The occiput is large with paired, dark-brown, bristle-like setae either close to the posterior margin of the compound eyes or the setae are lacking; the lateral margins are straight and tend to converge posteriorly, ending in a swollen, posterodorsal, blunt, dorsally directed tubercle (Fig. 10.189), and are beset with minute, dark-brown, spiniform setae; a low, rounded, small tubercle is on each side of the midline; the occipital lobes have a posteroventral row of large, stout, flat, scale-like spines, which continues longitudinally along the ventrolateral surface of the head. The posterior margin of the head is straight or sinuate. Mandibles have the formula L 1234 y- ab/R 1234 y ab or L 1234 0 ab/R 1234 y b; the tooth *a* in both mandibles is minute or absent. The maxilla's galeolacinia has seven teeth; the three dorsal teeth are slightly incurved, the two distal teeth are similar in length and robustness, and the basal tooth is reduced; the four ventral teeth are of different sizes and robustness, the basal tooth is the shortest, and the apical tooth is the largest. The ventral pad of the hypopharynx is subquadrate, creamy-pale, and set with a subapical, transverse row of stiff, anteriorly directed setae and long setae on the margins; the posterior margin is strongly concave.

Fig. 10.190 The head in ventral view showing the short prementum

Labium. The prementum-postmentum articulation almost reaches the posterior margin of the prosternum (Figs. 10.190 and 10.191). The prementum is yellow to yellowish-brown, subpentagonal, and very short and only 0.20–0.35× wider than long; the lateral margins are sinuate and divergent to the apex in ventral view (Fig. 10.190) and have a ventral, subbasal, transverse, sinuate carina; in dorsal view, a transverse row of 8–11 long premental setae plus 10–13 shorter setae of different sizes are on each side of the midline and are without any particular arrangement; a group of 10–13 small, sublateral setae is close to the palp's articulation, as well as a dorsolateral, full-length, longitudinal row of spiniform setae intermingled along the basal 0.60 with a dense fringe of long, stout setae. The ligula is poorly or well-developed (Fig. 10.190), widely convex or triangular, and rounded at the tip, its apical margin is finely serrulated, and a dorsal, submarginal row of long, piliform or claviform setae is along the entire margin. The labial palp is rather small and yellow to light-brown on the lateral surface, and the margins are reddish-brown to brown; the apical and ventral margins are finely serrulated and lack crenations on the apical margin (Fig. 10.189), and a submarginal row of short and long, piliform or claviform setae extends along the apical and ventral margins; the dorsolateral margin of the palp bears a row of small, spiniform setae along the basal half; 8–10 long palpal setae, with or without a subbasal setella, and a group of 9–10 short setae are on the basomesal surface of the palp. The movable hook is smooth, short, and thick along the basal half and slender and sharply pointed along the distal half; spiniform setae are lacking at the palp's articulation.

Thorax. The prothorax is scarcely narrower than the head (Fig. 10.188). The pronotum anterior lobe has an anterodorsal, thick, rugose ridge on each side, which ends dorsally in a swollen area anterior to the transverse midline, or this ridge ends in a large, rounded tubercle produced dorsally; the midlobe of the pronotum may or may not have a pair of anteromedian, low, rounded tubercles, and a large, depressed, transverse, irregular area behind the paired tubercles is on each side of the midline; the posterolateral corner ends in a large, rounded tubercle; the posterior lobe is very narrow and is produced in the middle as a posterior pull-flap. The proepisternum anterior margin is raised, thick, and beset with strong conical spines. The propleuron ventral margin is shallowly biconvex or strongly V-shaped and spiny. The pterothorax is rugose or minutely scaly. The metepimeron may or may not have a large, obovate, vertical tubercle. The legs are long (Figs. 10.188 and 10.191) (i.e., when fully extended, the hind legs far surpass the tip of the abdomen) and pale and have three wide, dark bands on the femora and tibiae; the dorsal (posterior) margins of the femora are granulose and fringed with stiff setae and have 2–3 regularly spaced tubercles of which the subbasal tubercle is the largest; the tibiae have short, stiff setae, and their apices bear abundant stout, thick, spiniform setae; the tarsi are the darkest and have ventral rows of stout, thick, spiniform setae; the pretarsal claws are simple, large, and hook-like, and each has a pulvilliform empodium.

Fig. 10.191 Ventral habitus of the nymph of *Zygonyx iris*

The anterior and posterior wing sheaths are subparallel or decidedly divergent (Fig. 10.188) and can reach at least the basal half and posterior margin of S6, respectively. The mesolaterosternites have a large tubercle densely covered with short, stout, scale-like setae, or these tubercles are lacking; the metalaterosternites have a smaller, low tubercle on the anterior margin, which is covered with small, scale-like setae, or these tubercles are lacking. The anterior margin of the metasternellum is shallowly concave and 0.20× wider than the metasternellum midline length.

Abdomen. The abdomen is mostly parallel-sided, widest at S6, abruptly narrowed posteriorly, strongly convex dorsally, and flat ventrally (Figs. 10.188 and 10.191). The posterior margins of S2–S4 or S2–S6 have short or long hair-like setae, which become short, truncate, scale-like setae on S5–S8 or S7–S8, and the posterior margins of S9–S10 are smooth.

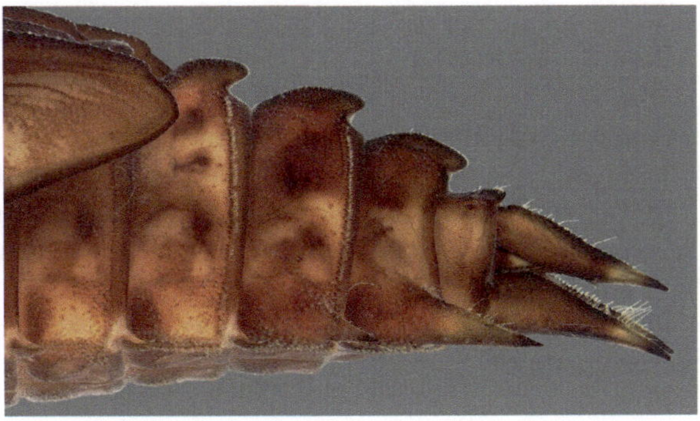

Fig. 10.192 The abdomen and caudal appendages in lateral view

The middorsal protuberances on S2–S10 are expressed as vertical spines on S2–S3; as strong, round-tipped hooks on S4–S9; and as a smaller hook on S10, and all are covered abundantly with small, scale-like setae (Fig. 10.192); the hooks on S4–S9 surpass the anterior margin of the next segment or beyond, and the hook on S10 slightly surpasses the base of the epiproct. The lateral margins of S2–S10 bear spiniform or scale-like setae; those of S7–S9 or S8–S9 end in short or large, strong, and sharply or roundly pointed spines. The spines on S7 are small, divergent from each other, and moderately incurved; those on S8 are short and roundly pointed or large and sharply pointed, divergent, and moderately incurved; and those on S9 are almost straight, divergent, sharply or roundly pointed, and the largest of the posterolateral spines and reach the posterior margin of S10 or nearly to the tips of the cerci. The sterna are covered with minute spiniform setae or scale-like setae. Sterna 3–5 are each divided into five sternites and sterna 6–9 into three sternites (Fig. 10.191). The female gonapophyses are vestigial and gomphid-like. The caudal appendages are pyramidal, long (i.e., as long as the dorsal length of S8 + S9) or shorter, and sharply (Fig. 10.192) or roundly pointed. The epiproct is at most almost twice longer than its basal width and shorter or slightly longer than the paraprocts; its dorsal margin is ridged along the distal 0.60 and has spiniform setae on the margins and ventral surface; the male's epiproct has a dorsal tubercle at the basal 0.40–0.60 of its length. The paraprocts are divergent in ventral view and slightly decurved or straight in lateral view; their mesal surfaces and dorsal and ventrolateral margins bear strong spiniform setae, and the ventral surfaces are mostly smooth or have sparse, minute, spiniform setae. The cerci are digitiform, slightly incurved at the tip, smooth, and 0.25–0.50× as long as the epiproct.

Measurements (in mm). The total length (including caudal appendages) is 21.2–29.7; abdomen (ventral, including appendages) 12.5–17.0; maximum width of the abdomen 6.4–8.0; maximum width of the head 5.8–7.1; and length of hind femur 6.3–8.5.

10.34.3 Remarks

Zygonyx is found in "swift clear streams in forest and open areas … with nymphs clinging to rocks in riffles" (Orr 2005). We have also found *Zygonyx* on the near-vertical rock surfaces of waterfalls. In Thailand, only two species have been recorded: *Z. iris* Selys, 1869, which is widely distributed, and *Z. ida* Selys, 1869, which is found only in the southern part of the country (Hämäläinen and Pinratana 1999). Of these, only the nymph of *Z. iris* has been described (Matsuki and Kitagawa 1986).

10.34.4 Distribution

Zygonyx has been recorded from the provinces of Chaiyaphum, Chanthaburi, Chiang Mai, Chumphon, Kanchanaburi, Krabi, Nakhon Nayok, Nakhon Ratchasima, Nakhon Si Thammarat, Phangnga, Phrae, Phuket, Prachuap Khiri Khan, Ranong, Saraburi, Satun, Songkhla, Tak, Trang, Trat, Yala (Hämäläinen and Pinratana 1999), Chiang Rai (Katatani et al. 2004), Kamphaeng Phet, Phatthalung, Phayao, Phitsanulok, Ratchaburi, Surat Thani (Ferro et al. 2009), and Nan (Katatani et al. 2010). ***New province records:*** Lamphun, Loei, Mae Hong Son, Mukdahan, Narathiwat, Phetchabun, Sakon Nakhon, and Uttaradit (Fig. 10.193).

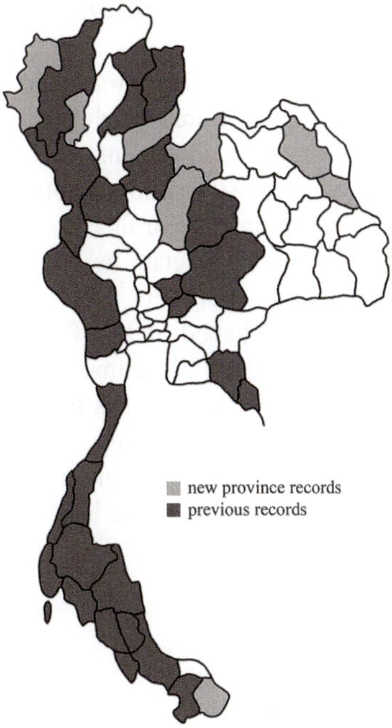

Fig. 10.193 Distribution of *Zygonyx*

10.35 *Zyxomma* Rambur, 1842

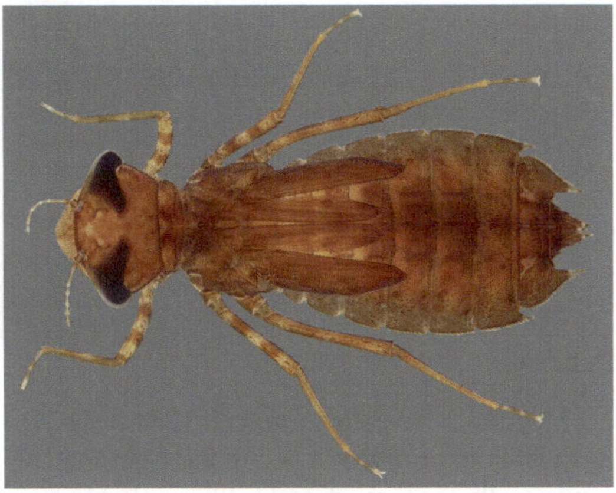

Fig. 10.194 Dorsal habitus of the nymph of *Zyxomma petiolatum*

10.35.1 General Appearance

The nymphs of *Zyxomma* are characterized by their small size (17.6–18.0 mm), with the body light-brown to dark-reddish-brown, legs relatively short, abdomen with low dorsal protuberances on S3–S10 and large posterolateral spines on S8–S9, S10 sunken into S9, and caudal appendages sharp (Fig. 10.194).

10.35.2 Description

Head. The head is subtrapezoidal, wider than long (Fig. 10.195), and granulose with the infraocular area roughly granulose. The labrum is yellowish-brown and setose on the distal half, and its anterior margin is shallowly biconcave ventrally. The clypeus is bare and mostly yellowish-brown, and a wide, dark-brown stripe is on the midline. The frons anterior surface is short, vertical, and flat; its anterior margin is straight and beset with stiff setae; its dorsal surface is yellowish-brown, mottled, and flat. The antennae are long (Fig. 10.195) and seven-segmented, the scape and pedicel are brown and barrel-shaped, and the flagellomeres are cylindrical, yellowish-brown basally, and yellow distally; sometimes antennomere 6 is the darkest and 3 is the longest. The vertex is convex and has three ocelli. The compound eyes are small, rounded, and prominent laterally (Figs. 10.195 and 10.197). The occiput is large (Fig. 10.195), moderately granulose, and covered with minute, spiniform setae and claviform, scale-like setae mainly on the lateral margins; the lateral margins are straight and strongly convergent posteriorly. The posterior margin of the head is straight.

Fig. 10.195 The head in dorsal view. Notice the small compound eyes protruding laterally

Mandibles have the formula L 1234 0 abb'/R 1234 y- abcd, with b > a in the left mandible and b = d > a > c in the right mandible. The maxilla's galeolacinia has seven teeth; the three dorsal teeth are slightly incurved, the two apicalmost teeth are of similar lengths and robustness, and the basal tooth is shorter and slender; the four ventral teeth are of different sizes and robustness, the basal tooth is minute, and the apical tooth is the largest. The ventral pad of the hypopharynx is wide and more or less kidney-shaped and bears short, regularly spaced setae on its surface and abundant long, reddish setae on the apical and lateral margins; the posterior margin is strongly concave in the middle.

Fig. 10.196 The head and prementum in ventral view

Labium. The prementum-postmentum articulation surpasses the anterior margin of the mesosternum. The prementum is subrhomboidal, yellowish-brown, and mottled on the ventral surface, and the lateral margins are smooth, sinuate, and divergent to the apex (Fig. 10.196); 8 + 3 or 10 + 3 premental setae are on each side of the midline. The ligula is well-developed, triangular (Fig. 10.196), and finely serrulated on the apical margin; a submarginal row of long, spiniform setae is on each side of the midline; and its tip is studded with 3–4 long, spiniform setae. The labial palp lateral surface is yellowish-brown and densely mottled (Fig. 10.197); the apical margin has 10–11 moderately developed crenations (Fig. 10.197), each bearing a graded series of 4 spiniform setae; the dorsolateral margin has a row of small setae; the ventral margin is finely serrulated and bears a submarginal row of elongate, spiniform setae, with those on the distal corner the largest; 8–10 long palpal setae and a row of minute, spiniform setae are present on the basomesal surface. The movable hook is relatively short, slender, sharply pointed, and smooth. One to three short, stout, spiniform setae are on the palp's articulation.

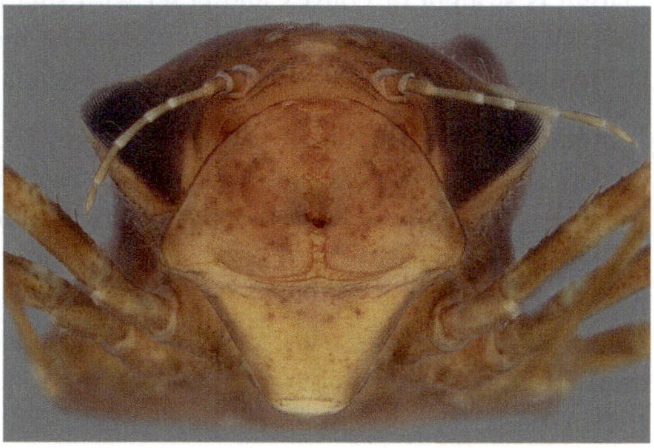

Fig. 10.197 The head and prementum in frontal view

Thorax. The thorax is narrower than the head (Fig. 10.194). The pronotal disc is light-brown and has a longitudinal, wide, brown stripe on each side of the midline; the anterior margin is shallowly concave, and the posterior margin is broadly convex and thick; the sides are rounded and granulose and bear spiniform setae. The proepisternum protrudes strongly as a conical tubercle beset with long setae. The pterothorax is mostly brown and finely granulose. The legs are relatively short (i.e., when fully extended, the tips of the hind legs slightly surpass the caudal appendages) (Fig. 10.198), pale, moderately setose, and spiny; three dark bands are on the femora and the fore and middle tibiae, whereas the hind tibia is dark-brown; the distal fourth of the ventral (internal) surface of the fore and middle tibiae and ventrolateral row of the fore tarsus have tridentate setae, and the remainder of the legs have simple, short and long setae; the pretarsal claws are simple and each has a pulvilliform empodium.

Fig. 10.198 Ventral habitus of the nymph of *Zyxomma petiolatum*

The anterior and posterior wing sheaths are parallel (Fig. 10.194) and reddish-brown, their costal margins are granulose, and they reach the posterior margin of S5 and basal half of S6, respectively. The mesolaterosternite is thick, and its anterior margin is straight or convex and beset with small, stout, bluntly pointed spines or with long, stiff setae. The anterior margin of the metasternellum is straight and 0.20× wider than the metasternellum midline length.

Abdomen. The abdomen is pyriform (Fig. 10.198), widest at S7, and truncate posteriorly. The terga are finely granulose or bear minute, spiniform setae, yellowish-brown to reddish-brown, and more or less concolorous or have a complex mottled pattern; S9 has a pale, star-shaped, submedian spot on each side of the midline. The posterior margins of S2–S4 or S2–S6 and the lateral margins of S2 or S2–S3 are bare, and the posterior margins of S5–S9 have a row of small, spiniform setae or they are minutely granulose on S7–S8, and S10 is bare; the lateral margins of S3–S10 or S4–S10 are fringed with small, spiniform setae, which increase in size and robustness posteriorly.

Fig. 10.199 The abdomen and caudal appendages in lateral view

The middorsal protuberances on S3–S9 or S3–S10 are low and spine-like (Fig. 10.199), and on S6–S10, they form a continuous ridge onto the epiproct (Fig. 10.194). The posterolateral spines on S8–S9 are large and sharply pointed, and their tips are parallel to each other; they are 0.5–0.6 and 1.0–1.1 mm in length, respectively, and 0.4–0.5× and 0.9–1.0× as long as the dorsal lengths of their respective segments (Fig. 10.199). S10 is short and sunken into S9. The sterna are yellowish-brown to reddish-brown (Fig. 10.198) and mostly bare or densely granulose; the laterosternites are more or less concolorous, whereas the mediosternites are densely mottled and have a basal, median, brown spot on 6–8 or a pair of median, brown spots on 2–8. The posterior margins of 2–7 or 2–8 are bare, and the remainder have a row of small, spiniform setae. Sterna 3–5 and 8–9 are each divided into five sternites and sterna 6–7 into three sternites. The caudal appendages are pyramidal, sharply pointed, and reddish-brown with the tips tricolored (Fig. 10.199). The epiproct is ≤0.2× longer than its basal width, and it has a full-length, middorsal carina, and the tip is straight; a row of long, stiff, bristle-like setae are on its lateral margins. The paraprocts are 0.1× longer than the epiproct; a row of strong spines and long, stiff, bristle-like setae are on the dorsal and lateral margins (Fig. 10.199); and their tips are straight and parallel (Figs. 10.194 and 10.198). The cerci are bare, digitiform, and the shortest of the caudal appendages and half as long as the epiproct, and their tips are straight (Fig. 10.199).

Measurements (in mm). The total length including caudal appendages is 17.6–18.0; abdomen (ventral, including appendages) 11.0–11.6; maximum width of the abdomen 6.5–6.8; maximum width of the head 5.0–5.2; and length of hind femur 4.2–4.8.

10.35.3 Remarks

Zyxomma is found in open habitats "in lakes, ponds, polluted drains and sluggish streams" (Orr 2005); the nymph is a bottom-dweller. Two species have been recorded from Thailand: *Z. petiolatum* Rambur, 1842 (Hämäläinen and Pinratana 1999), and *Z. breviventre* (Martin, 1921) (Sribal et al. 2018).

10.35.4 Distribution

Zyxomma has been recorded from the provinces of Bangkok, Chachoengsao, Chaiyaphum, Chiang Mai, Chon Buri, Kanchanaburi, Lampang, Nonthaburi, Yala (Hämäläinen and Pinratana 1999), Nan, Phrae (Katatani et al. 2010), Khon Kaen, and Phangnga (Day et al. 2012) and recently from Sakon Nakhon and Udon Thani (Sribal et al. 2018). ***New province records:*** Kalasin, Krabi, Phetchabun, Phuket, and Songkhla (Fig. 10.200).

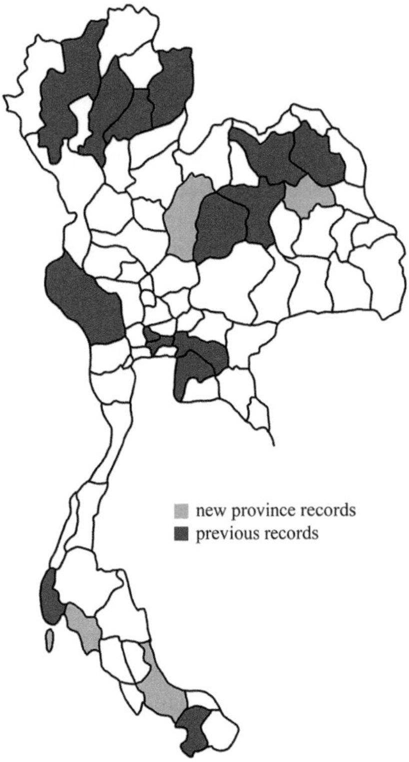

Fig. 10.200 Distribution of *Zyxomma*

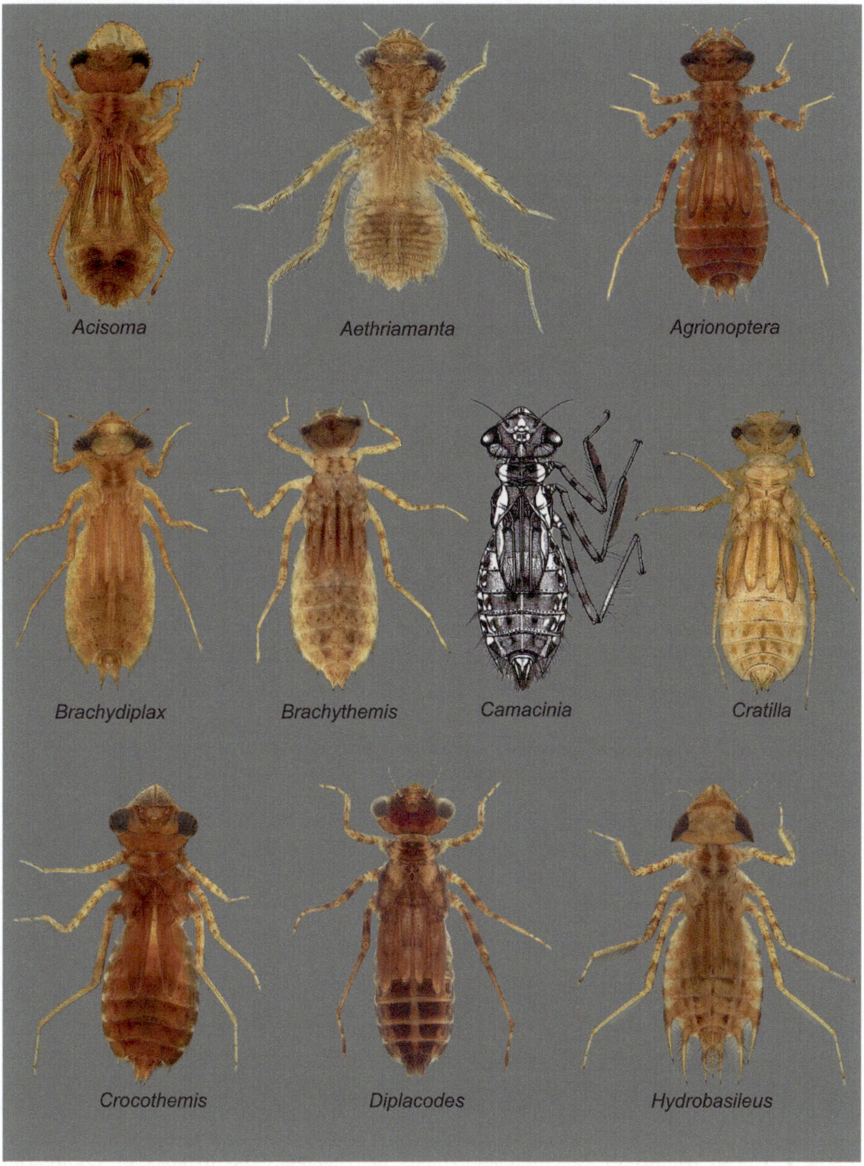

Plate 4 Dorsal habitus of Libellulidae nymphs of Thailand. The nymph of *Bradinopyga* is not illustrated. The nymphs of *Brachygonya* and *Chalybeothemis* are unknown

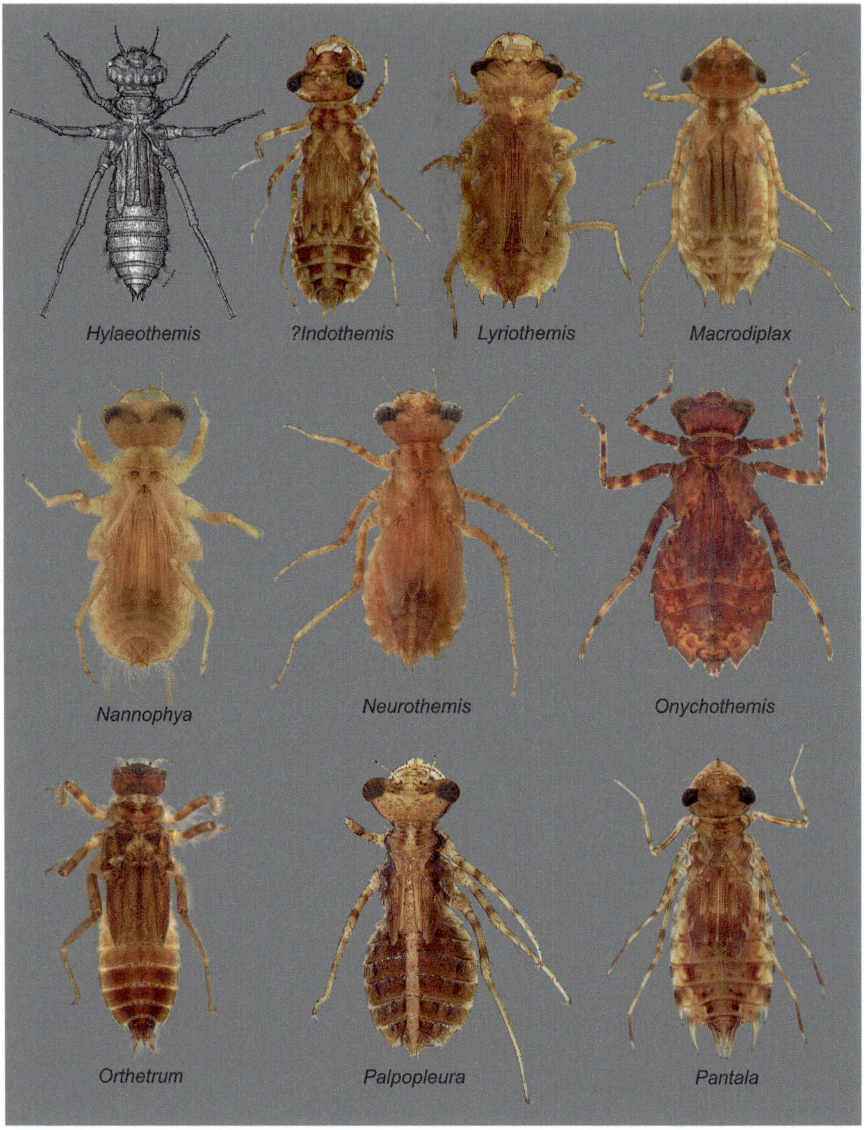

Plate 5 Dorsal habitus of Libellulidae nymphs of Thailand. The nymphs of *Lathrecista*, *Nannophyopsis*, and *Orchithemis* are not illustrated

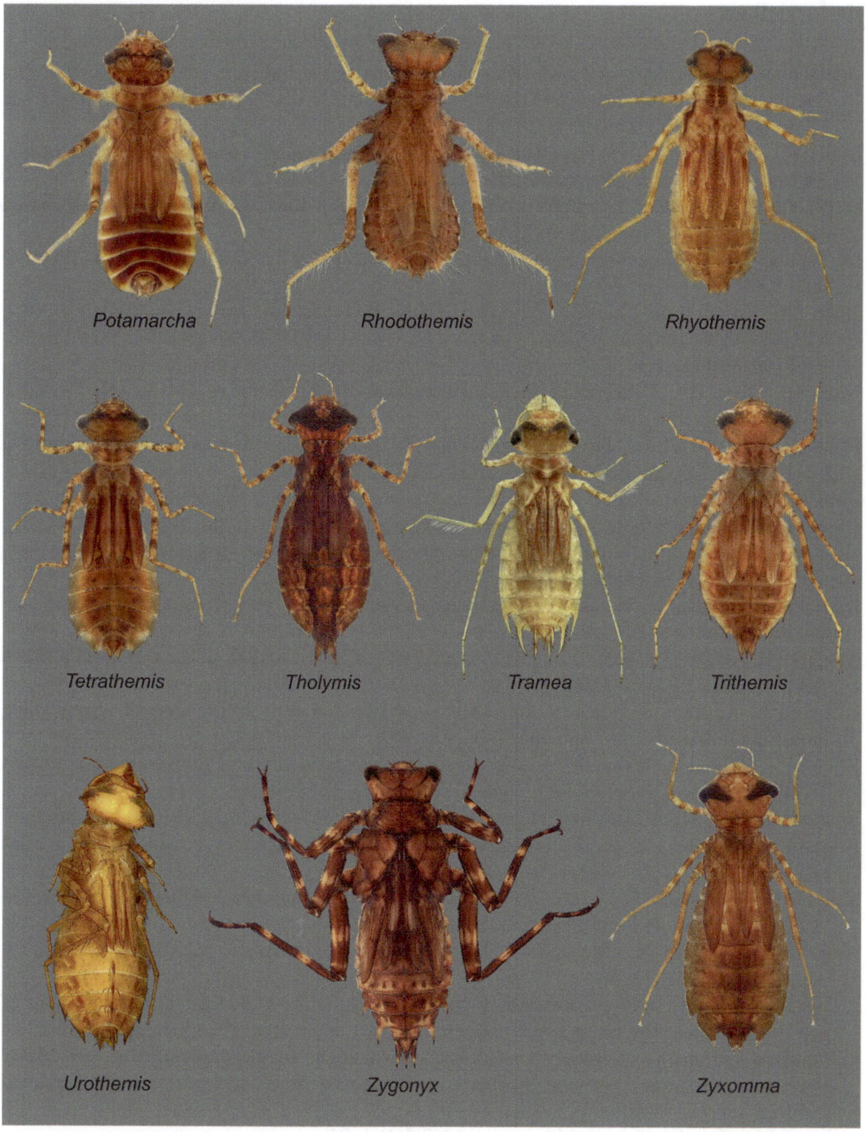

Plate 6 Dorsal habitus of Libellulidae nymphs of Thailand. The nymph of *Sympetrum* is not illustrated. The nymphs of *Amphithemis*, *Atratothemis*, *Nesoxenia*, *Phyllothemis*, *Pseudothemis*, *Pseudotramea*, *Risiophlebia*, and *Tyriobapta* are still unknown

References

Adambukulam SP, Kakkassery FK. Taxonomic studies of the last instar nymph of *Lathrecista asiatica asiatica* (Fabricius 1798) (Family: Libellulidae, Order: Odonata) by using its exuvia. J Entomol Zool Stud. 2013;1(5):103–9.

Butler SG. Description of the last instar larva of *Orchithemis pulcherrima* Brauer from Sarawak, Malaysia (Anisoptera: Libellulidae). Odonatologica. 2013;42(3):247–51.

Day L, Farrell D, Günther A, et al. New provincial records of Odonata from Thailand mostly based on photographs. Agrion. 2012;16(1):16–25.

De Fonseka T. The dragonflies of Sri Lanka. Colombo, Sri Lanka: WHT Publications (Private) Ltd.; 2000.

Dow RA. *Palpopleura sexmaculata*. The IUCN Red List of Threatened Species 2009: e.T163719A5641056; 2009. doi: https://doi.org/10.2305/IUCN.UK.2009-2.RLTS.T163719A5641056.en

Ferro ML, Sites RW, Vitheepradit A. Contributions to the faunistics of Odonata in Thailand. Insecta Mundi. 2009;104:1–24.

Fleck G, Legrand J. La larve d'*Aethriamantha rezia* Kirby, 1889 (Odonata, Anisoptera, Libellulidae). Rev Francaise Entomol (NS). 2005;27(1):17–20.

Hämäläinen M, Pinratana BA. Atlas of the dragonflies of Thailand. Distribution maps by provinces. Bangkok: Brothers of Saint Gabriel in Thailand; 1999.

Katatani N, Kitagawa K, Muraki A, Matuzawa T. Notes on Thai Odonata, Part 3. The Odonata from Chiang Rai. Aeschna. 2004;41:59–70.

Katatani N, Kitagawa K, Muraki A, et al. Notes on Thai Odonata, Part 5. The Odonata from Nan and Phrae. Aeschna. 2010;46:48–58.

Kitagawa K, Katatani N. Notes on Thai Odonata, Part 1. The Odonata from Chanthaburi in March 1998. Aeschna. 2002;39:33–42.

Kitagawa K, Katatani N. Notes on the Odonata of Thailand, Part 2. The Odonata from Trang Prefecture in March 1998. Aeschna. 2003;40:13–23.

Kompier T, Holden J, Makbun N. A new species of Lyriothemis Brauer, 1868 from Vietnam and Thailand (Odonata: Libellulidae). Zootaxa. 2021;4933(4):586–94.

Kosterin O. The journey of a dilettante. Agrion. 2006;10(1):5–7.

Kosterin OE. Odonata of the Cambodian coastal regions in late rainy season of 2011. J Int Dragonfly Fund-Report. 2012;45:1–102.

Kosterin OE. Dry season Odonata of the Cardamonean coast (Cambodia and Thailand) revisited in 2015. J Int Dragonfly Fund-Report. 2015;89:1–36.

Kosterin OE, Vikhrev N, Zinchenko V, Korshunov A. More Odonata records from Thailand. Agrion. 2010;15(1):12–6.

Lieftinck MA. A review of the dragonfly-fauna of Boeroe, in the Moluccas, with descriptions of new or interesting species, and an account of their larvae. Treubia. 1930;7, Suppl:305–30.

Lieftinck MA. Handlist of Malaysian Odonata. A catalogue of the dragonflies of the Malay Peninsula, Sumatra, Java and Borneo, including the adjacent small islands. Treubia. 1954;22(Suppl):1–202.

Lieftinck MA. Odonata. Insects Micronesia. 1962;5(1):1–95.

Lien JC, Matsuki K. Description of the larva of *Cratilla lineata assidua* Lieftinck from Taiwan (Libellulidae: Odonata). Tombo. 1988;31(1–4):35–6.

Matsuki K, Kitagawa K. Description of the possible larva of *Zygonyx iris malayanus* (Laidlaw) from Thailand (Libellulidae: Odonata). Gekkan-Mushi. 1986;183:24–5.

Orr AG. An annotated checklist of the Odonata of Brunei with ecological notes and descriptions of hitherto unknown males and larvae. Int J Odonatol. 2001;4:167–220.

Orr AG. Dragonflies of peninsular Malaysia and Singapore. A pocket guide. Kota Kinabalu, Sabah: Natural History Publications (Borneo); 2005.

Paulson DR, Schorr M, Abbott, J et al. (Coordinators). World Odonata List. Odonata Central, University of Alabama; 2023. https://www.odonatacentral.org/app/#/wol/. Accessed: 18 July 2023

Pinratana A Editorial. Malangpo 2003;20:195

Sangal SK, Kumar A. Studies on the taxonomy of larvae of Doon Valley Odonata 1. *Crocothemis servilia servilia* (Drury) and *Bradinopyga geminata* (Rambur) (family Libellulidae). J Nat Hist. 1970;4:33–8.

Sasamoto A, Kawashima I. Description of the last instar larva of *Hylaeothemis clementia* Ris from Laos (Anisoptera: Libellulidae). Odonatologica. 2009;38(4):369–74.

Sribal U, Paweenpermsuk Y, Thitiarchagul T, et al. New national records of Odonata from Thailand based mostly on photographs (Odonata: Argiolestidae, Philosinidae, Aeshnidae, Libellulidae). Agrion. 2018;22(1):30–6.

Suhling F, Martens A. Dragonflies and damselflies of Namibia. Windhoek: Gamsberg Macmillan Publishers; 2007.

Tang HB, Wang LK, Hämäläinen M. A photographic guide to the dragonflies of Singapore. Singapore: Raffles Museum of Biodiversity Research; 2010.

Theischinger G, Hawking J. The complete field guide to dragonflies of Australia. Collinwood: CSIRO Publishing; 2006.

Thipaksorn A, Kittayapong P, Milne JR, et al. Records of rice field Odonata from Thailand. Notul Odonatol. 2003;6(2):20–4.

van Tol J. An annotated index to names of Odonata used in publications by M.A. Lieftinck. Zool Verh. 1992;279:3–263.

Tze-wai T, Ka-king L, Kwan BSP, et al. The dragonflies of Hong Kong. Friends of the Country Parks. Hong Kong: Cosmos Books Ltd; 2011.

Yeh W-C, Lien J-C. Proven distribution of *Nannophyopsis clara* in Taiwan and morphological description of the ultimate instar larva (Odonata: Libellulidae). Tombo. 1995;38:24–6.

Incertae Sedis 11

Contents

11.1 Key to Incertae Sedis Genera of Thailand 472
11.2 *Idionyx* Hagen, 1867 472
11.2.1 General Appearance 472
11.2.2 Description 472
11.2.3 Remarks 476
11.2.4 Distribution 476
11.3 *Macromidia* Martin, 1907 477
11.3.1 General Appearance 478
11.3.2 Description 478
11.3.3 Remarks 481
11.3.4 Distribution 481
References 482

The genera *Idionyx* and *Macromidia* are small to medium-sized dragonflies distributed in the Oriental region (Wilson 1996), principally in Southeast Asia (Lieftinck 1971, Wilson 1996, van der Poorten 2012). Both genera traditionally have been considered by many authors as belonging to the family Corduliidae, subfamily Idionychinae (e.g., Lieftinck 1971, Asahina 1987, Hämäläinen and Pinratana 1999, Subramanian et al. 2013), whereas others consider them to be in the family Synthemistidae (e.g., Subramanian 2011, Payra et al. 2022). However, phylogenetic studies based on morphological and molecular analyses are inconclusive; thus, they are considered as *incertae sedis* (Dijkstra et al. 2013).

Adults of both genera have emerald green eyes and a metallic green body with yellow maculation. Males exhibit an acute anal angle on the hind wings, and small auricles are present on AS2. Wings can be totally hyaline or have an amber tint at the base. The adults are strong fliers and spend much of their time flitting over the water surface in fast-flowing streams with females ovipositing in pooled sections

R. Novelo-Gutiérrez, R. W. Sites, *The Dragonfly Nymphs of Thailand (Odonata: Anisoptera)*, https://doi.org/10.1007/978-3-031-33712-3_11

(Wilson 1996, Tze-wai et al. 2011). Nymphs live under stones in riffles, in underwater plant roots, and amongst algae.

Five species of *Idionyx* and two species of *Macromidia* occur in Thailand, and those of *Idionyx* have the widest distribution throughout the country.

11.1 Key to *Incertae Sedis* Genera of Thailand

1 Wing sheaths divergent (Fig. 11.1); 5–6 long premental setae on each side of midline; 5 palpal setae . . . . **Idionyx**

1' Wing sheaths parallel (Fig. 11.8); 9–10 long premental setae on each side of midline; 9–10 palpal setae . . . . ***Macromidia***

11.2 *Idionyx* Hagen, 1867

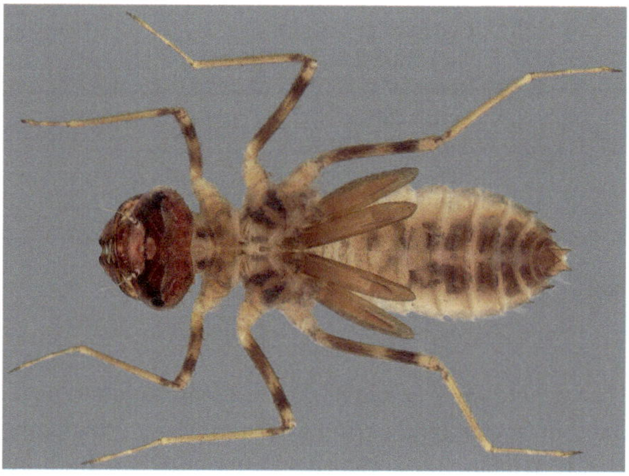

Fig. 11.1 Dorsal habitus of the fully developed nymph of *Idionyx*

11.2.1 General Appearance

The nymphs of *Idionyx* are characterized by their small size (15.0–19.5 mm), coloration of yellow-brown to reddish-brown, hirsute body mainly on the sides, long legs, divergent wing sheaths, and abdomen lacking dorsal protuberances (Fig. 11.1).

11.2.2 Description

Head. The head is relatively large, wider than long, and covered with scalelike setae (Fig. 11.2). The labrum is 0.7 mm in length, with long setae on most of its surface, including along the anterior margin, which is widely concave at the middle and not flattened ventrally. The clypeus is bare. The postfrons is produced anteriorly as a frontal shelf which is raised but not projected, its anterior surface is flattened and hirsute, and the apical margin is straight and is covered by fan-shaped setae dorsally and with some long-stalked, fan-shaped setae laterally. The vertex is reduced and has three ocelli. The antennae are short and seven-segmented; the scape is cylindrical, the pedicel is globose, and the third antennomere is the longest. The compound eyes are small and moderately protruding dorsally. The occiput is very large, bulging laterally, and the lateral margins are covered with scalelike setae intermingled with long-stalked, fan-shaped setae (Fig. 11.2); delicate setae are on the ventral surface, the anterior dorsal surface is mostly bare, and the posterior margin is straight and scaly.

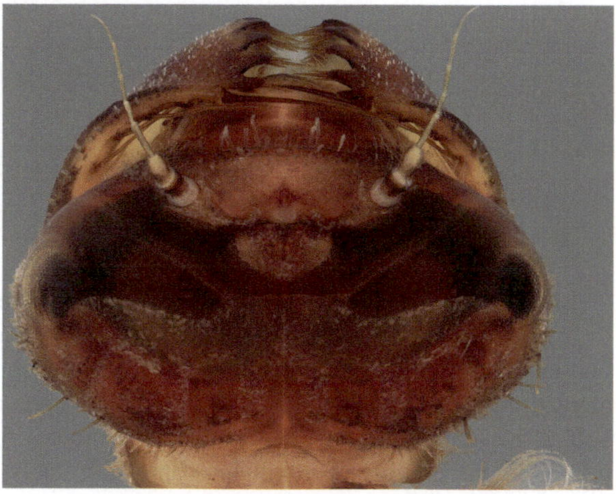

Fig. 11.2 The head in dorsal view

Mandibles have the formula: L 12'234 0 ab/R 1234 y ab; although there is no proper molar crest, in both mandibles, teeth *a* and *b* are connected to each other by a low, transverse ridge, as a reminiscent *m*. The maxilla's galeolacinia has seven teeth; the three dorsal teeth are slightly incurved and more or less of the same length and robustness, and the four ventral teeth are of different sizes with the apical tooth the largest and the basal tooth strongly reduced. The ventral pad of the hypopharynx is wide and subrectangular, with the anterior margin convex and the posterior margin strongly concave; abundant stiff setae occur on most of its surface, and the margins have long, robust, incurved setae.

Fig. 11.3 The prementum in ventral view

Labium: The prementum-postmentum articulation reaches the anterior margin of the metacoxae. The prementum is large and subtriangular (Fig. 11.3), with the lateral margins widely divergent and bare, but with a sublateral row of scalelike setae along the basal 0.75 of its length; a longitudinal, bare, central sulcus is in the basal half of the ventral surface; premental setae number is 5 + 5 or 5 + 4 on each side of the midline; the ligula is prominent, triangular, finely serrate on its apical margin and has a subapical row of stout setae. The labial palps are strongly developed and cover the anterior part of the head as a mask (Fig. 11.4); the apical margin of the palp is divided into 6–7 rounded lobes or serrulated crenations, with each crenation bearing 4–6 stout setae of different sizes on the mesal margin; the dorsal margin of the palp has a row of robust, scalelike setae; there are five palpal setae. The movable hook is long, slender, and sharply pointed.

Fig. 11.4 The head and palps in frontal view. Notice the large crenations

Thorax. The pronotal disc is trapezoid, with the anterior margin concave, the posterior margin convex (Fig. 11.1), the lateral margins straight, and all the margins covered with scalelike setae of different sizes; the propleuron has minute scalelike setae, and its margin is beset with long, white, recurved setae. The dorsum and lateral surfaces of the pterothorax are covered with minute, scalelike setae, and the ventral margins are hirsute. The legs are long (i.e., when fully extended, the hind legs far surpass the caudal appendages) (Figs. 11.1 and 11.5); the coxae and trochanters are setose, and the remainder of the segments are scaly; the femora and tibiae have sparse, stalked, scalelike setae; all legs are yellow or reddish-yellow, with two large, wide, transverse, dark bands on the femora (Figs. 11.1 and 11.5); the pretarsal claws are long and simple, and each has a pulvilliform empodium. The anterior and posterior wing sheaths are divergent and reach the posterior margin of S4 and basal half of S5, respectively (Fig. 11.6).

Abdomen. The abdomen is subcylindrical and parallel-sided on S1–6, and then convergent on S7–10, with a complex color pattern on terga 4–9; terga 2–10 are covered with minute scalelike setae; the posterior margins of S2–9 have a row of long, scalelike setae; and the lateral margins of S2–9 have stiff setae of different sizes. Dorsal protuberances are lacking (Fig. 11.6), and S8–9 each end in a short, incurved spine (sometimes minute on S8). The sterna are densely setose or scaly.

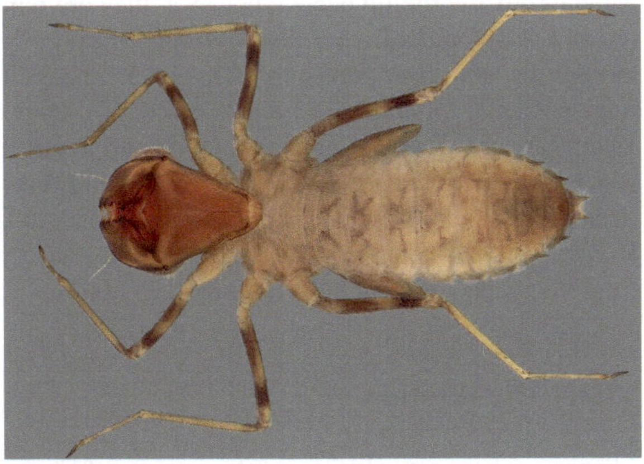

Fig. 11.5 Ventral habitus of the fully developed nymph of *Idionyx*

Sterna 3–9 are each divided into five sternites; sternum 10 forms a continuous ring with the respective tergum. Male and female gonapophyses are lacking. The caudal appendages are setose on the mesal surfaces, and coloration is yellow-brown, with the cerci sometimes dark-brown. The epiproct and paraprocts are pyramidal, covered with scalelike setae, acutely pointed, and their tips are slightly decurved (Fig. 11.6); the epiproct is basally wider than its length (Fig. 11.1); the cerci are digitiform; the size proportions are paraprocts 1.00, epiproct 0.95, and cerci 0.75.

Measurements (in mm). Total length including caudal appendages 13.7–19.5; abdomen (ventral) 9.7–11.8; maximum width of head 4.2–5.2; length of hind femur (lateral) 4.5–5.8.

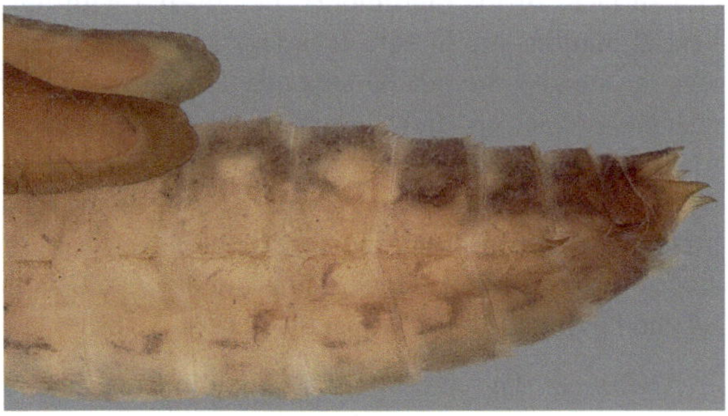

Fig. 11.6 The abdomen and caudal appendages in lateral view

11.2.3 Remarks

In recent molecular studies, the phylogenetic position of this genus was considered as uncertain (Dijkstra et al. 2013), although traditionally it has been included in Corduliidae (Lieftinck 1971). The nymph of *Idionyx* looks quite atypical as compared to other corduliid genera; however, they are reminiscent in general aspect of those of *Neocordulia* from the New World. Nymphs occur in "small streams flowing through lowland forest marshes" (Orr 2005; Zhang and Tong 2010). In Thailand, four species of *Idionyx* have been reported: *I. optatus* Selys, 1878; *I. selysi* Fraser, 1926; *I. thailandicus* Hämäläinen, 1985; and *I. iida* Hämäläinen, 2002, plus one undetermined species of the *yolanda* group and three more of the *optatus* group (Hämäläinen and Pinratana 1999; Hämäläinen, 2002). However, the record of *I. optatus* from Thailand has been questioned because the only definitive records are from Meghalaya state in India (Hämäläinen 2002).

11.2.4 Distribution

Idionyx has been recorded from the provinces of Chaiyaphum, Chanthaburi, Chiang Mai, Kanchanaburi, Loei, Nakhon Nayok, Prachuap Khiri Khan, Phrae, Ranong, Rayong, Saraburi, Tak, Trang (Hämäläinen and Pinratana 1999), Chiang Rai (Katatani et al. 2004), and Nan (Katatani et al. 2010). ***New province record:*** Trat (Fig. 11.7).

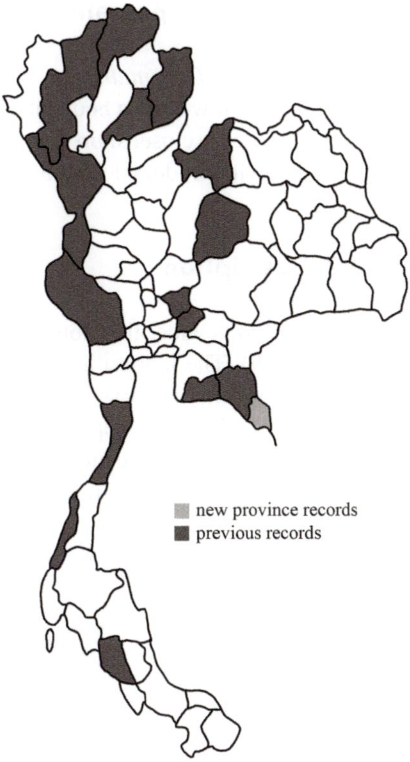

Fig. 11.7 Distribution of *Idionyx*

11.3 *Macromidia* Martin, 1907

Fig. 11.8 Dorsal habitus of the fully developed nymph of *Macromidia genialis*

11.3.1 General Appearance

The nymphs of *Macromidia* are characterized by their small to medium size (15–20 mm), yellow-brown body, libellulid-like feature (Fig. 11.8), appearing granulose and sparsely setose, short legs, broadly fusiform abdomen, and lateral spines on S8–9, but without dorsal protuberances.

11.3.2 Description

Head. The head is wider than long and with the lateral margins converging posteriorly (Fig. 11.9). The labrum is 0.6 mm long, setose, gently concave on the apical margin, and not flattened ventrally. The clypeus has minute, delicate setae. The frons, vertex, and occiput appear densely granulose due to the protuberant sockets of minute, fan-shaped setae and sparse, large, scalelike setae; the frons is projected as a high shelf and is widely rounded apically (Fig. 11.9).

Fig. 11.9 The head and prothorax in dorsal view

The antennae are seven-segmented; the scape is cylindrical, pedicel is subcylindrical, and antennomeres 3, 6, and 7 are subequal in length. The compound eyes are small and slightly protruding laterally. The occiput is large, with the sides produced to form broad flanges; the cephalic lobes are small and rounded; and the occipital margin is gently concave (Fig. 11.9). Mandibles have the formula: L 1234 0 ab'b/R 1234 y ab'b; there is no molar crest; in the left mandible, 4 > 3 > 1 > 2; in the right mandible, 1 > 4 > 3 > 2; in both mandibles, b > a. The maxilla's galeolacinia has seven teeth; the three dorsal teeth are slightly incurved and subequal in length and robustness, the four ventral teeth are of different sizes, and the apical tooth is the largest. The ventral pad of the hypopharynx is wide, soft, whitish, and setose on the anterior and lateral margins; it has three transverse rows of stout setae at the middle of the ventral surface, and its posterior margin is strongly concave at the middle.

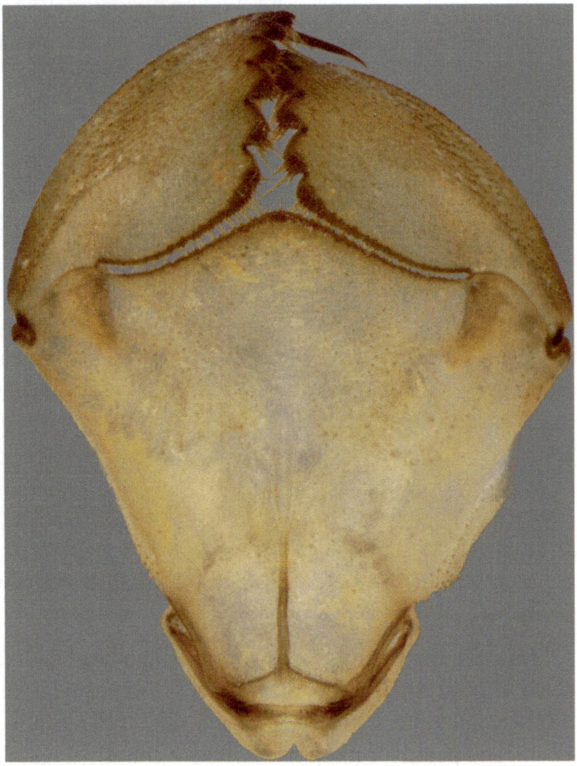

Fig. 11.10 The prementum in ventral view

Labium: The prementum-postmentum articulation reaches the middle of the mesosternum. The prementum is wider than long, and its lateral margins are strongly convergent basally (Fig. 11.10); a group of 23–25 spiniform setae are on the dorsolateral part at the base; a ventral, longitudinal, median sulcus is in the basal 0.40 (Fig. 11.10). Premental setae number is 9 + 6 or 9 + 7 on each side of the midline (Fig. 11.11); the ligula is well developed and rounded apically, and its margin is finely serrated; a dorsal subapical row of setae increase in size and robustness toward the center, and two large, isolated setae are at the extreme tip. The labial palps have 9–10 long setae and a group of 17–18 short setae close to the palp's articulation (Fig. 11.11); the distal margin has 7–8 widely rounded crenations, each of which bears a graded series of six strong setae (Fig. 11.10); the dorsal margin has a row of 21–23 spiniform setae, and the ventral margin is gently serrated; the lateral surface has minute, spiniform setae. The movable hook is sharp and slightly incurved at the tip.

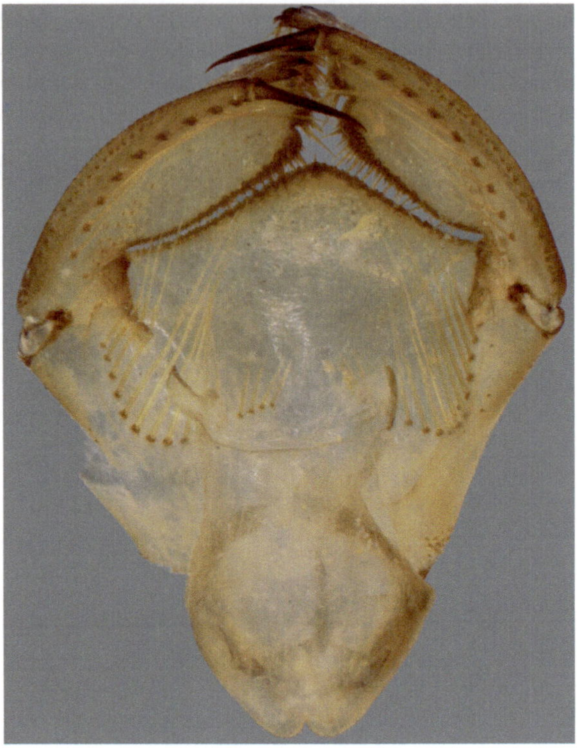

Fig. 11.11 The prementum in dorsal view, showing the premental and palpal setae

Thorax. The thorax is narrower than the head and has abundant small, scalelike setae (Fig. 11.8). The pronotal disc is subtrapezoidal, mostly yellowish, and it has a large, rectangular, bare area on each side of the midline (Fig. 11.9). The pronotum lateral and posterior margins are dorsoventrally thick, the posterolateral corners are slightly reflexed, and the posterior margin is widely convex at the middle. The ventral margin of the propleuron is divided into a wide V-shaped projection (Fig. 11.9). The pterothorax is mostly brown. The legs are short (i.e., when fully extended, the hind tibia reaches the posterior margin of S8); the femora are yellow with two transverse, grayish bands; the tibiae and tarsi are light yellow-brown; the pretarsal claws are simple, and each has a pulvilliform empodium. The anterior and posterior wing sheaths are parallel and surpass the anterior and posterior margins of S4, respectively. The sterna are light yellow and have abundant minute, scalelike setae; the mesosternum has a large, conical, roundly pointed tubercle close to the coxal articulation.

Abdomen. The abdomen is about 1.25× wider than the head, yellow anteriorly, and reddish-yellow posteriorly and has abundant small, scalelike setae on the surface and margins of the terga; it is flattened ventrally and moderately convex dorsally with the sides more or less parallel, reaching its maximum width at S6–7. Dorsal protuberances are lacking. The lateral margins of S8–9 each end in a strong, dark, incurved spine (Fig. 11.8). The sterna are yellow and covered with small, spiniform setae; sterna 3–8 are each divided into five sternites and sternum 9 into three sternites. The female gonapophyses are vestigial and evident as small, blunt, conical rudiments. The caudal appendages are sharply pointed, covered with small scalelike setae, and the tips are slightly decurved; the epiproct and paraprocts are pyramidal; the cerci are triangular in lateral view; the epiproct length is 0.80× its basal width; the relative sizes are epiproct 0.88, paraproct 1.00, and cercus 0.60.

Measurements (in mm). Total length including caudal appendages 15.0–19.7; abdomen 9–14; hind femur (dorsal margin) 5.0–5.1; maximum width of head 4.9–6.0; maximum width of abdomen (ventral) 5.0–5.6.

11.3.3 Remarks

In recent molecular studies, the phylogenetic position of this genus was considered as uncertain (Dijkstra et al. 2013), although traditionally it has been included in Corduliidae (Lieftinck 1971). Nymphs occur under stones in riffles in clear forest streams (Orr 2005). In Thailand, two species of *Macromidia* have been reported: *M. genialis* Laidlaw, 1923 and *M. rapida* Martin, 1907 (Hämäläinen and Pinratana 1999).

11.3.4 Distribution

Macromidia has been recorded from the provinces of Chaiyaphum, Chanthaburi, Chiang Mai, Nakhon Nayok, Nakhon Ratchasima, Tak (Hämäläinen and Pinratana 1999, 2000), Chiang Rai (Katatani et al. 2004), Phrae (Katatani et al. 2010), and more recently Ratchaburi (Day et al. 2012). ***New province records:*** Lampang, Kalasin, Prachuap Khiri Khan, and Sakon Nakhon (Fig. 11.12).

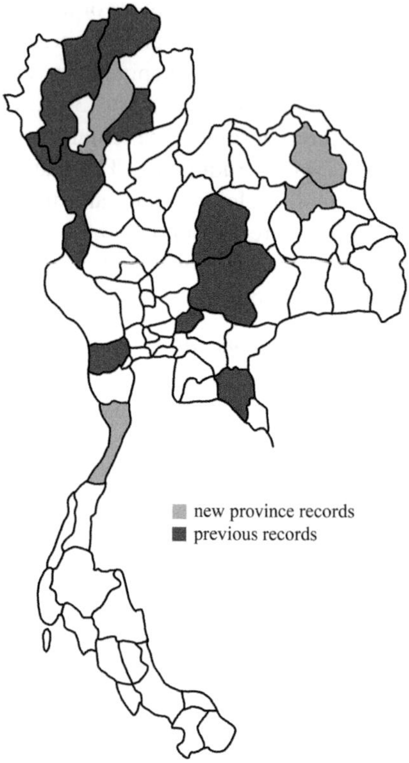

Fig. 11.12 Distribution of *Macromidia*

References

Asahina S. A list of the Odonata recorded from Thailand. 17. Corduliidae 2. Kontyû. 1987;55(4):699–720.

Day L, Farrell D, Günther A, et al. New provincial records of Odonata from Thailand mostly based on photographs. Agrion. 2012;16(1):16–25.

Dijkstra K-DB, Bechly G, Bybee SM, et al. The classification and diversity of dragonflies and damselflies (Odonata). In: Zhang Z-Q (ed) Animal biodiversity: an outline of higher-level classification and survey of taxonomic richness (Addenda 2013). Zootaxa. 2013;3703:36–45.

Hämäläinen M. The species list of Thai dragonflies increase steadily—an update. Malangpo. 2002;19:176–9.

Hämäläinen M, Pinratana BA. Atlas of the dragonflies of Thailand. Distribution maps by provinces. Bangkok: Brothers of Saint Gabriel in Thailand; 1999.

Hämäläinen M, Pinratana BA. Additions and corrections to dragonfly lists of five protected areas in Thailand. Malangpo. 2000;17:156–7.

Katatani N, Kitagawa K, Muraki A, Matuzawa T. Notes on Thai Odonata, part 3. The Odonata from Chiang Rai. Aeschna. 2004;41:59–70.

Katatani N, Kitagawa K, Muraki A, et al. Notes on Thai Odonata, part 5. The Odonata from Nan and Phrae. Aeschna. 2010;46:48–58.

Lieftinck MA. Studies in Oriental Corduliidae (Odonata) I. Tijdschr Ent. 1971;114(1):1–63.

Orr AG. Dragonflies of peninsular Malaysia and Singapore. A pocket guide. Kota Kinabalu, Sabah: Natural History Publications (Borneo); 2005.

Payra A, Despande A, Koparde P. Up and above: northernmost records of *Macromidia donaldi donaldi* (Fraser, 1924) and *Merogomphus longistigma* (Fraser, 1922) from the Western Ghats of India (Odonata: Synthemistidae: Gomphidae). Rev Soc Entomol Argentina. 2022;81(2):63–9.
Subramanian KA. *Macromidia donaldi*. The IUCN red list of threatened species 2011: e.T175184A7118824; 2011. https://www.iucnredlist.org/species/175184/7118824
Subramanian KA, Rangnekar P, Naik R. *Idionyx* (Odonata: Corduliidae) of the Western Ghats with a description of a new species. Zootaxa. 2013;3652(2):277–88.
Zhang H, Tong X. Descriptions of the final instar larvae of three Chinese *Idionyx* species (Odonata: Anisoptera: Corduliidae). Zootaxa. 2010;2716:53–63.

Glossary

Anal pyramid Epiproct, paraprocts, and cerci as a whole.
Annular Ring-shaped.
Anteclypeus The ventral part of the clypeus where the labrum articulates.
Antennomere Any segment of the antenna.
Burrower Aquatic organism that commonly resides in the bottom substrate.
Burrowing hooks Thick lateral spines at the distal end of fore- and middle tibiae.
Cephalic capsule The exoskeleton of the head.
Cercus (pl. cerci) Paired appendage of the 11th abdominal segment.
Climber Aquatic organism that commonly resides on aquatic plants, filamentous algae, or mosses.
Clinger Aquatic organism that commonly resides on underwater objects.
Crenation A low, rounded lobe on the margin of the labial palp and/or ligula.
Crenulation A smaller rounded lobe than a crenation.
Cuticle The noncellular outer layer of the integument.
Denticle A small tooth.
Ecdysial line A line of weakness on the head and notum where the cuticle splits during ecdysis.
Ecdysis The process of shedding an exoskeleton.
Emarginate Indented or notched.
Empodium A single pad-like structure of the pretarsus beneath and between the claws.
Endophytic oviposition Egg-laying strategy in which eggs are inserted into plant tissue.
Epicranial Situated on the top of the head.
Epimeron The posterior sclerite of the thoracic pleuron.
Epiphyte Plant that grows on the surface of a tree and derives its moisture and nutrients from the air, rain, or debris.
Epiphytic oviposition Egg-laying strategy in which eggs are attached to the outside of a plant.
Epiproct The median dorsal sclerite of the 11th abdominal segment.
Episternum The anterior sclerite of the thoracic pleuron.

R. Novelo-Gutiérrez, R. W. Sites, *The Dragonfly Nymphs of Thailand (Odonata: Anisoptera)*, https://doi.org/10.1007/978-3-031-33712-3

Exophytic oviposition Egg-laying strategy in which the eggs are dropped into the water.
Exoskeleton A skeleton on the outside of the body.
Flagellum The part of the antenna beyond the pedicel.
Frons The part of the cephalic capsule between the clypeus and vertex.
Fusiform Spindle-shaped.
Galeolacinia The fused distal lobes of the maxilla.
Glabrous Having a surface without hairs or projections.
Gonapophyses Midventral processes on abdominal segments 8 and 9 of females forming the first and second valvulae of the ovipositor.
Granulose With abundant granules.
Hemimetabolous Development by gradual metamorphosis.
Hypopharynx The tongue-like structure anterior to the labium.
Instar Developmental stage between two successive ecdyses.
Integument The outer layer of the insect comprising the cuticle and cellular epidermis.
Keel A sharply projected ridge.
Labium The posteriormost of the mouthparts functioning as the lower lip and modified in nymphal Odonata for prey capture.
Labrum Sclerite articulating with the clypeus that serves as the upper lip.
Lanceolate Spear-shaped.
Larva Immature stage in the life cycle of holometabolous insects.
Ligula The distomedial portion of the prementum in Odonata larvae, which comprises fused glossae and paraglossae.
Mandibles The first pair of feeding appendages.
Maxilla (pl. maxillae) The second pair of feeding appendages.
Molting The process of shedding an exoskeleton.
Notum (pl. nota) The dorsal sclerite of each thoracic segment.
Nymph Immature stage in the life cycle of pauro- and hemimetabolous insects.
Palp Segmented appendage on the maxillae and labium.
Paraproct Paired ventrolateral sclerites of the 11th abdominal segment.
Pedicel The second segment of an antenna.
Pharate Condition in which, before ecdysis, the new cuticle has become separated from the old cuticule.
Phytotelmata Epiphytes such as bromeliads and tree holes filled with rain water.
Pleuron (pl. pleura) The lateral part of a thoracic segment.
Postclypeus The dorsal part of the clypeus.
Postmentum The basal segment of the labium.
Predator Organism feeding exclusively on living prey.
Prementum The distal segment of the labium bearing the labial palpi and ligula.
Pretarsus The distal leg segment bearing the claws and empodium.
Prothorax The first segment of the thorax.
Pterothorax The wing-bearing (2nd and 3rd) segments of the thorax.
Pyramidal A three-dimensional shape forming a pyramid.
Scape The first segment of an antenna.

Sclerite A hardened plate of the cuticle.
Semi-terrestrial Organisms living out of the water but in terrestrial environments saturated with water.
Serrate Sawlike, with notched or toothed margins.
Serrulated With finely-toothed margins.
Seta (pl. setae) A hair-like unicellular external process from the integument.
Setose With abundant setae.
Spine A multicellular, rigid outgrowth of the integument.
Spiracle The opening through which air enters into the tracheal system.
Stadium (pl. stadia) Length of time of an instar between two successive ecdyses.
Sternite Any one of the subdivisions of the sternum of any segment.
Sternum The ventral region of a body segment.
Subequal Approximately equal.
Subquadrate Almost square.
Subrectangular Almost rectangular.
Subrhomboidal Almost rhomboid in shape.
Sulcus (pl. sulci) A linear furrow in the cuticle.
Suture A seam, line, or furrow in the cuticule (*sensu lato*); a line representing the fusion of two segments of the body (*sensu stricto*).
Tagma (pl. tagmata) Any region of the insect body (i.e., head, thorax, and abdomen).
Tarsomeres The subdivisions of the tarsus.
Teneral The condition of an insect immediately after ecdysis when the new cuticule is almost colorless and unsclerotized.
Tergite Any one of the subdivisions of the tergum of any segment.
Tergum The dorsal region of a body segment.
Truncate With a square margin.
Tubercle A multicellular, small, usually nipple-shaped protusion of the integument.
Vertex The top of the head bearing the ocelli.
Wing sheath External covering of the wing rudiment.

Index

A
Acisoma, 292, 303–306
Acrogomphus, 23, 109, 117
Aeshnidae, 9, 34, 35
Aethriamanta, 293, 307
Agrionoptera, 299, 311
Agrionoptera insignis, 311, 313
Amphigomphus, 112, 121
Amphigomphus somnuki, 121, 123
Anaciaeschna, 44, 51
Anaciaeschna martini, 44, 51, 53
Anax, 48, 49, 55
Anisogomphus, 116, 125
Anisoptera, 33, 34
Anotogaster, 35, 250
Anotogaster gregoryi, 250, 253
Asiagomphus, 115, 129

B
Brachydiplax, 300, 316
Brachythemis, 299, 321
Brachythemis contaminata, 320, 323
Bradinopyga, 300, 325
Burmagomphus, 116, 134

C
Camacinia, 294, 295, 328
Camacinia gigantea, 328
Cephalaeschna, 49, 50, 59, 60
Chlorogomphidae, 35, 239
Chlorogomphus sp., 32, 35, 239, 240
Clypeus, 60, 73
Conservation, 4–7
Cordulegastridae, 35, 249
Corduliidae, 36, 271
Cratilla, 295, 296, 331
Crocothemis, 301, 336
Crocothemis servilia, 336, 337

D
Damselflies, 2
Davidius, 109, 110, 138
Diplacodes, 302, 341
Dragonflies, 1

E
Epiophlebia, 2
Epiproct, 67
Epophthalmia sp., 35, 259
Euthygomphus, 116, 142

G
Gomphidae, 10, 34, 105
Gomphidia, 112, 146
Gomphidictinus, 113, 151
Gomphidictinus perakensis, 113, 151, 153
Gynacantha sp., 34, 44–46, 64

H
Heliaeschna, 45, 48
Heliaeschna simplicia, 47
Heliaeschna uninervulata, 45
Heliogomphus, 114, 115, 156
Hemicordulia sp., 37, 272, 273
Hydrobasileus, 25, 37, 298, 346
Hydrobasileus croceus, 25, 37, 298, 346
Hylaeothemis, 292, 351
Hylaeothemis clementia, 351

R. Novelo-Gutiérrez, R. W. Sites, *The Dragonfly Nymphs of Thailand (Odonata: Anisoptera)*, https://doi.org/10.1007/978-3-031-33712-3

I

Ictinogomphus, 113, 162
Idionyx, 472
Incertae sedis, 15, 36, 471
Indaeschna, 47, 48, 71
Indaeschna grubaueri, 47
Indothemis, 298, 355

L

Labial palp, 42, 44, 45
Lamelligomphus, 24, 30, 112, 167
Lathrecista, 298, 360
Leptogomphus, 109, 171
Libellulidae, 36, 287
Life cycle, 2, 3
Ligula, 71
Lyriothemis, 300, 362
Lyriothemis pachygastra, 362, 365

M

Macrodiplax, 300, 367
Macrodiplax cora, 367, 369
Macrogomphus, 114, 175
Macromia, 36, 258, 264
Macromia arachnomima, 36
Macromidia, 472, 477
Macromidia genialis, 477
Macromiidae, 35, 257
Megalogomphus, 111, 179
Microgomphus, 114, 115, 184

N

Nannophya, 301, 372
Nannophya pygmaea, 372, 375
Nannophyopsis, 294, 377
Nepogomphus, 110, 111, 188
Neurothemis, 32, 302, 379
Nihonogomphus, 111, 192
Nychogomphus, 112, 196
Nychogomphus duraricus, 196, 198

O

Occiput, 60, 65, 69, 74
Odonata nymphs, 23
Onychogomphus, 111, 200
Onychogomphus louissiriusi, 200, 202
Onychothemis, 294, 385
Orchithemis, 296, 297, 390
Orchithemis pulcherrima, 297
Orientogomphus, 110, 204
Orientogomphus minor, 110, 204, 206
Orthetrum, 27, 29, 295, 296, 392
Orthetrum sabina, 392, 395

P

Palpopleura, 302, 398
Pantala, 293, 403
Pantala flavescens, 403, 406
Paragomphus, 110, 208
Paragomphus capricornis, 25, 110, 208, 210
Paraprocts, 67
Periaeschna, 25, 45, 46, 73
Petaliaeschna, 42, 43, 77
Phaenandrogomphus, 112, 212
Phaenandrogomphus tonkinicus, 212, 214
Planaeschna, 31, 49–51, 83
Polycanthagyna, 46, 48, 87
Polycanthagyna erythromelas, 87, 89
Potamarcha, 28, 34, 295, 296, 408
Potamarcha congener, 28, 34, 408, 411
Procordulia, 272, 277
Prothorax, 44, 89
Provinces, 8

R

Rhodothemis, 293, 414
Rhodothemis rufa, 414, 417
Rhyothemis, 299, 420

S

Sarasaeschna, 42, 43, 91
Sarasaeschna pryeri, 91, 93
Sieboldius, 113, 216
Sinictinogomphus, 112, 113, 221
Sinictinogomphus clavatus, 221, 222
Stylogomphus, 114, 225
Somatochlora, 272, 280
Somatochlora uchidai, 280, 282

T

Tetracanthagyna, 26, 27, 29, 42, 95
Tetrathemis, 298, 425
Tholymis, 299, 431
Tholymis tillarga, 431, 434

Tramea, 300, 301, 436
Trithemis, 299, 442

U
Urothemis, 297, 447
Urothemis abbotti, 447, 450

Z
Zygonyx, 291, 294, 453
Zygonyx iris, 453, 456
Zygoptera, 33
Zyxomma, 459, 462
Zyxomma petiolatum, 459, 462

GPSR Compliance

The European Union's (EU) General Product Safety Regulation (GPSR) is a set of rules that requires consumer products to be safe and our obligations to ensure this.

If you have any concerns about our products, you can contact us on ProductSafety@springernature.com

In case Publisher is established outside the EU, the EU authorized representative is:

Springer Nature Customer Service Center GmbH
Europaplatz 3
69115 Heidelberg, Germany

Batch number: 10755214

Printed by Printforce, the Netherlands